I0474137

Cystic Fibrosis & the Brewer's Yeast: A Microbiology Tale
by David Wooster, Ph.D.

~ ~ ~

~ ~ ~

~ ~ ~

TABLE OF CONTENTS

PART 1: FALL SEMESTER

PART 2: WINTER SEMESTER

PART 3: SPRING SEMESTER

PART 4: SUMMER SEMESTER

PART 1: FALL SEMESTER

THE OTHER KIND OF DOCTOR

Almost everyone gets it wrong so I should explain that I'm not the kind of doctor that practices medicine but rather the "other kind". We're usually professors most Ph.D.'s, which is why I teach as well as do research for a living. To my students I'm Dr. Ketchum, one of a handful of microbiologists in this small town along the foothills of the Rocky Mountains, but to most everyone else I'm just plain Ben. Lots of folks get the two types confused, though, and as I often explain more out of desperation than anything else when someone still doesn't get it is that, if I were to practice medicine, I'd go to jail just like anybody else without a license. But what bothers me even more these days is when my family reminds me that I'm pushing 40 and still not married. Maybe it's just that I've just never found the time, I'm not sure, but if it is an excuse it's one I've been using for some time now.

And as a microbiologist what interests me most is figuring out all the different ways these clever little "bugs" have found to get along in the world. Extreme microbiology, you could say, is my specialty. We're all supposed to have one these days. Well it turns out these overlooked creatures have had more than three billion years of experimenting with different strategies for survival and there seem to be an endless number of ways they've come across, too, almost as many as there are microbes.[1]

On the other hand, we humans are the relative newcomers, an extensive collection of cells – around ten trillion of them – all grouped together and working side-by-side doing different jobs in vast colonies we call "tissues" and so we live our lives on a grand scale compared to single-celled life forms like bacteria and as a consequence tend to take for granted that rain, or wind, or most anything else that Mother Nature can throw at us on a regular basis doesn't have any real effect, except for perhaps having to pop open an umbrella or throw on an overcoat once in a while.[2] Our size and complexity for the most part renders us indestructible to these everyday forces. But for microbes life can be a horse of a different color.

Changing conditions, maybe it's something as simple as a mud puddle drying up in the heat of the midday sun, or perhaps the danger lies closer to home, in the form of the sticky foot of some passing insect, either of which

---

[1] And many are good at hitching a ride. About 10% of the weight you carry around every day is bacteria. Each of us has around 200 trillion microbes living in and on our bodies (10,000 different species).

[2] Single cells staying together to become a multicellular organism was such a good idea, it happened throughout evolution several different times.

can spell the difference between life and death to these economical little beings and unlike us with our specialized tissues – muscles, bone, and skin for instance – microbes are for the most part stuck where they are, without the elaborate defenses against nature.   Well, most of them are anyway, which brings me to my most recent setback.

The anthrax grant I had applied for, the one I had been counting on to fund my research for the next several years, was turned down by the National Institutes of Health, the largest granting agency in the US government when it comes to doling out the loot in science.   It probably won't come as a surprise to you to learn that, on the whole, money is valued above most everything else in research because it's funding that keeps labs like mine rolling along just like money greases the wheels in other walks of life, so NIH is about as close to a higher power as you can get in science, the ones charged with handing out the money and keeping the research establishment honest, deciding who gets what and who doesn't and on days like this one I often wish I'd gone into my second area of interest – history.   But apart from the financial side of things, disease-causing bacteria like *Bacillus anthracis* are of interest to me because of this amazing ability they all have of enduring some of the harshest conditions that nature – and indeed even some scientists – can throw at them.[i]

And it's only within the last century that anthrax has been considered a rare disease.   In fact, getting anthrax is one of the few things George Washington and Karl Marx probably had in common.[3]   And one of my favorite writers, Ernest Hemingway, got anthrax when a few spores entered a cut on his foot while swimming in a river in France on his honeymoon.

Notoriety isn't anything new for this passive-looking bacterium as it had already secured its place in the "Biological Hall of Fame" long before 9/11 came along.   Anthrax's rod-shaped cells were the very first microbes ever proven beyond a shadow of a doubt to cause a disease,[4] all by a country doctor in Germany (the MD kind) by the name of Robert Koch way back in the 1870's.   There were no Petri dishes in Koch's lab in those days so he had to improvise by nurturing his anthrax cells deep within the sterile confines of an ox's eyeball.   And as for an incubator, the good doctor made due with ordinary sand he heated with a kerosene lantern.[ii] [iii]

---

[3] Anthrax was in the New World before Columbus arrived.   Washington got anthrax from his horse, while Dr. Marx (the Ph.D. kind) got it from some horsehair furniture.   Well into the 1940's, horsehair was used as a binder for plaster in buildings, as workers discovered when they got anthrax in London's subway while remodeling the Kings Cross Station in 1998. More recently the Pentagon building burned for 3 days after 9/11/01 as firefighters tried to reach the fire.   Water from their hoses was being blocked by horsehair used as insulation in the Pentagon.   Jet fuel had penetrated all the way to the wood beneath.   Some of the strongest memories of survivors are of burning horsehair.

[4] *Bacillus anthracis*.   Technically, I should be calling it as an *endospore* since an endospore forms *inside* the cell rather than *around* the cell like a true spore does.

Whenever *anthracis's* natural surroundings become hostile and it starts to run low on vittles – animal flesh in this case – in order to survive it has this inner set of software, "Extreme Biology 2.0" I guess you could call it, a combination of genes that it can switch on (or off, depending on the gene) to transform itself from a run-of-the-mill rod-shaped bacterium into an even smaller, rounder, heat and drought-resistant spore all in the space of about a dozen hours, almost like going to seed, yet this is a single celled microbe we're talking about here.[iv]   This cell's ability to form these impervious little time capsules seems miraculous to me no matter how many times I see it unfold under the microscope or think about it while I'm drifting off to sleep at night, yet what often wakes me up in the middle of the night is the thought that this same talent for survival is also what terrorists find so interesting. For me, though, the payoff for working with these clever little bugs is the opportunity to understand life better at one of the more extreme ends of the spectrum.[v]

Bacteria like anthrax are survivalists that would put the human kind out here in Montana to shame and I'd be willing to bet there are few, if any, chemicals underneath your kitchen sink, including all the detergents and disinfectants, that could affect an anthrax spore in any observable way. While it's true boiling them in water can kill most of their spores, it takes about an hour to accomplish the task.   You can torture them with strong acids or expose them to harsh ultraviolet light to your heart's content…all to no avail. They'll simply shuck you off as if you were a mild cold.[vi]   But put these same intractable little beings inside a Petri dish with just the right amount of water and nutrients, stand back and watch, because their spores will come right back to life before your very eyes.

And I have no doubt that, out in the middle of some rancher's field not far from where I'm sitting right now, hidden within the topsoil of the sweet grass on the prairie, anthrax spores have lain dormant for 50 or even 100 seasons, just waiting for their chance to spring back, almost like Lazarus but they're not really dead as it's just *anthracis's* way of waiting around for conditions to change, its chance to steal back into the natural order of things after being taken up by the "walking around type of incubator," a grazing animal like a cow, of which my family owned several hundred head when I was growing up not far from here.

Anthrax has the talent to kill quickly, too, as researchers in Zimbabwe discovered when they came across a hippopotamus in the middle of the jungle. It seems anthrax had gotten there ahead and killed the poor critter so quickly it never even had the chance to hit the ground.   They found the hippo's carcass still up on all fours.[5]   And when two of the postal workers in Washington D.C. first began feeling sick during the letter attacks following

---

[5] In a warm, blood & broth-filled Petri dish, *Bacillus anthracis* will divide to form two new cells about every two hours.

9/11, symptoms like a sore throat and difficulty breathing came on so rapidly that anthrax appeared even to a trained medical observer as if it were just a bad case of the flu. One of the postal workers was even sent home from the emergency room with some Tylenol. Needless to say, he was dead by nine the next morning.

It's because anthrax employs what we sometimes refer to as the "smash and grab" strategy. Unlike some bacteria, anthrax doesn't rely on getting spread around by a semi-healthy, walking around patient who's coughing and sneezing all over the place. Nope, anthrax plays hardball from the get go, dispatching its victim as quickly as it can, using this strong cocktail of toxins it brews up just for the occasion, then devours what's left of the body by making use of its digestive enzymes, in the end forming hardy little spores again, then waiting down in the soil for a rainstorm or some wind to bring it back up to the surface so the whole ancient process can begin all over again. And unlike most people these days, waiting is something *anthracis* does surprisingly well.[vii]

It helps to be tiny, too, even by microbial standards, when it comes to getting spread around. You'd have to line up about 50 anthrax spores just to span the thickness of a single piece of paper; in fact they're so small they can penetrate right through the walls of an ordinary envelope, which is why it cost the EPA 30 million dollars to get rid of anthrax from the US Post Office's Brentwood sorting facility even though the total amount released was probably less than would fit on the tip of your little finger.[viii]

A good sign of a microbe's success in nature is when its lineage goes back a long way. The ancient Greek doctor Hippocrates described what *anthracis* could do to grazing animals, and it's likely this bacterium is what caused the 6th and 7th plagues in Egypt's cattle as described by the Old Testament.[6] And it was the philosopher Aristotle who recognized over 2000 years ago that anthrax came and went season after season in the same exact fields... but for reasons he could only guess at. Not knowing anything about microbes or spores (a way to see them wouldn't be invented for another 2000 years) Aristotle passed the whole thing off as being caused by rodents and lizards nipping at the cow's feet.

When I was growing up on my family's ranch – the Triple K – I used to listen to stories my grandfather would tell around the campfire about how, back in the days of the great cattle drives down in Texas, back when the cowboys would bring their herds to the nearest railroad depot for shipment back east, how anthrax was so common back then that, even to this very day the spores that got carried along with the cattle back in the 1800's still reemerge from time to time... infections reappearing as if echoes from a

---

[6] Hippocrates is considered the first true *physician* because he didn't rely on supernatural forces to explain disease.

distant past. Fortunately, we were far enough north that we never saw anthrax to that extent, and also fortunately for us, in order to make anthrax dangerous on a truly terrifying scale, their spores probably need to be treated with a powder so they can get dispersed into the air, or "weaponized" as we say, and that isn't easy either because being spread through the air is not part of anthrax's normal lifecycle.[ix]

Even the most experienced infectious disease experts in the world, some of which I had the opportunity to work with back when I was a graduate student, are at a loss to explain all the peculiarities of *anthracis*, like why it is cats and dogs will get only a mild case of the disease, perhaps resembling a cold (if they get any symptoms at all), while these same spores can be so devastating if they find their way inside the gut of a cow or a horse, all of which are, of course, mammals. No one knows for sure even to this day.[7] In fact, one of the surprising things about bacteria is how boring they all tend to look under the microscope, even to a trained biologist like myself. Anthrax cells resemble little, rectangular boxcars all lined up bumper-to-bumper in a train yard but with nowhere to go and yet what these tiny microbes can do with their metabolisms is truly remarkable. A saying among microbiologists is that bacteria are interesting not so much because of what they look like, but because of what they can do. Some bacteria have metabolisms that will even allow them to dine on oil spills and monuments made out of the hardest stone.[x]

But I don't want to waste all your time with this small talk, which is why I need to get back to the main problem that was on my mind that evening – how it was I came to lose out on the anthrax grant in the first place. I could shade the truth some by telling you that I was only interested in anthrax for patriotic reasons after 9/11, but the fact is, that was only part of it. You see, I saw anthrax as my best chance to gain tenure here at the university. And an assistant professor in a tenure-track position without achieving that brass ring after 4 years is…well you're pretty much out of a job is what it is.[8] Contracts like mine are only temporary, that is until we either gain our tenure, or we don't and have to go someplace else.

Unfortunately for me, the granting agency didn't seem to care about this small detail of mine. They seemed more concerned with the fact that my institution didn't have a BSL-3, or bio-safety level 3 lab, a highly secure area set aside for working with pathogens like anthrax or West Nile virus, microbes that can be spread through the air and so used as a weapon. Of course I realized the importance of containment and addressed the issue in my grant

---

[7] The largest number of people ever killed at one time by anthrax happened in Sverdlousk, Russia in 1979 when a worker removed a clogged air filter without telling anyone. As a result, 68 died when anthrax spores were released from the bioweapons facility the following day.
[8] In fact, I was officially classified as *contingent faculty*.

proposal as a footnote, which may have been a mistake looking back. The smallpox virus has been eradicated from nature since 1980 yet the last person to die from the disease was when a few of the virus particles got loose and floated out the window of a laboratory at a medical school in England, killing a photographer who was also working in the building.

BSL-3 labs are designed to contain germs like anthrax, preventing their escape with ingenious yet often surprisingly simple techniques like unidirectional airflow and workers having to walk through two sets of doors just to get in and out of the lab and things like that. I was aware of the containment problem and had planned to collaborate with a colleague at a nearby facility who happens to have a BSL-3 lab, but unfortunately the panel that reviewed my grant must have seen things differently. Maybe they thought 45 miles was too far away, I never asked. I have to laugh at the thought of 45 miles being a long ways away for government types back in Washington D.C. because even a couple hundred miles out here in Montana is considered "right next door" and if NIH had only known my high school football team used to travel twice that far on Friday nights just to play a single game, well maybe they would have reconsidered I'll never know.

But even two hundred years ago when the explorers Lewis & Clark passed this way they had backup plans for there was a bit of good news that came my way that day too. You see, the second grant I had applied for – my back up grant so to speak – was approved by NIH and due to this bit of good news I was able to keep my entire staff of two technicians and two students on, at least for the time being.[9]

Yet I still couldn't help thinking that maybe the real fly in the ointment that evening was that I wasn't all that fired-up about working with this second project, for this newer one involved the brewer's yeast and about the only thing I knew about brewer's yeast was that they could make beer and wine out of sugar and bread dough rise when you add water to flour and that's about it. Brewer's yeast aren't anthrax by any stretch of the imagination. But as my dear old grandmother still likes to remind my older brothers and me from time to time: "No person should ever turn down milk just because they can't get any cream."

I should also mention that I had picked up on a few interesting details about the brewer's yeast while researching the grant proposal, things I hadn't come across in any undergraduate biology class, like how alcoholic fermentation was the very first chemical reaction ever written on a piece of paper back in the late 1700's by a French chemist named Lavoisier, or that the yeast is both a microbe and a fungus, meaning it's a lot more closely

---

[9] Lewis & Clark tried to have a backup plan in case the Native Americans turned hostile by camping on river islands whenever possible and anchoring their keelboat backwards to beat a hasty retreat upriver if & when the time came.

related to you and me than it is to any bacterial cell it might associate with down in the soil. In fact the unicellular yeast presents some of the same arrangement of molecules to the outside world that a bumblebee does on its surface. But in all honesty, I just wasn't all that interested. In fact, applying for the yeast grant was really just a way of keeping money coming into the lab if something should happen to the all-important anthrax grant, which, like I said, it did.

There were some other things on my mind too that day, like how missing out on all that anthrax money would affect my teaching load,[xi] and how it's possible I might even lose some of my bench space to another scientist who brought in more money than me this time around, but then I realized I'd probably manage somehow. We Ph.D.'s tend to be a resourceful lot and what a lot of us won't tell you is that we all have our bread and butter projects in the lab, the tried and true ones we were brought up on because we learned them from our mentors back in graduate school bringing in money and publications almost guaranteed.[xii]

And then on the side, in some out of the way corner, there may be a pet project or two on the back burner, the long shots we secretly hope will win us a Nobel Prize someday and these second projects are the ones we hardly ever talk about yet may have a student or two tinkering away on in the hopes of gaining enough preliminary data for a grant application in the future, but I need to get back to the brewer's yeast since the yeast is what precipitated my most recent adventure.

In spite of the fact that both are microbes, switching from a bacterium like *Bacillus anthracis* to a fungus like the yeast isn't all that simple. It's not true that all microbes are alike, in fact there's a surprisingly wide range of them in your mouth and under your feet and some that live in ponds and are so large they can even be seen with the naked eye. And if you could somehow gather up and weigh all of the living organisms here on Earth, the ones you can see with just your eyes, like trees and dogs, people and elephants, we would all actually get outweighed by everything you'd need a microscope to observe, microbes are that numerous here on Earth.

Well in nature the yeast – whose real name is *Saccharomyces cervisidae* by the way – is a loner, kind of like an amoeba or a paramecium, yet it's still a fungus, in fact it's more like a mushroom in the way it reproduces. The brewer's yeast even has a compartment like our own cells do called a nucleus to gather up its DNA inside of.[10] I like to think of the yeast as the "buzzards of the soil" because like other fungi they eek out a living waiting around for things to die, things like berries falling off of a shrub or leaves off a tree, and then they consume them. But instead of claws & beaks, fungi come

---

[10] Bacteria are classified as *prokaryotes* because they don't have a nucleus.

equipped with powerful enzymes, useful for digesting their food from the outside (unlike we humans who, of course, digest our food on the inside), and which is also why you might hear yeast and other fungi being referred to as "nature's recyclers" and while both are microbes and good at what they do, the similarities between bacteria and yeast don't extend far beyond that and no one's ever died of a brewer's yeast infection as far as I know.

Anthrax, on the other hand, is glamorous and lethal and scientists are people too so given the choice...well you're probably beginning to see my predicament already. Not only wasn't I very interested in the yeast, but my staff was even less interested, including the graduate student, and graduate students tend to get worked up about most anything involving science. And it's hard to keep the best and brightest at a university working for you when all you have to offer them is something that makes alcohol out of grape juice or bubbles rise in champagne when what they were really expecting was a BSL-3 level pathogen complete with plane tickets to seminars in far away places to give poster presentations about their work to really important people.

So I spent the rest of what would turn out to be a rather eventful night wandering the halls of my department, alone down in the "dungeon" as we sometimes call our basement on account of how the only two places natural sunlight can penetrate is through two small windows at the stairwells on either end. I should also mention that we have these painfully long winters up here at this latitude and not being able to see the sun a lot doesn't help morale either.

Well I spent the better part of that evening feeling lower than a snake's belly in a wagon rut, trying to figure out the upside to all of this, and it seems that part of being a good scientist I've learned over the years is being a good salesman so I knew that if I could somehow interest my own group in the idea that working with the brewer's yeast is important in its own right, then I might just have a chance of keeping them on. But how could I if even I didn't believe it myself? By midnight I was no closer to a solution and since everyone had already gone home for the evening, I was all alone. I should also mention that not many microbiologists wear cowboy boots so that night I made so much clomping around noises that the janitor came down to see what all the commotion was about and was visibly relieved to see it was just me.

So after reassuring Charlie and more out of boredom than anything, I found myself reaching for a packet of brewer's yeast – the kind you can find in any supermarket, the freeze-dried vacuum-packed variety[xiii] – and I added about a teaspoonful of water and a pinch of sugar, just like the directions tell you to do, to a few of the crumbs so I could have a look at them under the microscope, hoping for some kind of inspiration I suppose. The name *Saccharomyces* literally means "sugar fungus" in Latin and in fact they do love their sugar – wherever they can find it. Almost a third the weight of a freeze-dried yeast cell is going to be sugar, stored mostly in the form of glycogen, which are long chains of glucose molecules strung together for convenience. They keep this sugar in reserve for the chemical energy it contains, not so different than the way I keep gasoline in my pickup truck for its chemical energy, except that the yeast doesn't move around under its own power or need oxygen to burn its fuel like we do.

I always like to start off with the clunky microscope, the heavy steel "boxy" one with the two wide eyeholes on the top so I don't have to do so much squinting when looking through it and, like I said, yeast are larger than a bacterial cell is so it doesn't take that much squinting just to catch a glimpse of them.[11]

It wasn't easy being alone in the lab that evening thinking about my missed opportunity with the anthrax project and I have to admit that few things are as uninteresting as a microbe that just sits there when you add water to it – rather ungrateful when you think about it – and just sit there is pretty much what the brewer's yeast cell does. Lying in wait is part of its strategy, whether the yeast finds itself in the soil or on the skin of a ripening grape or the petals of a nectar-secreting flower attempting to bribe a passing insect into providing a free ride for its pollen onto the next flower. [xiv] Still, knowing all this didn't get me excited one iota.

And after a few minutes, every last one of the yeast had settled on their own accord down to the bottom of the droplet in front of me on that glass slide, which is kind of amazing in its own right because water to a microbe would be like us falling into a vat of maple syrup, water's that viscous to microscopic things.[xv] There's no worrying about brewer's yeast swimming off on you, though, no need to constantly refocus the microscope just to keep track of them because believe me...they aren't going anywhere. You won't see any twitching or tumbling around, no gliding from place to place like an *E. coli* does inside your intestines whenever it whips itself around with its flagella to snag what's left of your breakfast, no crops of cilia along its outside like a *Strombidium* has to twirl itself around until it gets where it's got to go.[12] Nope, they won't even ooze about like an amoeba does on a rock. Instead these little round specks just sat there that night imbibing sugar, as if mocking me, as immobile on the bottom of that glass slide as my own career had suddenly become on land.[xvi]

The very first person to ever see bacteria lived in Holland and his name was Antony van Leeuwenhoek. Back in the 17th century Leeuwenhoek was

---

[11] Indeed, the average microscope throughout the late 1800's was stronger than the one I was using that night – but mine is more comfortable.

[12] Some *Vibrio* bacteria can travel 100-times their own body length in under a second using a rotating flagella for a propeller (it would be like us trying to run 300 miles an hour on land).

also the first to catch sight of a yeast cell, too, and it turns out that even the "Janitor from Delft" wasn't all that impressed. In fact, he didn't even think yeast were alive. Like me, Leeuwenhoek was partial to his bacteria, once referring to those he examined from his mouth as "little beasties" because they were so swift while twirling about "like a top" and "prettily a-moving" as he put it. Yep, that Dutchman sure had a way with words.

And even though it's relatively gangly by microbial standards (50 yeast cells could be lined up end-to-end just to span the diameter of the dot above the letter i) I kept thinking to myself how far I was from where I'd expected to be when I first left graduate school just a few short years ago. Here I had been trained to work with bacteria, and not just any bacteria but the kind that cause devastating diseases. And pathogens are infinitely more interesting than a yeast cell is to someone with my background. Yeast don't secrete toxins of any sort, certainly not the kind some bacteria can use to crack open a red blood cell to steal its iron, or make someone cough up phlegm from their lungs. Nope, being a non-pathogen like the yeast means having none of these strategies at your disposal. Well the fact was, I just couldn't get excited about the yeast in the least and I had to...soon...because fall semester was getting ready to start and I was in danger of losing my entire staff.

THE E-TICKET RIDE

It's a funny thing how eyestrain can set in a lot sooner after you've had a rough day, and so after a few minutes of squinting I gave up on the yeast and logged onto the Internet in the back room of the lab, the small place we have set aside for two computers, a large incubator, and a sterile hood for streaking microbes onto the surfaces of agar dishes. Unfortunately, the computer didn't do much for my eyestrain other than make it worse, but the consolation was that I now had the whole world at my fingertips – kind of like a shot of adrenaline when you live this far out in the middle of nowhere – so I clicked on it more out of boredom than anything, first to my e-mail, then I found myself at that giant encyclopedia in the sky, the one with just about everything in it if you care to look, and with two fingers typed in the words "brewer's yeast".

All I wanted was to put my mind at ease so I could get some sleep. Well at first nothing much happened, no catchy photos of any kind, so I followed a couple links that lead nowhere in particular, then gave up and did an image search. It was then that I came across something able to hold my attention and it was unmistakable, in fact it didn't even need a caption underneath to explain it. I had found the most amazing photograph of the pyramids I have ever seen...all three of them, too...looming high above the desert floor with an orange, puffy sun setting in the background. When I scrolled over it, the caption read simply, "Brewer's yeast floating in the Nile gave the world its first leavened breads around 2600 BC."

Well like I said, history was going to be my second choice of study, so I lingered there for a spell just to see if I could figure out what the connection was between the yeast and the pyramids until it dawned on me I had stumbled onto a website devoted solely to the history of civilization. Still, I couldn't help wondering: what did the dawn of civilization have to do with a lowly fungus like the yeast? I got about a paragraph into it when this plane trailing a banner flew right over the top of the pyramids and so I clicked on it.

Now I consider myself a person of science even though I still hold out hope of a higher power having a hand in things here on Earth, and as it turns out this plane belonged to some travel agency way off in New York City. And my vacation was fast approaching yet I was in no mood to spend it like I had the previous three summers – working down in the dungeon again – all the more reason to hit the back button on the browser and resume reading, this time looking for an excuse to go someplace interesting.

The website explained how the ancient Egyptians were the first people to turn the leavening of bread dough into an industry. In fact, Moses and his followers when they left Egypt missed their leavened bread so much that one of the first things they did after wandering around in the desert for so long was to start growing wheat so they could bake bread again just like they'd learned to back in Egypt. Apparently this process had been perfected by the Egyptians, who then went on to invent some 40 different kinds of bread.

So the Egyptian knack for raising dough seems to have rubbed off on people from time to time. I'm always teaching students new techniques in the lab myself, whether it's how to purify bacterial toxins using an affinity chromatography column or something as simple as the proper way to store glass beakers on a shelf without collecting any dust (upside down, if you care). This was all interesting and yet the failed anthrax grant kept drifting in and out of my mind that evening. The travel agency had mentioned something about a limited time offer, so I clicked on the plane again as it buzzed by a second time, this time a little more wide-awake and determined to flesh out the details.

For most folks there comes a time during their formative years when they run across a book they can still remember reading. For me it was *The Innocents Abroad* by Mark Twain.[13] I'll just mention how this thin little book is basically a compilation of stories written by Twain back when he was traveling through Europe and the Holy Land and that, even though it's a century and a half old, still manages to transcend time in some interesting ways. I had missed out on my chance to go to Europe and see all the great ruins while

---

[13] I would learn later that Twain was no stranger to bacterial diseases, having lost two children to them, one to meningitis, the other diphtheria.

everyone else my age was going mostly because I was too busy volunteering in a lab to get some experience I could put on an application for graduate school (science is competitive).   The farthest I'd ever ventured was up to Canada, and that was only one time to cheer on one of my brothers in a rodeo.   Come to think of it sitting in my chair that night, I realized I had never gone off anywhere on my own that was very far away in my whole life except to graduate school.   How nice it would be to have the stories to tell once in a while, I imagined, just like my grandfather used to do for all of us.

Mark Twain arrived in Egypt back in 1867 when he was about the same age I was now, on a Civil War era steamship called the *Quaker City*, among the very first to take part in such a journey back to the Old World – the very first American tour group you could say – and Twain wrote about it for a newspaper back in San Francisco.[14]

I scanned the fine print on the screen, hungry for more details.   The ad claimed the ticket was just $550.   A measly few hundred bucks for a round-trip flight all the way to Egypt and back?   Just as sure as a goose goes barefoot, there had to be a catch.   I could feel my heart make an unscheduled appearance in my throat as I started to consider the possibilities.   But what were the add-ons, I kept wondering?   There are always add-ons these days...a hidden fee of some sort.   Nope, none that I could detect.   It even promised tax was included.

I thought about the travel guides scattered among the microbiology textbooks gathering dust on my shelf across the hall in my office.   One was a guidebook for Nepal, a place I never expected to see.   I just liked to take it down and look at the glossy photos once in a while. I rubbed my eyes again, unable to blame my worsening eyestrain on the yeast, for they had only managed to resurrect what I'd already gotten from grading so many final exams the previous week.   It turns out it's practically impossible to design a biochemistry exam in multiple-choice format, which is why they always end up looking more like essay tests, and essays by definition require close-up scrutiny to grade.   College students like partial credit, as you probably knew already.

This was getting sillier by the minute, I thought.   What was a middle-aged professor whose chances for tenure had just been diminished by a couple orders of magnitude going to do wandering around Egypt, I kept asking myself?   But human nature being what it is I just ignored all of this and kept on going, entering some dates off the top of my head, fully expecting the airfare not to be there anymore.   Then I clicked on the "Begin Search" button

---

[14] The Alta California, which also sponsored his trip.   Specifically tailored tours of Europe were not new, however.   Thomas Jefferson, for example, took an English garden tour in 1786, which would have an influence on the design of his own gardens back at Monticello in Virginia.

and within a few seconds the screen changed to black, everything except for that little hourglass thing, and so I waited...and waited...and then waited some more until finally a computer-generated image of a plane ticket materialized right in front of me. It even had its own bar code, and most amazingly of all, the fare was still $550. All that was left for me to finalize the agreement, it promised, was to enter my name and a valid credit card number along with that 3-digit code on the back.

My fork in the road that night had materialized in front of me in the form of an airplane with a cartoon face, one that was now standing on its tail, winking, and using its wings like hands to prompt me for more details. Window seat or aisle? The most I'd ever bought online before was a country & western CD so I chose the window seat, took a deep breath, wiped the accumulating sweat on my palms off onto my jeans, and then clicked the accept window seat icon. The screen went blank again, as if commiserating with some higher power for advice on the matter, and then after what seemed an eternity the plane flew back across the screen trailing a banner that read, "Congratulations, Dr. Ketchum. You have just purchased a round-trip nonrefundable ticket to Cairo, Egypt. Please print a copy of this for your records. Would you like a rental car when you arrive?"

I wish I could say I had decided at that moment to go, but the truth is, I didn't know for sure until somewhere between school and my apartment on the long walk home that evening when it occurred to me that in less than a week's time a jet bound for Salt Lake City would be taking off from Missoula, Montana with or without me on board, then a second seat assigned to yours truly and already paid for would arrive in Frankfurt, Germany and finally a third would venture all the way to Cairo, with someone else sitting in it if I backed out, that I decided to go. And I decided it all without having the foggiest idea of just how it would help me with the new project. The final nudge came with the realization that a two-week trip might be just the thing to get me caught up on some pleasure reading, but without the eyestrain that always seems to accompany grading exams.

While shuffling through the leaves alongside the road that evening I had plenty of time to think about just where it was I found myself at this point in my life, which wasn't good by hardly anyone's measure. Feeling the cool dampness of the leaves as they worked their way up above the tops of my boots, my mind began wandering again, recalling how fascinating Egypt had always been to a high school student in Thistle, perhaps because it lasted so much longer than any other civilization. We Americans tend to be fascinated by things that are old, and Egypt is...well...pretty darn old by anyone's measure.

The ranch I grew up on was less than an hour's drive from campus, yet it may as well have been a world away these past couple months. Somehow I had slipped into a place I never knew existed, an unfamiliar place I hadn't grown up in or knew hardly anything about, a place I felt more and more an outsider in every day. Not many ranchers' sons spend all their free time thinking about tenure. At the very least Egypt offered a chance to find a new perspective, like standing on higher ground to survey the surrounding terrain better maybe. Egypt was the ancient world's granary. Thistle, the town I grew up near and home to my small university, is the granary for the entire Lewis Valley. But unlike here where the valley is walled off by mountains on all four sides, Egypt was protected by deserts. And if it wasn't sand and lack of drinking water blocking the way for invaders back then, it was the Mediterranean Sea to the north.[15]

Yet when Jesus walked the earth some 2000 years ago, Egypt was already ancient, even more so to Christ's time than his time would be for us today. The great pyramids had already been standing for 20 centuries before the Roman Empire began running things, and to Greek travelers like Antipatros, the mystery of how the pyramids were built had already been lost to time, and Antipatros was no slouch either, in fact he walked the Earth in the year 200 BC.

After opening the door to my small apartment I was greeted by about a feedbag's worth of affection launched into my midsection, his two front legs draped over my shoulders, the momentum of his body sandwiching me up against the wall while a sandpaper tongue was scouring my face. Barney is a white boarder collie I'd raised as a pup back on the ranch. The black patch over his eye made him look like a pirate, and he was definitely getting older, which is why I sometimes called him a senior pirate, and he had less energy than he used to and so was more likely to be found taking long naps by the furnace grate on the floor in the kitchen when I arrived home than looking out the window for me, but mostly I never noticed. Dogs, not horses, were the very first animals ever domesticated by humans and I've never doubted as to why. They've been by our side since before we took to agriculture even, there to protect us and so we take care of them in return. After walking Barney once around the block we both gladly welcomed our sleep.

BIOPROSPECTS

---

[15] Egypt's relative isolation also explains why it didn't need a standing army to defend itself during the Old Kingdom period, which lasted about 500 years, and why hieroglyphics would someday die out as a writing system. For nearly 2,000 years not a single person alive could decipher them, not until after the Rosetta stone was discovered by Napoleon's traveling army of scientists in the late 18th century.

The next morning, after discussing it with some of my colleagues, a consensus had emerged down in the dungeon that the reason I had gotten such a good deal the previous night was that I was willing to fly last minute. Apparently I had stumbled onto the fact that some airlines would rather sell tickets at greatly reduced prices online anonymously, than to publish lower prices, which could trigger a fare war among the airlines.

The next several days went by slower than a hay wagon hitched to a hog as the whole thing had taken on an air of incredulity to it, as if it were happening to somebody else. Of course my family was surprised and it helped bring the reality home some when I phoned them with the details. Even so, I still had to take the ticket out once in a while just to see my name on it. Thoughts of the anthrax grant were fading from my mind, replaced by things I hadn't considered since I was a kid, back before I knew what a grant application looked like, and since it's possible to get a visa at the airport in Cairo upon arrival, I had only to figure out a way to keep my lab up and running, as well as decide between the two backpacks in the hardware store window, either the one with the strap across the waist for added security, or the burlap one with no strap but was at least lighter in color and so better equipped to reflect the Saharan sun. In the end, I decided on the more secure pack, which probably says more about me than anything.

My last day before leaving the butterflies were still coming and going in spite of the fact that, twice a week another professor had promised to check on things like the temperature of our incubators and the levels of liquid nitrogen in the cryogenic tissue culture jugs, requisitions for the new yeast project would be placed in their appropriate wire baskets in the office upstairs by my two technicians, plus I knew I had some of the best in the department working for me, so what could go wrong? As someone wise once said, you know you have a good horse underneath you when you can forget all about the saddle. Or maybe it was that a good horse is seldom spurred. I'm not sure which applies best here so you can go ahead and take your pick.

As lunchtime rolled around I found myself brushing up on some ancient history, ham sandwich in one hand, guidebook in the other, eager to take down some notes, looking for any clue that might get me started in the right direction with the project. I had forgotten how the pyramids were built by some 20,000 farmers fed a steady diet of bread and beer delivered to them on site, in fact the pharaoh even had an entire city raised near the pyramid just for providing his workers with these necessary staples. I'd also forgotten how wine containers were found alongside King Tut's mummy to quench his thirst after death, complete with information describing the wine's vintage and place of origin.[16]   How strange to think that the same microorganism that once nourished a nation and helped raise their pyramids was what my own future could very well depend on now.[xvii]

I laid the book down on my lap and considered how little time I had left now.   I was no kid anymore; in fact I wasn't even the youngest assistant professor in the department.   Thomas Jefferson authored the Declaration of Independence when he was just 33 and Napoleon was all of 29 when he led an entire army off to invade Egypt.   And the Frenchman who would someday decipher the Rosetta stone, Champollion, already knew Greek and Latin and was busy studying Coptic, Hebrew, and Arabic...all by the age of 12.   Tenure committees like mine are made up of other scientists and in a year's time these eight would either recommend to the provost that I be made an associate professor and given all the rights and responsibilities associated with that honor, including my tenure, or that I should be let go.   There was, in all practicality, no in-between.[17]

My last afternoon before leaving was spent moseying around the lab, trying to put things back in order, looking for anything I might have missed, wondering if Twain had felt this nervous before he left for Egypt, but then I recalled how Twain had already been to Hawaii and back by then (folks called them the Sandwich Islands in those days).

A quick glance through any microbiologist's freezer will turn up all sorts of curious artifacts: projects that had once been at the forefront of one's thoughts but were now relegated to plastic cryo-vials not much bigger than a pinky finger, the kind with screw-top lids and rubber gaskets for keeping out the air. Or in the refrigerator there might be other forgotten about microbes growing on Petri dishes, their once-separate colonies now having merged, becoming one continuous lawn of microbes, no longer content to live as separate little spots; other dishes having dried out in spite of the wax paper stretched around their outsides in a vain attempt to keep the air out and the jelly from losing its water and then shrinking, forming a network of little cracks, like those you see at the bottom of a dried up mud puddle.   Another shelf was filled   with   glass   test   tubes   waiting   to   be   inoculated...or contaminated...whichever came first, perhaps by some fungal spore from a distant mushroom taking advantage of the day's local air currents.

We microbiologists are all collectors at heart, not so different than the way an entomologist collects bugs and then displays them on pins with their names written in impossibly tiny script on small paper flags, or the way a coin

---

[16] Some of the bread in King Tut's tomb still had fruit in it...tangible evidence that pharaohs craved variety in the afterlife. / Ancient Egyptian doctors sometimes used moldy bread as a poultice to prevent wound infection.

[17] Sacajawea was just 16 when she helped guide the Lewis & Clark Expedition to Montana. Sitting Bull – Custer's rival at the Little Big Horn – would count his first coup in battle at 14. Custer was promoted to brigadier general at age 23 during the Civil War.

collector is always on the lookout for rare specimens in his or her change at the supermarket, we microbiologists are never that far away from our next find either.

It might be something from right under our feet, about a hand's depth in the soil, or floating around in the air above our heads, or even lingering on the outside of our bodies, or inside the gut of a cow; maybe even on the surface of a fern. Microbes are everywhere you turn in nature and insulated kettles of liquid nitrogen are the mantles above the fireplace we display our trophies in; perhaps a species of bacteria we worked with several years ago on a project that never went anywhere, the microbes themselves now lying in suspended animation, as if waiting for the next idea to hit us before being thawed back to life again.

As for me, having cut my teeth bio-prospecting in the thermal features of Yellowstone National Park, it's only natural that my first inclination was to learn all I could about hyperthermophiles, the kind of single-celled life forms with names like *Aquifex* and *Thermotoga,* inhabiting waters so hot you wouldn't dare stick your hands into,[18] bacteria once attached to pebbles on the bottom of a hot spring, now stored silently away in a plastic vial and labeled with a thin black sharpie complete with, not only the name and date they were found, but also the substratum they were collected on, in geologic formations like "Whirligig" or "Steamboat Geyser" or "Grand Prismatic Spring". This part of the American West is different from the rest of North America for a lot of reasons, not least of which is that the crust is thinner here and so the liquid magma from down below has a chance to rise up closer to the surface than it normally would, nearer to where the rainwater and snowmelt from the mountains can get at it. This water then becomes super-heated underground, the Earth's thermal energy making it expand and begin its rise back up to the surface. It's all part of Mother Nature's perfect, time-honored recipe for a geyser.

On another shelf in a plain white freezer was a bacterium I'd come across while working in an abandoned sub-surface mine near Butte, Montana, as inhospitable a place if there ever was one for a living creature, with water pools 1000 times more acidic than what's inside a car's battery, strong enough to dissolve metal tools left inside them overnight by some overworked graduate student.[19] Yet the microbes I was holding in my gloved hand would have preferred this harsh acidity to the balmiest beach in Hawaii. Extremophiles are a reminder of how even the scientific literature can be

---

[18] Near-boiling at over 80° C (176° F).
[19] Most bacteria are killed by acid and some historians have speculated that the reason Lewis & Clark's men suffered from dysentery for much of their journey was that they didn't add vinegar, a weak acid, to their drinking water even though it was a common practice in the early 1800's. Why they didn't is still a mystery.

biased from time to time because to a microbe like this one, it is *we humans* who inhabit the extreme environments. The oxygen we take into our lungs whenever we inhale and that bathes our skin continuously is a toxin to so many different microscopic bacteria, most of which we'll never know about.

A feeling of regret washed over me upon recalling that bio-prospecting would be out of the question on this trip. Things had gotten too difficult in the wake of 9/11. Prior to this it wasn't unusual for microbiologists to carry our finds around, distribute them at meetings in ordinary envelopes stashed away in our top pockets, freeze-dried and harmless like so many grains of coffee lying there lifelessly. I once sat through an entire talk by the biologist Lynn Margolis – in fact I can even remember her topic, which was on the evolution of mitochondria from bacteria – with an envelope filled with bacteria collected from rustsicles on the hull of the Titanic. This particular microbe had kept its appetite for iron over millions of years...and the person I'd gotten them from wanted me to help his company find better ways to remove dissolved iron from drinking water, so he had the microbes lyophilized (freeze-dried) into a coarse dry powder which was by then resting harmlessly in my top pocket. No one was ever the wiser.[xviii]

It's not hard to revive microbes usually. Once back in the lab just a small amount of sterile water and the right choice of carbon for an energy source, usually glucose, some ammonia salts so it can manufacture amino acids, that's about all it can take, that and a modicum of patience. Microbes make a living as opportunists,[xix] not so different than scientists do these days come to think of it. We have to be with funding the way it is. Today, the red tape alone would be enough to stop all but the most foolhardy from bringing back a microbe in their top pocket from overseas, and with my tenure on the line I sure wasn't going to be accused of carrying anything suspicious back from the Middle East. I had enough to worry about as it was.

It just dawned on me that I never did explain how I came by the yeast project in the first place, the reason I was awarded the grant and was now on my way to Egypt. Why would NIH give an assistant professor like me – and a tenureless one at that – $250,000 to study yeast if it wasn't to make a better jug of wine or loaf of bread?

Well the answer is simple enough. The brewer's yeast – like the bacterium *E. coli* that lives inside our gut – has become one of the workhorses in modern microbiology labs over the years, useful for making all sorts of human proteins inside of.[20] They're miniature fermentation vessels, each yeast cell its own test tube, delineated not by glass but by a fatty cell membrane, tiny cells that double as fermentation factories for the assembly of

---

[20] I knew yeast were being used to make non-yeast proteins because of a paper I'd come across by some structural biologists growing anthrax toxin (which normally functions as a set of 3 different bacterial proteins) inside yeast cells.

valuable medicinal proteins. Microbes like the yeast can be coaxed into making all sorts of human proteins like insulin, interferon, and antibodies for example, which is a lot cheaper and easier than trying to acquire these same proteins from human tissue. A whole heck of a lot easier, believe me.[21] [xx]

The downside is that, because these proteins have to be purified away from the bacteria or yeast, often times pesky chemicals can slough off these microbes and co-purify along with the human proteins you're after. And if these stray contaminants get injected into a patient then all hell can break loose. They will get recognized by special cells we all have whose job it is to constantly go around looking for things that don't belong in the body and then set off inflammatory reactions when they find them, complications that can be serious, even deadly on occasion.[xxi]

My grant involved making a trial human protein of my choice, then genetically modifying the yeast so it wouldn't produce so many of these stray yeast molecules in the first place. Pharmaceutical companies spend millions to purify human proteins away from bacteria and yeast to guarantee they are free of contamination and if we could genetically engineer a brewer's yeast cell that made even less of its own molecules to begin with, well this would be of benefit. A tall order, humanizing the yeast, but the amazing thing about the brewer's yeast, as I would soon come to discover, is that it has been so amenable to change over the years.

It's one of the few microbes in the lab where you can actually add a stretch of human DNA, say the gene for insulin, and the microbe will take up this gene and incorporate it into its own chromosome. Chemically, DNA is DNA throughout all the kingdom of life, whether it's DNA from a bacteria that causes anthrax or an amoeba, a bird or a dinosaur. It's the *sequence of the bases* that makes DNA different between different organisms, and in fact microbes will make the corresponding protein the gene codes for just as if it were making the protein for its own self. And that's about all you have to do to get your foreign DNA inside a yeast cell, just add it to the yeast culture along with a quick jolt of electricity, which is pretty amazing when you think about it.

But this kind of research is more along the lines of filling-in-the-blanks than actually discovering something new and completely different and therefore it's never as exciting. A lot of research is like this, though. Few of us have ever won a Nobel Prize just by filling-in-the-blanks in science. Nobel Prizes are usually given out to those not afraid of going out on a limb once in a while. And so I was following an inner voice to Egypt that, deep down told me if I could just learn something new and interesting about the brewer's yeast, uncover something special, then it might give us an edge with the new

---

[21] Most human insulin used by diabetics today is made inside genetically engineered *E. coli*.

project and light a fire underneath all of us to accomplish something important.

A RIVER IN EGYPT

"Egypt," as the wise man said, "is the gift of the Nile." Or maybe it was "Egypt is the Nile," I couldn't recall exactly as the cabin crew began its preparations for landing. Merely crossing the Atlantic had been the longest trip of my life so naturally I was feeling the excitement that always builds whenever you begin your decent someplace new, plus I had the added knowledge that this trip would, at the very least, achieve a lifelong ambition. As the plane began its trek over northern Africa I couldn't help wondering if I'd been able to make out the triangular shape of the delta down below, assuming the sun hadn't already set. The ancient Greeks were first to call where the Nile empties into the Mediterranean a delta because their letter "delta" was shaped like a triangle except that the triangle below me probably delineated the first place humanity ever used irrigation for crops.

I watched the flight attendants collect all the leftover drink containers as our plane droned on towards the place where perhaps the oldest civilization in human history began, having begun my journey in one of the newest. Up until the Eiffel Tower, Egypt's Great Pyramid was the tallest structure ever built by humans, a feat that was to remain unrivaled for four thousand years. In fact, the pyramids are all that's still standing of the original Seven Wonders of the Ancient World.[22]

Without warning, the plane veered sharply to the right and then dipped as I was beginning my way towards the rear of the plane to take one last look in the mirror. I continued risking my neck because I was to be met by a woman doctor named Tahany Hassan (the MD kind) from the American University in Cairo. And as I'd already seen a photo of her on the Internet I expected she would be attractive. While trying to find my seat again, a new wave of anxiety surged through me, not because of Tahany but for my own safety, as the plane dipped yet again, then a third time, finally leveling out just low enough to make out the long columns of lights marking Cairo's runway.

On the one hand I knew that the dishes rattling away on the shelves were simply due to the randomness of the air molecules the plane's wings were slicing through, yet I still couldn't help looking across the aisle into the flight attendant's eyes to see if she was as concerned as I was. At least I had no baggage to claim, I reassured myself, having stuffed everything I could think

---

[22] There was a brief period of only 60 years when all Seven Wonders of the Ancient World were standing at the same time.

to bring inside my new backpack.   I'm still not sure why I thought the markets in Cairo wouldn't carry toothpaste.[xxii]   With my feet touching the pack, I leaned further back into my seat and reminded myself that I too was a doctor, even if it wasn't the MD kind. That should at least count for something in her eyes.

To get a visa involved a simple cash transaction and a cursory glance at my virgin passport; my first welcome sign coming in the form of an official stamp and a smile.   Feeling a sense of kinship towards my fellow passengers, I followed the ones in front of me through a turnstile, backpack slung over my shoulder, turned sideways to make it through the rotating stall and was suddenly greeted by my first real surprise in Africa...a gust of sticky, warm air.   Even deserts can be humid.   My excitement continued as I pushed through a gauntlet of turbans and beards – locals holding up handwritten signs advertising hotels and tours, some with passenger's names scrawled on them – and for a brief moment the uncomfortable image of myself as a prisoner shuffling through a lineup went through my mind as each one's attention seemed trained on me.

I wiped the sweat from my forehead and used it to dampen down the place where my hair always has a habit of standing straight up from in every photograph I've ever taken.   Tahany was probably in the crowd somewhere and she didn't know what I looked like yet.   I deliberately hadn't gotten around to putting a photograph of myself up on my own department's website. Like I said, I've never taken a picture I've ever liked.

Cairo, like so many ancient cities, has accreted over the centuries, reminding me of the way a snowball does while rolling down a hillside. Newer portions were added on as more recent arrivals saw fit, and when the Moslems began their turn running Egypt in AD 641, they founded the walled city of Al-Fustat, what is today Islamic Cairo.   A little further to the south, Egypt's Coptic Christians – here since the days of St. Mark – had already built their own version of a city with some buildings surviving into the present, as does the overall layout.   Coptic Cairo today picks up where the ancient remains of an even earlier fortress built by the Romans left off, and I had plans to visit it the next day if my jet lag allowed.   But even in the Middle East things change, albeit more slowly compared to a lot of places, and by the mid-1800's an even newer version of Cairo had emerged, one like what you'd expect to see in Europe today, in fact it's why Cairo is sometimes called the "Paris on the Nile".   The solution turned out to be straightforward enough. To make room for the new addition, Ismail the Magnificent simply had some unused swampland near the Nile northwest of the old city drained away, and because of his decision this is where New Cairo and my hotel sit today.

With a population of just over twenty million, Cairo is easily the capital of the Arab world.   In fact, few countries are so dominated by a single city as

Egypt is by its own capital.   And as I was discovering on the way to my hotel that evening, one in every two cars in Egypt is also taking up space on one of Cairo's congested streets somewhere.

I gazed at the exotic script on the billboards advertising cigarettes and Coca Cola, still looking for more tangible evidence of being someplace far away during the drive south from the airport when my driver – a young university student named Mustafa – pointed out the Arab League Building to my left.   I should probably have mentioned that it was one of Tahany's medical students, Mustafa, and not Tahany herself, who was the one holding up the sign with my name on it, and Mustafa who felt the need to apologize for his boss's absence, while I looked down at the ground in an effort to hide any disappointment that might have shown.   He had been given the task of explaining why his mentor had been called away at the last minute.   One of her patients was in the midst of having a baby, yet I still couldn't help wondering if maybe she hadn't found an old photo of me while doing a Google search.   I went through a period where I was a lot heavier than I am now back in graduate school.

During the drive Mustafa explained that, further out in either direction from the Nile, are situated Cairo's suburbs with names like Giza and Heliopolis and Shubra, the last one being where Tahany and her extended family lived. Mustafa, it turns out, was unusual in that he lived alone in a small apartment in Giza, a suburb that stretches almost the entire ten miles from the outskirts of Cairo all the way down to the pyramids to the south.   With fully ninety percent of her population living within just a few miles of her banks on either side, Egypt is still very much a gift of the Nile.

As we neared the pulsating heart of New Cairo we came upon Tahrir Square, which looked more like a circle from my vantage point, and the smell of exhaust actually became palatable through the open window.   My pulse quickened as the carbon monoxide made its presence felt in my veins.   I was getting a taste of what it must be like to be thrust into a time machine of sorts, of being very far away from one's comfort zone, of what an acrobat must feel when he looks down and suddenly recalls that he's been working without a net the whole time.   Adrenaline produced above my kidneys entered my bloodstream, teaming up with the carbon monoxide already coursing through there as we rounded the bend that thrust us into a giant beehive of activity. We found ourselves in the midst of what can only be described as frozen chaos.

All around was a life-sized, 360-degree panorama of faded busses, passengers hanging out of open doorways, motorbikes with extended families squeezed onto gas tanks as if so many condiments of a sub sandwich, a specter of smoke hanging over everything and reflecting the headlights, in the background some Arabic tunes emanating from some place indeterminable.

I'm still not sure how many lanes there were because there were too many vehicles converging from all directions to see any lines, motorbikes impatient with the pace of traffic squeezing between busses and dodging pedestrians taking their chances, the exhaust causing my eyes to water over, me looking out the window and wondering how anyone could go through this ritual every day. A feeling of vulnerability sent me reaching for my backpack to make sure it was still down beside my foot, as if it could have walked off on its own. The fact that Mustafa didn't seem fazed made the scene seem all the more surreal, but after a good ten minutes the gridlock broke apart like so many ice sheets do on the Yellowstone River in springtime and Mustafa located an opening and rushed down a side street leading to my hotel.

The Bedouin had come highly recommended by Tahany in one of our string of e-mails, perhaps because her brother ran it, or owned it, I still wasn't sure which. When she'd mentioned how it had once been used by Egyptologists back in the 19th century, I made the reservation and besides, it was supposed to be a stone's throw from one of Cairo's most important thoroughfares, the Sharia Talaat Harb,[23] a section that experienced a building boom back when the Suez Canal opened in 1869, two years after Twain and his party rode the train here from Alexandria. I was promised "a single shared balcony overlooking a bustling avenue." The room had this, as promised, but otherwise turned out to be rather plain, accompanied by a single wooden table and chairs, a musty odor, a double bed and that was about all I can recall now. The only other concessions to the world I left behind were a ceiling fan, a telephone that never seemed to work when I needed it and a western-style bathroom I wasn't required to share. Egypt, I was learning, has a noticeable lack of hotels in the mid-price range, as the tourist industry has grown up for so many decades around cost-conscious backpackers.[xxiii] These days at the other extreme are the newer five-star extravaganzas with names like Nile Hilton and Sheraton Cairo. Still, having traveled halfway around the world I wasn't going to allow a heated swimming pool and exercise room into my recollections when I'm old and grey sitting around the fireplace with my own grandchildren someday.[xxiv]

After checking in I laid on the bed face-up, the balcony door wide open, staring up at the ceiling fan that was making inroads at replacing the musty smell, soothing my exhaust-stung eyes as they followed the fan blades twirling fast enough to remain a blur, barely able to convince myself I had actually made it…and all in one day. Twain wasn't the first tourist here and of course I wouldn't be the last. Egypt would come to remind me of one of those flowers that manages to emerge from the crack of a busy city sidewalk and bloom against all odds. The entire civilization sprouted out of the world's

---

[23] My guidebook described Cairo's older streets as "living museums".

largest desert – above a narrow patch of land fed and watered annually by the Nile – yet Egypt was also a tourist destination as far back as the ancient Greeks and Romans. In fact, it was the Greeks who first paid homage to Egypt by referring to her as their own "cradle of civilization", a complement coming from a people that considered all other outsiders barbarians. In much the same way someone from my own world might take a trip to Europe after finishing college, the future movers and shakers of ancient Athens and Halicarnassus – young and wealthy men like Plato and Herodotus – paid a visit here to take in some of Egypt's wisdom and bring it back to enrich their own corner of the Mediterranean.[24]

Egypt was where the ancient arts of medicine, astronomy, and mathematics had already been incorporated into daily life long before a Greek traveler named Pythagoras "discovered" the formula for describing a right triangle here. More recently Napoleon's three-year sojourn into Egypt beginning in 1798 – part military and part adventure – would soon reveal Egypt to an astonished Europe. Napoleon took along 167 artists, scientists, and scholars to record not only the major finds but the details along the way, all of which, the terrain and ecology included, were dutifully depicted.[25] However, the spike in tourism Egypt has most recently experienced can be traced to a single event that took place on November 16, 1922, which is when King Tut's tomb was discovered by an artist with no formal education in archeology, a painter by the name of Howard Carter, and he and the boy king would generate headlines around the world for months. [xxv]

I laid the guidebook back down on the bed beside me and looked at my watch. It was almost midnight and only now were the first signs of fatigue invading my frame. I took another sip of my tea...then cradled the cup back in its saucer on the bed, imagining what it must have been like for Twain's party in the 1800's. The first two weeks onboard their floating hotel – a side-wheel steamship – had been spent just getting from New York City to the Azores west of Portugal, where they had to stop and recuperate from seasickness. Their next destination was Gibraltar and eventually Egypt. The *Quaker City* while crossing the Atlantic had taken an enormous pounding from storms I could barely imagine while I, on the other hand, had made the entire journey in just under thirteen hours and two plane changes, and was only now succumbing to indigestion from my umpteenth bag of smoked almonds taken on an empty stomach. In the mid-1800's there were several

---

[24] Herodotus, the world's first historian/tourist, wrote about Egypt and as a result, Greeks that came to Egypt after him (during Aristotle's time, for instance) wanted to see the same sites Herodotus had described. / Many ancient Greeks were convinced their own civilization had its beginnings in Egypt.
[25] He was in Egypt hoping to cut Britain's trade route to India, and besides, being Napoleon he probably enjoyed the idea of traveling in the footsteps of Alexander the Great two thousand years earlier.

guidebooks Twain and his party could have chosen from, most offering more in the way of romance than practical advice, always somber in tone a few even instructed the reader on how he or she was supposed to feel while staring up at the great ruins of antiquity.[26] To his credit, Twain made up his mind early on to use his own experiences, one of the first travel writers to interject personal humor this way into a travel guide. I couldn't help thinking that, in spite of the gulf in time between us, that I had come here for many of the same reasons Twain and his companions had... adventure... and a chance to look at my everyday world in a new way after returning home.

Just as I was drifting off, the sense of helplessness that sometimes accompanies sleep startled me back awake. It was the crowdedness of it all.[27] Cairo at one time had been a wealthy city situated at an important point along the Silk Road, home to a wide assortment of traders dealing in rich spices and perfumes and other novelties from the Far East.[28] It was no wonder perfumes were important back then, I thought. What with all the heat and humidity things had to have gotten rather smelly, still it was strange but in all that 21st century gridlock Mustafa and I had come upon on our way to the hotel, I didn't notice a single face that looked angry or upset in any way.

AS WIDE AS THE MISSISSIPPI

I only realized I had left my balcony door wide open when my sleep was being replaced by the Moslem call to prayer. I slept through the traffic noises as the sun was busy making its appearance, nudging itself above the red tiled rooftops, the call to prayer enveloping the streets and narrow alleyways below me, filling what looked from my vantage point like numerous capillaries branching off its main artery feeding the Sharia Talaat Harb with an endless stream of cars, busses, and motorbikes. The melodious voice of the call to prayer sounds to the unaccustomed traveler like a lonely man singing but is in fact the recitation of passages from the Koran spoken over a loudspeaker and in spite of my jet lag I was still able to recall my appointment with Tahany later in the afternoon. But for the moment I was all too happy to have the better part of a day free to wander around in Twain's footsteps through an ancient city, my first, a place I had only seen pictures of but never dreamed I'd be standing in just a short week ago.

---

[26] According to historian Stephen Ambrose in his book *Undaunted Courage*, the journals Lewis & Clark kept of their trip across the continent can be considered the first guidebooks written west of the Mississippi.

[27] Only 4% of Egypt's land is inhabitable (about the size of Maryland), and yet this is where 80 million live today.

[28] It was the rise of the Ottoman Empire and their monopoly on the spice trade that led the Italian explorer Columbus to set sail for India in 1492.

Mustafa had suggested I leave the Egyptian Museum for later as Tahany had a surprise in store, so with 21st century guidebook in hand and what was left of my complementary breakfast rolled up in a napkin in the other, I opened the door to my Funduq and walked out onto the sidewalk, experiencing Cairo's sights, sounds, and smells for the first time in full daylight. Oddly enough, what caught my attention first wasn't the traffic as I had been warned so often of by the guidebook, but the peacefulness of the Nile, flowing in silent contrast to all the human activity around it.

Bazaars are what most often come to mind when thinking about the Middle East and so the Khan al-Khalili, described in my book as "an oriental bazaar of fable not to be missed", had easily made my to-do list. Everything else I came across, including what Tahany and I would see at the museum, I reasoned, would be icing on the cake.

I walked over to the sidewalk's edge, staring down at what was a more swiftly flowing river than I had first realized. Like all big rivers, the Nile near its shore appears alive and animated and yet towards its center almost eternal, which is one of the things I've always admired about rivers: they seem so preoccupied with themselves. In 1867, Twain couldn't help drawing comparisons either, writing that in Cairo "the Nile at this point is muddy, swift, and turbid, and does not lack a great deal of being as wide as the Mississippi."

A few feet away a vendor had made his own pyramid, stacking a few dozen oranges atop a folding table in front of a juice stall on the sidewalk, an older gentleman apparently intent on selling drinks, which I now felt obliged to investigate. Reaching into my blue jeans I emerged with a wad of Egyptian bills hastily exchanged back at the hotel. Egypt never got around to using money for nearly all of its long history, perhaps not so hard to imagine in this day and age of plastic credit cards and electronic banking. At various times the brewer's yeast made all the currency anyone needed, staples like bread and beer, which could be exchanged for goods and services.[xxvi] [29] Modern Egyptian currency is colorful, reminding me of play money and my first purchase was for a glass of juice – a mango squeezed together alongside half a stalk of sugarcane split lengthwise and fed through a metal vice. After finishing it I handed the glass back and hailed a taxi, which was soon whisking me the mile or so south to Islamic Cairo.

I did know already – having learned it in graduate school – why Robert Koch was in Cairo in the late 1800's. The German doctor, having wrapped up his work on anthrax, was now hot on the trail of another microbe, this one responsible for the disease called cholera. Koch was eager to apply some of

---

[29] The English word *salary* comes from *salt*, a reminder that salt was sometimes used as payment. / Beer, unlike wine, is made from grain and since grain is easier to store than grapes, beer could be made more often during the year.

the same techniques for cultivating microbes he had been so successful with for *Bacillus anthracis*.   With the opening of the Suez Canal, travel between India and Europe had become faster in the 1870's and cholera – always the opportunist – was able to make its way from Asia to Europe that much faster, hidden within the holds of ships.   Koch's competitor, Louis Pasteur, had sent a French team to Cairo because he too was looking for the microbe during that same outbreak 140 years ago.

My guidebook offered advice on the best way to cross a street in Cairo, something I doubted Twain would have required.   It suggested that I use others as a human shield.   An amazing thing books are though, I thought as I walked, no less so than the printing press used to produce them, developed by Gutenberg, an innovation that would help bring an end to Europe's Dark Ages; making books more widely accessible and in quantities that, for the first time in human history, allowed knowledge to outpace religious authority.[xxvii]   It was Gutenberg who realized he could combine the high pressures of the winepress along with small metallic stamps useful for imprinting designs onto leather, thereby creating a practical way to transfer print onto paper.   He also took advantage of Europe's newly invented oil-based inks (the Chinese had already invented movable type, but they used wooden or clay stamps).[30]

Construction of Islamic Cairo was ordered by a General Jawhar as soon as the planets were aligned properly, and even though they were designed with pedestrians in mind, the layout of the old city is still much as it was centuries ago, back when Cairo was an important stop along the Silk Road from China to Europe.[xxviii]   Caravans rested not far from where I found myself that morning, and in the 1300's the Bedouins came there looking for exotic goods, forming the nucleus for what would over the centuries grow into the Khan al-Khalili bazaar, my final stop before meeting Tahany at her university. Unfortunately for Cairo's merchants, in 1497 the Portuguese explorer Vasco de Gama rounded the southernmost tip of Africa, circumnavigating the continent and in the process relegating Cairo to another unnecessary stop along the old Silk Road, whereupon the city soon fell into a prolonged period of decline.

After leaving the taxi and asking yet again for directions to reassure myself I was walking in the right direction I rested at an outdoor food stall, ready for my first taste of Fuul – Egypt's national dish – consisting of small brown beans soaked overnight in water, then boiled and topped with olive oil, or as I would discover a few days later in Luxor, a fried egg worked well in the mornings too.   By the end of my trip I was even having the vendor mash it all together and stuff it inside a piece of pita bread so I could enjoy it while exploring Cairo's endlessly fascinating side streets.   The Fuul I had that

---

[30] The guidebook called this part of the city "Islamic Cairo", while the tourist map I held from Tahany's brother preferred the term "Fatimid Cairo" for some reason.

morning would have been recognizable to any pharaoh in ancient times; in fact bread is about as popular now as it would have been when King Tut was commandeering his chariots. But bread was a staple just about everywhere, it seems. Even Meriwether Lewis described it in his journal after coming across some Native Americans making a kind of bread out of ground sunflower seeds.

I should confess that I had ulterior motives that morning while visiting Cairo's Citadel. My primary goal at the ancient fortress built by Saladin (but currently a military museum with three mosques) was not so much to learn the history of Cairo as to catch my first glimpse of the pyramids, which my guidebook assured me would be visible off in the distance. And as it turns out, Cairo is full of surprises for I soon discovered I didn't have to go all the way to Giza to touch them, either. The walls of the Citadel are themselves made from many of the stones taken from the pyramids in the 1100's at the instructions of that same General Saladin.

After reaching the Citadel I climbed the stairs – in hindsight perhaps a little too enthusiastically – full backpack resting on my lower spine, then once at the top I gazed across a broad horizon spiked by minarets, many in pairs appearing to guard the domed mosques which rival professional football stadiums in size back home.[31] With the help of a woman from Belgium, we soon located the pyramids off near the horizon, so small they could have been mistaken as a staircase carved out of desert floor.[32] The area beyond them was so sparse that a woodpecker flying over would have to pack its own lunch, as my grandmother would have said.

I rested a few minutes, this time lingering on a section of the guidebook that caught my attention back on the plane, a page explaining the business of antiquities. I looked out at the Nile while digesting what I had just read. If not for the river and its periodic inundations every year, all of Egypt would have remained a desert and the material making up the pyramids would have stayed an ancient seabed, just as the vast majority of the nation is resting on today. Instead – complements of the Nile – Egypt was to become the most stable civilization in human history...until oddly enough not a single person alive would be able to read or write hieroglyphics.

Periodic flooding in the delta is why ancient artifacts, especially wooden ones, tend to be a harder find here than in Upper Egypt. The ancient capital of Memphis is still under much of this mud, which is why information about Lower Egypt comes mainly from writings found near Luxor... introducing an obvious bias for Egyptologists to contend with. Contemporary farmers in the

---

[31] One of them – the Al-Azhar mosque & university – was founded in AD 970.
[32] The largest workforce in history up until then – 10,000 to 20,000 – was assembled to build the Great Pyramid.

Delta sometimes plow artifacts right into the ground without realizing it.[33] The Upper Nile, in contrast, has gotten so little moisture over the centuries that sandals, baskets, and even paper have been found in excellent condition, manufactured 500 years before the birth of Christ.

With the help of a sidebar I became adept at estimating the age of Cairo's buildings. As a rule, the less ornate and intricate the pattern on the exterior of a mosque, the older the mosque tends to be. In the 1300's, the stone domes were plain. But over time, zigzags and floral arabesques began to appear as ornamentation. Minarets, which rise from the city's haze like needles from a pincushion, first began dividing up the skyline around AD 673, eventually lending Cairo another nickname, this one the "City of a Thousand Minarets". [xxix]

After visiting the Al-Azhar mosque, the bazaar was, mercifully, only a short distance down a narrow alleyway and so it wasn't long before I was elbow-to-elbow with locals thicker than cats at a fish fry, some barely old enough to play Little League back home yet hoisting wide trays above their heads with cups of tea and flat bread piled high in the middle. I looked down to see one youngster still too small to be carrying a tray walking beside me, looking up and smiling. It took another two blocks before I realized he had appointed himself my tour guide and another to learn that his name was Mohammad. His presence seemed to be having a deterring effect on the other vendors and for the first few minutes it's fun to be held in such high of regard, but after a while you begin to feel like a walking ATM machine in a place like this. In Twain's day, tourism was still so new that his party was often mistaken by the locals as foreign agents intent on upsetting the carefully balanced power structure already in place for centuries. At other times they were accused of bringing cholera and so quarantined.

As Mohammad and I made our way down the increasingly wider alleyways towards all the activity, the air grew saturated with incense and carob, smells which gradually transitioned into eye-watering tobacco as the anthrax grant was fading more from my mind with each step.

My third purchase (I was actually still keeping count) in Egypt was for two cups of tea, after which my guide and I pressed on. It was about then that I began getting the uneasy feeling I was being led. Mohammad always seemed to be ahead of me, turning around and waiting patiently whenever I was unable to shuffle sideways through the thick crowds due to my backpack. I couldn't slip through like he could and at times I imagined we were making our way through a dense jungle, except that, rather than vines blocking our

---

[33] Over the last two thousand years, the Nile Delta has covered entire cities like Damanur in silt, a site still being uncovered. / Some farmers, continuing a tradition begun in ancient times, still grind up nitrogen-rich bricks from ancient monuments in the Nile Delta as a fertilizer.

path I had to contend with belly dancer costumes and colorful rugs slung just far enough overhead to require a duck.

I tried explaining to Mohammad with hand gestures what it was I was looking for, the kind of souvenir I wanted, delineating with my fingers in the air the outline of an Egyptian papyrus, but I still hadn't quite gotten the message across even though he nodded as if he understood me. At one point the boy even used his own sign language to enrich our tour.

Using his index fingers he made a sweeping motion across first one wrist and then the other. "Teiff....cot," he explained while looking up to see if I understood. Not until his second hand had been severed did I realize he was trying to tell me that the bazaar once had laws requiring the loss of a limb for stealing. We weaved back through the crowds for another ten minutes and just as I was beginning to lose confidence in him, we came upon a vegetable stand where I was instructed to "Come ein" by an older, graying gentleman pulling out a wooden stool, as if he'd been expecting me. He had materialized out of the shadows sporting a white beard, turban, and the same smile I'd come to expect in Cairo. It seemed the vegetable stand was performing double duty as an antiques shop. Grinning like a butcher's dog, he turned and patted the boy on the head, handing him something I didn't catch sight of, after which the boy disappeared into the alleyway before I had a chance to thank him.

"He is my son," the man explained, a toothpick hanging limply from one corner of his mouth. He was speaking in hushed tones, as if comforting a small bird he was trying hard not to scare off its perch. "So tell me something. What brings you here this fine day if I may be so bold as to ask? Did you come alone my friend?"

Relieved to be able to speak in complete sentences again since leaving the hotel, I explained how the papyrus I was looking for needed to be a genuine replica, not the kind Mustafa had warned me of, the kind where the paint flecks off every time you roll it up. I also didn't want one of the knockoffs made of palm fronds or bamboo leaves in place of genuine papyrus. The man seemed to be listening but unfortunately was no longer smiling. Undaunted I continued on, describing how I wouldn't settle for anything less an authentic scene copied from the walls of a temple or a tomb, preferably of ancient Egyptians baking bread or brewing beer. Unfortunately, his English wasn't fluent enough that I could explain why I needed it – that it was because I wanted something of the yeast to bring back home for inspiration.

His next move was to stage a disappearing act, venturing into a square room made of hanging rugs, after which he reemerged with a tray full of carefully balanced sheesha pipes, all copper and crowded together, each identical except in a bewildering assortment of sizes. Still stooping, he selected the pipe nearest me and pretended to take a puff from it, apparently deciding I needed to know how one worked. Feeling somewhat obliged I selected the smallest and feigned an interest, more out of politeness as I still had ten more days to fill up my pack, yet his son had been so helpful, if indeed Mohammad had really been his son.

"Do you have anything else?" I asked as I handed him back the pipe and stood to make my way over to the fruit stand, back towards the foot traffic, already putting into place my escape plan. His face drooped and his overall posture changed to disappointment. Then, just as quickly he pulled out a business card, explaining how there was money to be made if I could sell some of his fine pipes back in America. It was a game of chess and we both knew my turn was coming.

"How much for grapes?" I asked, plucking one off and holding it up to the light filtering in between the cloth awnings. Turning it over gently between my thumb and forefinger, I couldn't help admiring the dark, rich surface, as well as the white patches on its exterior, so white in places it stood in brilliant contrast to the purple of the skin, skin so laden with antioxidants that in the dim light the grape appeared black. Except for wild honey, the grape is perhaps the best surface in nature for capturing the wild brewer's yeast.

The Egyptian took out a 1-pound note to indicate how much he expected and then, curiously, began reciting a verse, only part of which I caught through the roar of voices. "Drowsing through the climbing vine, the bee hints at the promise of sun-kissed wine."

I reached into my pocket as he went to get a plastic bag, giving me another chance to examine the grape. It was what I had come half way around the world for. The yeast was in my hand, a part of the grape's natural flora, awaiting its opportunity to turn the grape's sugars into more yeast. One of the things that makes life different from the rest of the natural world – a place of rocks and wind, mountains and rivers – is its ability to reproduce, and most microbes do it whenever, wherever, and however they can. I recalled one sermon back in Thistle about how, after Moses and his people had fled Egypt and while still wandering around in the Wilderness of Paran, the prophet had sent his scouts ahead to Canaan, after which they brought back symbols to show the land was fertile, one of those being a handful of grapes. There was no sign of disease anywhere on its surface.[34] I took my change and used what was left of my afternoon to hunt down the surprisingly elusive papyrus.

Papyri were the first portable books of the ancient world, the Ebers papyrus alone containing some 800 spells, one describing the proper use of opium to keep babies from crying, another on how malachite could be used to

---

[34] Today, more grapes are grown than any other fruit. / Pliny the Elder described 91 varieties of grapes in the first century AD.

cover an open wound. Thanks to the scientific method we know today that malachite contains copper, an effective antiseptic for killing staphylococcus bacteria.[xxx] The Egyptians had other uses for this mineral, sometimes mixing it with animal fat and using it as a green make-up. Other heavy metals like lead would have been applied around the eyes to darken them and, like copper, lead's antibacterial effects would have offered protection from infections initiated by the constant exposure to dry winds and sand.[35] If Mark Twain had visited Egypt during Roman times and his ship had anchored in the harbor of the world's first cosmopolitan city – Alexandria – the *Quaker City* would have been searched and its captain required to hand over every last papyrus on board so it could be copied and deposited in Alexandria's magnificent library.[xxxi]

As I made my way from the center of the bazaar back towards its perimeter where a taxi should be waiting, I encountered less of the alabaster pyramids and stuffed camels and yet still no sign of the papyrus. I was beginning to wonder if they existed. The guidebook promised if I walked north I could find just about anything, which I was sure I had. I was raised on a ranch so directions come second nature; the sun is great for telling direction when you have it. And yet the book hadn't let me down yet. Growing tired again, I found another bench, this time outside a coffee stall, and sat while reading in a sidebar about how the ancient Egyptians used beer as a solvent to suspend onions and salt in to induce vomiting if someone was unfortunate enough to be bitten by a snake and that Egypt has over thirty poisonous snakes, making the legend of Cleopatra using an asp to poison herself more likely than not true. Yes, I'd picked a good guidebook, I reassured myself while rubbing my lower back with my fingers to ease the growing pain. At least I'd had the foresight to do that.[xxxii]

---

[35] Several heavy metals have antibacterial properties, and I was reminded of how Meriwether Lewis would administer a mercury salve to his men whenever they had bouts of venereal disease they acquired from the Mandan women. The use of this treatment inspired the 19th century adage "A night in the arms of Venus leads to a lifetime on Mercury." As a Montanan I grew up learning about Lewis & Clark and as an adolescent boy this bit of exotic information didn't escape my notice either.

INVASION OF PRIVACY

In the rear of the cab on my way to American University, I couldn't help looking in the mirror again, smoothing my hair down while trying to remember what it was that I was doing, exactly. On the one hand, I was going at Tahany's invitation because she wanted me to know more about her work. But what if something more came of it? I squeezed the handle on the door tighter as the driver took yet another chance I wouldn't have, after which he slammed on the accelerator and we darted down what I took to be the garment district, catching random glimpses out the window of vendors trying to scratch out a living selling tennis shoes on blankets almost touching the tires as the taxi soared by.

I was an assistant professor coming up for a full committee vote...my last chance at tenure. I had signed a four-year contract when I joined the university, in other words I was a temporary hire. If I didn't get my tenure I'd be out of a job in a year. It was that simple. And if that wasn't enough, all of it was during a time when there were some alarming cutbacks taking place in science, unprecedented ones in fact. The Iraq War, Hurricane Katrina, deficit spending, Afghanistan, all had come together like the perfect storm after 9/11. I could barely support myself if I lost my job and had to go back to being a postdoctoral fellow in someone else's lab, let alone take care of a wife on the other side of the world.[36]

I took the guidebook out and plotted my position on the insert map with my pen, drawing a circle and then studying the details inscribed within it. Not bad for my first day on what was for me a new continent, I reassured myself. The book claimed the university was in the same direction as my hotel, yet nothing looked familiar, even as we rounded Tahrir Square, maybe because we came at it from the opposite direction. As we veered around a corner I caught sight of some paunch spilling out of an unfastened gap in my shirt, my belly jiggling as if dancing to the rhythms of the road's potholes. Maybe I'd walk back to the hotel afterwards, I thought, and besides, getting some exercise might be a good way to take in the city after dark. I considered how, back on the ranch I never had time to put on extra weight. But then I went away to college and everything took on a more sedentary tone working in a lab twelve hours a day.

Once we got to the seven-story red brick building I was struck by how it could have been on any campus I'd seen before. A handful of students were standing around a fountain, chatting like young people do everywhere,

---

[36] In 1970 the average age of a scientist in the United States when awarded his or her first research grant was 32. By 2009 it would climb to 42 and in 2012 it was reported that, due to budgetary constraints, NIH was funding only 1 in 7 grant proposals: the lowest in its history.

discussing friends or upcoming tests most likely. Just inside the door on the wall was a directory and I scanned the names behind the glass. The sign above the door in front of me identified it as lab 123. Tahany's lab was 723. I looked up through the high stairwell, then down again at my gut. We had missed our first meeting and now she was directly above and the butterflies began hovering en masse again. On the way up the stairs my pack seemed heavier, which struck me as odd since I hadn't added anything to it.

After a few more minutes I had to take off the pack and sit down to rest. Tahany, like many MD's, was working on a project that had the potential to yield more immediate benefits than basic research like Ph.D.'s tend to work on. Ph.D.'s are usually more concerned with uncovering basic theories and mechanisms underlying a particular phenomenon, it's how we advance, get our work into print, gain tenure, etc, while MD's tend to be more focused on helping their patients, even if there's no particular explanation for how something works, a potential new drug perhaps, or a test to diagnose a disease faster and more accurately, as Tahany and her group were working on.

An important problem with schistosomiasis – Tahany's field – continues to be that the worms causing the disease not only invade the body, bypassing its primary defense, the skin, to lay its tiny eggs wherever it can once inside, but these eggs clog the circulatory system and set off inflammatory reactions in the process. Even worse, the worms can go undetected for years, thus aiding their silent journey back into the water supply. It seems a person can become infected and carry the worms without ever knowing it. Called asymptomatic carriers it's why one in ten Egyptians still harbors the worms.[37] In fact, worms cause more socioeconomic damage worldwide than any parasite other than the single-celled protozoan responsible for malaria.

The number of people living with these tiny worms is staggering, even to someone who studies diseases, something like 200 million globally, but no one knows for sure. The worms are so common in some villages that symptoms like blood in the urine (called hematuria) can be seen as a rite of passage for males entering adulthood, not so different than the way menstruation is for young women elsewhere. Tahany was trying to attack the problem of treatment selectively. Her idea, I seemed to recall, was that if a test could somehow identify non-symptomatic carriers of the worms, then mass treatment of entire villages with strong medicines like praziquantel could be made obsolete. All I knew as I struggled up what remained of the stairs was that it had something to do with finding new ways of detecting the elusive worms.

---

[37] Typhoid Mary was a famous carrier of the *Salmonella* bacterium.

By the time I managed the 5th floor I had come to the conclusion that, what with teaching and grading, not to mention writing one too many grant proposals the previous semester, it had all finally caught up with me. Standing on that stair half out of breath meant there was no longer any denying it; I had officially entered middle age.   The same piece of luggage I had been so proud of just a few hours ago was no longer a source of convenience but had at some point turned on me, intent now on carving its way ever deeper into my shoulder blade.   The pain was excruciating, burning even, as if someone had stuffed rocks into my bag while I wasn't looking.   I sat down again to catch my breath, remembering how, on their way to the Pacific, Lewis & Clark would walk alongside the Missouri River for 30 miles or more...in one day! They were both about the same age I was now. Looking for any reason to stall longer, my eyes caught on one of the research posters hanging on the wall nearest me. May as well have a look, I allowed myself, after all, early 19th century explorers didn't have to sit around grading papers all day.

With posters detailing work presented at past meetings, Tahany's department was like any I had seen before. The names were in Arabic, but everything else seemed to be in English, this being the world's default language when it comes to commerce and science.[xxxiii]   Many of the posters were either about West Nile virus or Rift Valley Fever; only one was focused on cholera.[xxxiv]   Robert Koch would have been pleased to know his work hadn't gone unfinished, even if his name wasn't listed on any of the posters. Apparently the 5th floor was where the virologists and at least one bacteriologist were located.

Back home we often hang onto our posters detailing work (used as visual aids at seminars) because it can lend a sense of accomplishment to an otherwise dull existence in academia.   And besides, it beats just throwing them in the wastebasket. Despite what a lot of TV shows try to convince you of, doing science isn't what you might expect most days. [xxxv]   I often find myself walking out into the hallway of the dungeon just to have a look at my old posters whenever experiments go awry or take too long, usually late in the evening when negative results seem that much more ominous for some reason.   Mustafa had mentioned on the ride from the airport what Tahany's budget was for her entire lab and I calculated that it probably amounted to less than what I paid one of my two technicians in a full year, making me all the more curious to see what her lab would look like. She must be pretty resourceful is what I remember thinking.

On the day I trudged up those stairs, I had been in research for three years as an independent investigator, and before that working as a postdoctoral fellow in someone else's lab to gain more experience, and even before that for seven long years as a graduate student, and I have yet to step foot inside a lab that actually impresses me, aesthetically, when I first walk in. Even the nicer ones aren't much to talk about, more like your uncle's basement, the one with the pool table that has become a catchall for every stray item in his house.

Research labs like mine (so-called "wet labs") can actually be pretty scruffy places, stale-smelling incubators that could use a good cleaning out, stationed next to refrigerators which substitute as benches for empty test tube racks on their tops, floors sticky here and there with dried broth containing fossilized footprints no one has had the time or inclination to mop up yet, the kind of places you're more likely to see a mouse than some model with spotless teeth and lab coat holding onto the latest digital pipette and clipboard.   There might be screws lying around belonging to things you wouldn't dare touch because someone is probably planning to use them for something they were never intended for, columns for separating (i.e. purifying) proteins for instance, pipes that look like PVC plumbing underneath someone's kitchen sink, complete with lubricated fittings.

The first time I was given a tour of a research lab, I was pretty disappointed to be honest.   Back then I was still expecting to see something impressive.   Too many books and movies, I suppose.   Most benches in laboratories are long tables no more pleasing to the eye than those of a commercial kitchen, but instead of mixing pasta or cutting up vegetables on they're for doing science.   The challenge of modern molecular biology is mostly an intellectual one, and sometimes it can be aesthetically pleasing if you're lucky, but not as often as you might expect.

Hot, newly poured Petri dishes have a peculiar and unappetizing odor to them, maybe because of the volatile ammonia salts and amino acids in their hot broth, and they require a flat surface for the agar to solidify on as it cools, otherwise you won't have nice flat surfaces to separate your colonies of bacteria onto after the gel hardens.   The excitement, if there is any, may come afterwards when you're in an office pouring over your data, scouring photomicrographs and you happen to notice some detail that didn't catch your eye all those other times you looked at the same data, something you somehow missed back when you were doing the experiment.

In graduate school, the task I didn't mind the most (I'm kind of ashamed to admit) was washing the lab's glassware.   It was mostly physical, not mental, and it took no real effort and I was guaranteed of seeing progress, not so different than watching dirt pile up alongside a posthole for a new fence on a ranch.   Sometimes a person needs to see progress, any kind, especially when a single experiment can last more than a week before you get any results to share – good or bad, but most of the time somewhere in between – meaning that you have to repeat the whole experiment all over again because the answer just wasn't clear enough this go around, changing only one of the many possible variables to see if that was the real problem or not.[38]

Tahany's lab would be no different than other labs I'd seen except that it was smaller, with just two rows of benches dividing it up into three claustrophobic workstations. Some of the desks were built right into the benches, making it look at first glance more like a library. There were books and catalogues piled high as if waiting for the slightest tremor to set off an avalanche. I cupped my eyes to reduce the glare from the hall as I looked in through the window, my attention settling upon two figures, a man and a woman about the same age, perhaps 30 or so. The woman was taking samples from a rack of test tubes using a slender glass pipette while holding each up to the light and I wondered if this exotic creature with the dark hair gathered up into a bun could be Tahany. I had seen photos of her and yet I still couldn't be sure. Her hair had always been down around her shoulders in the photos.

My heart cheated me out of a beat or two as I twisted the knob and then pushed on the door just enough to creak it open. Then I stuck my head inside, feeling a bit apprehensive remembering that Tahany still didn't know what I looked like. Like I said, I've never taken a photo I've ever liked. The young woman turned and smiled, pipette still in hand. I was beginning to think everyone in Egypt smiled; maybe it was some sort of a defense against the heat.

"Right on time, Benjamin. But you look so tired. Come in…sit down…please," Tahany offered up a chair that had some books on it as she finished transferring what was left of the yellow broth into an empty test tube. "So how was your first day? What do you think of Egypt so far?" Her tone seemed to match those of her e-mails, and I found myself relaxing a bit more.

"Fine," I answered, "Having a wonderful time. Still no papyrus, though. And I've had a heck of a time finding any sunscreen over SPF 8. Who would have thought? Egypt being in a desert, I mean." That's great, I thought. Now I must be sounding like an old woman to her.

Tahany indicated with her shoulder where she wanted me put the books, on top of what turned out to be a desk. The giveaway was the ceramic cup performing double duty as a paperweight, necessary due to the twin ceiling fans whirling away above our heads.

"This is one of my students. His name is Kenan. He's from Turkey." American University has an eclectic group of students and faculty I would come to find out that day; a reflection of the Cairo I'd seen since arriving, little

---

[38] I've often felt Meriwether Lewis would have been a better scientist than me. As Jefferson described him in a letter once "…of sound understanding and a fidelity to truth so scrupulous that whatever he should report would be as certain as if seen by ourselves…"

different than New York or London is that way I suppose. "Benjamin is from Montana."

"The States?" Kenan asked, eyes widening. He asked it with the same tone I'd heard everywhere that day whenever someone wanted to know where I was from whether it was in the bazaar or walking down one of the side streets, or even sitting in the back of a cab. I had been surprised to find on this, my first day in Egypt that, even in the capital of the Arab world, Americans are not hated like I'd believed for some reason, quite the opposite really. America is still viewed as a shining city on a hill by many in the Middle East, a place always just over the horizon. It reminded me of the way someone from Iowa or Kansas who has never been to California might look forward to visiting Hollywood someday.

I laid my pack on the floor, sliding it with my foot up against the bench, yet for a brief moment I was unable to budge it; then consoled myself with the knowledge that at least Twain had the use of a donkey to carry his belongings around while in Cairo.

Tahany's complexion was as smooth as corn silk, giving away the fact that she spent the bulk of her time indoors. It was as if her skin should have been darker, like her hair and eyes, yet it was closer to the shade of milk, and I realized I was looking at the underlying part untouched by the sun. She was prettier than a tub of freshly churned butter and her face had all the graceful features of someplace far away. Still apparently avoiding my gaze, Tahany scooted the stack of manuals over to one side with her arm but without losing any of the liquid in her test tubes.

"Are you prepared to learn about Bilharzia? I hope so. It's about all we think of in here these days," she wiped some beads of sweat from her forehead with her sleeve while I thought about how I'd seen Egyptian women taking pains to avoid the sun all afternoon, perhaps to look more like Tahany. Her accent was noticeable, but not enough to hinder any conversation, and I stopped even being aware of it at some point. I could tell she was taking pains to pronounce her words carefully and I flattered myself into believing she was doing it to impress me. The more she spoke, the more obvious it was she had gone to school in England. It was then that it finally dawned on me I was being expected to shift gears onto a disease I was unfamiliar with, wishing I'd taken more time to read up during the long plane ride. Bilharzia, I did know, was another name for schistosomaisis. Some others in use depending on where you live are Katayma fever, swimmer's itch, snail fever, and blood fluke.

"Actually, I've been ready," I lied, then added, "Like coming full circle."

Tahany's work with the schistosome worm and its lifecycle had been what initiated our long string of e-mails my first year at the university. That morning back in Thistle I had been lecturing my "sunrise clinical microbiology class", a small group of pre-med students who met out towards the edge of campus, and I remember it so well because those mornings always began early enough that the frost was still on the grass and the student's invariably picked it up on their tennis shoes, causing them to squeak on the floors of the hallway and, like a deer caught in the headlights, I made the mistake of taking an obscure question from one of my more wide-awake students in the back of the room. Every morning class has one, except that they usually sit closer to the front. Well, having learned in graduate school that, with college students it's always best to admit when you don't know something because this also seems to increase their respect for you, I answered truthfully and got out of it by promising I would get back to him the next day. As soon as I made it to my office I flicked the computer on and began searching for an expert, anyone who could answer his question, one that had been about vaccinations. I fired off perhaps a dozen e-mails and Tahany was the only one gracious enough to return hers, which explains how I came to be sitting in her lab at that moment looking rather perplexed.

So Tahany knew I already had *some* knowledge of Bilharzia but apparently wanted me to know more so she could explain her project in detail, the one she and Kenan were no doubt working on when I wandered in. In some ways, I got the feeling she was explaining it as much to him as to me. Maybe Kenan had just joined and needed to hear the lecture again, to reinforce it in his mind. I often do that with my own students; reinforce things every chance I get. A lot of details in science need to be explained in different ways before they finally stick. But once they do, you often get a good student out of the deal.

"The schistosome worm even when fully grown, is barely visible without a magnifying glass so you're probably getting some idea how small her eggs must be. Here Benjamin. Can you find one?" she smiled again but still avoiding my gaze, picking up a glass bottle filled with clear alcohol that looked like water, only the sharp aroma giving away the liquid's identity. Tahany put the bottle and a magnifying glass down on the bench beside me. I picked them up and began searching for her eggs, wondering if it was Kenan or I who was being tested. I quickly reasoned that the worm and her eggs must have settled to the bottom, water and protein being denser than alcohol, so I began by looking for them there, using just the magnifying glass. Kenan set a cup of steaming hot tea down in front of me, then without turning due to the lack of space, retreated back the way he came.

"Bilharzia is an ancient disease," Tahany began, "in fact my ancestors never did get the connection...that it was caused by a worm, probably because of their small size, but also because, even in ancient times, with their respect for the dead, dissections were actually frowned upon. Bilharzia was only known about by the symptoms it caused. The worms themselves weren't even seen until the 1850's. That's when Theodor Bilharz, a professor of anatomy, saw them during an autopsy at a medical school here in Cairo."

As Tahany talked, Kenan began pointing towards an old portrait, a black and white photo on the wall at the far end of the lab just above the door. At first glance, Bilharz looked like one of those Civil War generals with his flowing white beard, except that rather than a uniform Bilharz had on a tuxedo.

"In 1910 the calcified eggs of the worm were noticed by the careful eye of Sir Armand Ruffer[39] while dissecting the kidneys of two 20th dynasty Egyptian mummies.[40] I sometimes wonder what my people would have thought if they had known the same medicine – antimony disulfide – would remain the most effective treatment for schistosomiasis over the next three thousand years."
xxxvi

I set the magnifying glass down, my way of admitting defeat, still unable to find any of the worms even after letting them settle to the bottom where they should have been easier to spot. Maybe it was the heat I consoled myself. The only time we get any real heat and humidity in western Montana is for about a month in the middle of summer.

"Antimony is a heavy metal," I explained as confidently as I could in an effort to keep the conversation from drifting into awkwardness.

"It's not easy," Kenan reassured me as he plucked one of the worms out with a pair of tweezers and, due to the lab's confinement, handed them off to his boss, who in turn handed them to me. Apparently, I was getting another chance.

"You see? They're so small. But the real problem is that they are such prodigious egg layers." I still couldn't see the worm so I took off my glasses. Being nearsighted I use them mainly for distance. It was then that we finally came face to face. The worm was extremely tiny, narrow, and snow-white. It was no wonder I couldn't see any of her eggs. They were microscopic.

"Different types of worms have their own preferences as to which areas of the body they invade and within a single mated pair hidden within the walls of the bladder, she may lay hundreds or even thousands of eggs in a single day," I nodded as Tahany spoke into my ear. It was the closest we had been and I could now smell her perfume. It's a strange sensation having the scent of someone you've just met up inside your nose, kind of like a pleasant invasion of privacy. Tahany was more beautiful than her photo let on and the more I studied her the more I had to remind myself we'd only just met.

---

[39] A pioneer in the field of what would someday be called *paleopathology*. Ruffer was a professor of bacteriology at Cairo Medical School and he also identified TB, atherosclerosis, and various dental conditions in mummies.
[40] The 20th dynasty was during the New Kingdom period, about 1200 BC.

"Well, the real problem isn't her size, but the fact that only about half her eggs will ever find their way back out…to be excreted. The rest will get shunted…build up inside the vital organs, like the liver or the brain, triggering an immune response," she sighed and then forged ahead using a more resigned tone, "The body can do more harm than if it hadn't done anything at all, had simply ignored her eggs."

"Isn't someone in your department trying to find a better drug?" No sooner had I gotten the words out than Kenan, who had worked his way back around again, was removing the lid off a cardboard box and sliding a large, clunky, dissecting microscope in front of me. It was the same make of microscope I had used to see the yeast just a few days before. I looked through the twin eyeholes but could see only a few specks of dust, then felt foolish once it became clear Kenan hadn't loaded the slide into the bay and I wondered if Tahany realized what kind of a disheveling affect she was having on me.

"Oh, we're not looking for new drugs anymore," Tahany explained, "These days the emphasis is not on treatment but prevention. Some of us are trying to rid the snail that the schistosome worm spends part of its lifecycle in. Almost anyone can get bilharzia if they're not careful, even Napoleon's army got it. Backpackers and Peace Corp volunteers have sometimes picked up the worms just by using river water to shave with. You see…the adults can penetrate right through skin. They chew their way in by secreting enzymes, digesting proteins. These enzymes work quickly, too. The whole journey from water into someone's bloodstream can happen in minutes."

"Not the kind of souvenir you probably want to take back home," Kenan added with a grin.

"The greatest obstacles," Tahany explained, "are as ancient as Egypt herself; poverty and lack of knowledge," she then frowned, "Eradication is difficult because the worms can hide inside of horses and dogs whenever humans are unavailable. The worm has one weak spot, though. It needs a snail. It has to spend part of its life inside this one snail. Not just any kind will do." Her words made me think about smallpox and how, even in this day and age of recombinant DNA technology we're all so proud of, how the smallpox virus is still the only human pathogen that's ever been completely eradicated from nature. Smallpox was exiled to the deep freeze mainly by vaccination and international cooperation, and of course money, but also because the virus had an Achilles heel: it could never hide inside another animal. Smallpox is strictly a human virus and so it infects only us.

"The schistosome worm has relatives, over 300, inhabiting the bodies of a billion people at any given moment. Dams, irrigation projects, they all have their place in improving the standard of living here, yet they come with a price because they create new habitats for the snail. My country has used crayfish to eat the snails, which has met with some success. Yet nature always seems a step ahead. There is even some evidence the worms have evolved to take advantage of human activity, the times of day we bathe in rivers coincide with the times the worms release themselves from the snails."[xxxvii]

I looked over at Kenan, who seemed as if he might have been waiting for a break in the conversation. When he saw my looking at him he pushed the microscope closer. Loaded into the bay was a glass slide with a pinkish film smeared onto it. As I began my search, the medical student explained how I was looking at a tissue slice of a former patient's bladder. Clearly visible through the lens were the large, flat plates – skin cells being impossible to mistake – along with dozens of tiny pink specks, which turned out to be the eggs scattered throughout the tissue, some equipped with grappling barbs that looked like fishhooks. They had been subjected to a synthetic dye because nothing in nature is that bold of color, not even red blood cells are that brightly pink. Few things in nature's realm can match the loud colors that synthetic stains manufactured in a chemist's lab can. Next we examined a drop of urine from the same patient. Looking for the worm's eggs in the patient's urine under a microscope is how laboratory technicians still find them to make a schistosome diagnosis. It's tedious work, the kind that reminds me of trying to coax a hot mule out of a cool barn, the kind of work I'm glad I don't have to do much of these days.

The eggs in the urine were few and far between. "They seem rather sparse and difficult to find," I said. "Which must be why we have graduate students." I was relieved when Tahany smiled too. Her eyes caught mine for a moment and then, like a butterfly, flitted away again. Maybe she wasn't so put off by my appearance, I reasoned. We seemed to be hitting it off. If she was disappointed in my physique, at least she was kind enough not to mention it.

"And this sample is from a patient who shed a lot of eggs."

"There must be a better way," I added, "Didn't you say something about DNA? Weren't you...?"

"Yes. I was just coming to that, Benjamin. We've been developing a PCR test."

"So you're doing an amplification, then?"[xxxviii]

"The same technology that has allowed the sequencing of DNA from ancient mummies works well for worm eggs in a patient's urine."

"So you want to use PCR to sequence the worm's DNA?" I ventured.

"No. Not sequence it. Just detect its presence."

"We only need the DNA from one of her eggs," Kenan explained.

Tahany finished his point for him, "In theory, only one egg is needed to confirm the worm's presence, and since all her eggs contain the same DNA then...", there was a brief silence as she considered her words. "That's the reason PCR works so well for crime scenes and why you only need a drop of blood, and if you can amplify it, that cell's worth of DNA means you can have as much as you need, even for DNA fingerprinting and sequencing if you care to do it."

Tahany reached above her head to the shelf and opened one of her notebooks. A dozen or so dark sheets of film cascaded onto the bench. They were negatives with images of what looked like ladder rungs but without any railings, as if the rungs had been suspended in midair all by themselves...defying gravity. I knew that each rung on the negative represented a different length of DNA fragment amplified enough times to make the spot visible on the film. The worm's DNA was discernable because, during one of the steps of PCR, all the fragments were labeled with a radioactive molecule, a probe, tagging and identifying the DNA like a lighthouse can a shoal, labeling it as coming from the worm. PCR is a bit like amplifying a page from a book simply by making Xerox copies of it, and then making copies from the copies using an additional machine and so on, which is where the amplification part comes in. PCR is all about making DNA copies of very specific pieces of DNA, just like when you want to make a copy from a book you select only the pages you need, not the whole book. When you use a PCR test to probe an organism's genome, you usually look at only one of its unique genes to identify it.

"This lane," she ran her fingers down the surface of the film, "it shows up fairly well. Don't you agree Benjamin?" I nodded and a warm feeling flooded me whenever she called me by my full name. "But it's not as reproducible as we need it to be. This is DNA from one single egg cell. But sometimes the signal blends in with the background and that's the problem now."

"Noise is always a problem. In every experiment," I sympathized.

"When you can see the worm DNA, it amounts to a red flag, a tag identifying a patient who's been infected, someone who can now be put on the drug and treated. No more shedding of eggs back into the environment," Tahany must have noticed I was impressed because she next tried to temper my growing enthusiasm using a more somber tone, "There are some problems, though."

"Like what?" I demanded, "These look really promising, Tahany. What's the matter?"

"Like the background noise on this negative over here," she ran her fingers across one of the fainter rungs on the negative, "and how sometimes the test says someone is infected when they are not really. Perhaps these are rare instances of cross-contamination. Or maybe the patient is infected...but just not producing any antibodies yet. We don't know. It takes only a single egg to contaminate an entire sample...and if it's an egg in the wrong place in the lab... and I forgot to mention how these eggs do like to stick to things... then your test can lose all credibility."

Kenan, who it was obvious now wasn't new to the project, added, "While we believe the test can work, we are still left with the same old regulations. The government wants to cling to the standards for diagnosis set in place decades ago."

"And in the meantime," Tahany added, "others will become infected. More work needs to be done to convince the authorities.   But I just don't know how..." she looked away from me with sad eyes.   "Things move so slowly around here.   It's probably not like that where you come from, is it Benjamin?"

"But it will work," I said, unable to answer her because parasitology and government regulations were pretty far out of my area.   "Eventually. If you don't give up."

We continued looking at the negatives and then studied the statistics her group had compiled from the data, checking and rechecking the numbers, until we were interrupted by a door slamming shut.   We all looked up in time to see a small, slender, well-dressed man in three-piece suit entering the lab. The closer he got, the older he looked, as his salt-and-pepper hair was receding noticeably at the temples.   I looked at Tahany who was smiling.

"Oh, I almost forgot.   The surprise.   This is Gamal.   Gamal is a microbiologist like you.   I asked him to show you around the museum tomorrow."

We shook hands as I tried to hide any disappointment that might have been obvious.   A slight, bespectacled man with soft hands grasped mine... definitely the hands of a scientist.   He was older than Tahany, perhaps 50, I guessed.   His English would be turn out to be as capable and yet his delivery lacked the same authority Tahany's had.

"Gamal works in the department, part time with me," she explained, while I resisted the temptation to look around her lab to see where it was he kept his belongings in such a confined space.

"But I mainly specialize in the brewer's yeast of which I hear you are quite interested at the moment," Gamal said.

"Gamal is an expert.   He works at the Stella brewery," Tahany added, apparently feeling the need to give him a reference, "He knows all about yeast genetics."

I had already been surprised on this, my first full day in Egypt, to find not only toothpaste, but beer and wine available in grocery stores, but perhaps I shouldn't have been.   Ten percent of Egypt isn't even Moslem... it's Christian. In fact, the coming of Christianity after the crucifixion of Jesus is the reason Egyptians stopped mummifying their dead.   These strange newcomers insisted that the body was a shell meant for carrying around the soul, a soul that ascends to heaven after leaving behind something no longer needed.[xxxix] And as I learned that first day, many Moslems in Egypt drink alcohol, even though the Koran forbids it.   It's because in life few things are as simple as you might first imagine and much the same way many Christians don't necessarily feel the need to go to church every Sunday just to be good Christians, there are Moslems in Egypt who will allow themselves a glass of wine or a beer occasionally.   It's near where it all began after all, brewing and winemaking.   In fact, Egypt has no less than two major breweries in addition to a domestic winemaking industry.

"And you're in luck. He's also an Egyptologist," Tahany said with obvious pride.   As for Gamal, he didn't say much of anything, contenting himself with studying me.   I was studying him too, but trying not to be as obvious, wondering if he and Tahany were only coworkers, or whether there was something more between them.   There was affection in her voice and I could see the two of them together, if not physically, then at least in an intellectual sense.

"It's not luck, really," Gamal added, "Most Egyptians are Egyptologists." Everyone laughed except for Kenan and I got the impression he would have preferred to return to his work.   From the look of things on his desk, he had a lot to do before calling it a night.   I started looking around for my backpack, hoping my search would be taken as a sign I was getting tired but Gamal was still curious and wanted to know more about me.

"So what made you become a microbiologist?   Everyone has a story, it seems. More often than not I find it quite interesting…if you don't mind my asking."

I had my stock answer prepared beforehand, before I even climbed the stairs.   And I delivered it the same way I always do.   "I was raised on a ranch, but for some reason never wanted to be in the rodeo," I began, "Maybe because I'm not all that tall…at least by Montana standards.   Instead, I just left the roping and calving up to my brothers.   I was more interested in finding out what was living inside the bull than riding up on top of him."

Tahany smiled and I could see my story was having its intended effect of lightening the atmosphere while deflecting some of Gamal's scrutiny from me, except that Gamal hadn't changed his expression much, in fact it seemed as if he was hanging on my every word, weighing each one carefully in his mind to find out which might have the hidden clue inside.   God, I thought to myself, he sure is the studious one.   And I also thought about how I had learned over the years that people as they get older tend to lose the rough edges that once defined them as a child.   Like a rock in a mountain stream that slowly becomes rounder, approaching perfection given enough of a chance, as an adult I had gotten better at evening out my own edges, the ones I didn't want anyone to notice, as well as turning something sad into something not quite as bad as you might have first imagined.

I rarely let on even to those I knew well just why it was I had been drawn into the strange world of microbes and it's kind of a mystery to me even to this day how a single afternoon can stand out as if to the exclusion of all the others. One day after having recently turned thirteen, my family's prize steer – the one my father was planning on breeding – suddenly rolled down into the bottom of a ditch beside the road, as if toppled by a strong breeze. I happened to be the one to see it while on my way home from school.

Our family's ranch is situated where the nearby mountains transition into foothills as they come down to meet the Lewis Valley below. Our valley is broad and flat for the most part and named after the explorer Meriwether Lewis, one of the leaders of the Corps of Discovery. The town of Thistle, where the university is and where I had hopes of gaining my tenure, is in the same valley but at the opposite end. Thistle's one of those towns in the American West that remind me of charms from a bracelet that somehow manages to dislodge and roll, not behind a refrigerator or a stove, but between four mountain ranges, the kind of place where farmers follow the river down into town every Sunday in their pickup trucks to go to church and stock up on supplies while catching up on the latest gossip as they get their hair cut in the town's only barber shop, not the fancy hair salons the tourists on their way down to Yellowstone use, and they wear plain blue jeans without any extra rivets in them, the kind with the straight boot legs that hug the lower portions of their body so tightly, like saran wrap, that they have to walk a bit sideways no matter what the style happens to be in the big cities.

Brucellosis was the vet's pronouncement, which sounded like a death sentence. And it is...at least for the animals. Brucellosis is caused by a tiny bacterium, one so small, even by bacterial standards, a pest new to North America[41] and of course so tiny I would never see it on that day, or any day for a long time to come, not until I entered my second year of graduate school, and yet I was still struck by the ferocity of the thing, the rapidity of something so small as to be invisible, and by the fact that it could bring down something so many times larger than even my grandfather was. I was completely fascinated by how quickly this once noble beast had been rendered helpless by something so tiny. And to a 13-year old boy with an overactive imagination anyway, the whole thing probably seemed like the extinction of the dinosaurs must have been.

---

[41] It was brought west from Europe. The bacterium makes an attractive bioweapon (like anthrax) because it can remain alive outside the host and can even be aerosolized (spread through the air). It replicates only after gaining entry into the host's cell – a macrophage – and is why once a cow gets the microbe, the animal either dies or it stays with it for life.

After the vet left, my father realized it probably began back when the herd's milk began drying up a month or so earlier, which meant my father had missed it, failed to recognize something every rancher is brought up knowing about, the early signs of brucellosis…a gradual decrease in the herd's milk production.

"Must have gotten it from a bison or an elk lookin' around for food," is how the vet explained it as he scuffed a small hole into the dirt road with the heel of his boot, trying hard not to look my father in the eye, and he was right, it had been a harsh winter, worse than usual, with plenty of snow on the ground, which meant not as much exposed vegetation. The vet's younger brother and my father had gone to school together as boys.

"Everyone has to eat, even bacteria," he muttered in that slightly ashamed tone people often use in the valley when they don't know what else to say. It was all clinical the way the vet and my father talked, but after he left that day I was sure of what it was I wanted to do with the rest of my life. I only explained to Gamal and the others in Tahany's lab how my parents had decided at some point that I was too different from my brothers, and that college might be a better choice, but I've always thought it was when I refused to chew tobacco that guaranteed me my tuition money. Something else I didn't mention to the group that evening was how aware I had been that my father suffered in silence a full two years after we had to cull the rest of the herd and my mother always said I got most of my traits from him anyway. I guess finding out ones parents aren't perfect, as is inevitable at some point in every young person's life, is difficult and yet strangely liberating at the same time, for it frees one from the responsibility of being perfect too.

Gamal offered to show me around the department, which meant an interesting, if sore, hour more of retracing our steps up and down the staircase, Gamal discussing the various projects being conducted at the school, periodically taking time to summarize the more interesting posters that caught my eye along the walls. I knew from my own experience that each poster represented the culmination of several months – if not years – of tedious work done by a team of scientists and technicians. Whoever said genius is 99% perspiration and 1% inspiration may have lived before the invention of automatic pipettes and disposable plastic Petri dishes, but he was still right about microbiology in the 21st century.

We wound up outside Tahany's lab as darkness was bringing with it the vulnerability one often feels upon finding oneself in an unfamiliar place as nighttime arrives. Just as I was getting ready to leave Tahany became more serious all of a sudden, guiding me aside by my shirtsleeve.

"Did you work out a time to meet tomorrow?" she whispered.

"Around nine," I whispered back but without knowing why, "The King Tut exhibit is supposed to be less crowded in the afternoon. His earthly goods take up a whole wing of the museum it seems."

"You don't have to worry about him being there," Tahany explained while looking intently into my eyes, "Gamal is my fiancé."

I'm pretty sure my heart cheated me out of another beat as she dropped the f-word, and I desperately hoped it didn't register on my face.

"Oh I'm not too worried," I stammered back, "Really, Tahany. It's been an amazing trip so far. If it ended tomorrow I'd say it was well worth it." I attempted a smile, still not sure if I succeeded, which made my way down the stairs an interesting contrast to my trip up. This time the trip left me winded not so much physically as emotionally. And I was also thinking about how if Tahany's work could be realized in her small lab on the top floor of a plain brick building in the middle of Cairo, how it would deliver so many benefits far beyond even Egypt. Bilharzia, I knew, is a problem in so many places around the world.

I gently refused Kenan's offer of a ride as it was only a few blocks back to the hotel, at least according to the guidebook, and I even managed it without too much delay and without feeling too sorry for myself, either. As I relaxed on the balcony back at the Bedouin, looking out over the three-story tiled roofs, I tried imagining what 1870's Paris must have been like, my stocking feet propped up on the railing, muscles still tingling, then wondered if it was offensive should anyone happen by, showing your bare feet that way. I changed my position to be on the safe side, considering how I'd forgotten to ask the most important question, yet here I was trained to ask questions...all sorts of them. But not those kind apparently. There are no classes in relationships in graduate school and I had managed to get blindsided yet again. I sure was a good one for wishful thinking. She was a catch and Gamal had to know it.

Strangely enough, I wasn't as jealous as I might have been in my younger days...progress of an unexpected sort. No, I was more resigned to being alone than I ever had before.[xl] I leaned further back, consoling myself with the notion that perhaps my own disappointment had less to do with Tahany and more with my own expectations.

One of the reasons I bought the ticket to Egypt in the first place was because of the book I'd come across as a teenager. When he boarded the *Quaker City* in New York, like me Twain was also a bachelor in his 30's. And while on the trip he fell in love, not so much with a woman as with her image. A fellow passenger, Charles Langdon, had brought onboard a photograph of his sister and Twain later claimed that after seeing Olivia Langdon's photo he knew right away they would be married someday. He'd already decided before they'd met. How's that for confidence, I thought? And when they did go out together for the first time, it was to hear another writer, one who had come all the way from London to speak. They would marry two years later, with the first royalty check for *The Innocents Abroad* arriving the same day and the book would soon make Twain famous.[42] I've often wondered whether they shared fond memories of their first encounter the night they heard Charles Dickens speak on his American Tour.

CHRONOLOGICAL ORDER

Blaming it on the jet lag, I awoke with the sun already having beaten me, which will always be a point of contention to a rancher's son no matter which hemisphere he hangs his hat in. I hadn't forgotten to ask for a wakeup call, even going so far as to give the attendant some baksheesh, then remembered that my phone didn't work. I must have slept through the call to prayer too.[43] It was 8:25, meaning I had just enough time to get dressed and get to the museum, sunburned, muscle cramped and all. Unlike that first day, I'd be leaving my backpack behind regardless of what the guidebook warned about theft in big city hotels. If a microbiologist could take time away from his job at the brewery to tell me about the yeast, then the least I could do was to be on time.

A road twice traveled is never as long, so it was a liberating jaunt down to the square without my backpack on. When I got to the same vendor and after buying another glass of juice, I turned north. From there, the top of the rectangular two-story museum was within sight. I bypassed the long line of buses and headed for the front gate. Not surprisingly, Gamal was already waiting. I looked at my watch – 9 o'clock on the dot – not bad considering I was ten time zones away from my own comfort zone.

"So what do you think of her, Ben? Is she old enough for you?" Gamal was craning his neck to look up at the building's roof. He barely came to my shoulders, and I'm not all that tall. "However it depends on your perspective. When this building was first constructed, King Tut was still slumbering in his tomb...just as he had been for 3,000 years, so perhaps it is not as old as you might first have thought. It holds more than one hundred and twenty thousand relics. Just on King Tut's body Howard Carter found one hundred and forty items." Dusty and crammed and "a bit like going through someone's attic" was how my guidebook described it.

---

[42] *Innocents Abroad* became an American bestseller, with numbers approaching that of the Bible. Twain's book was sold by door-to-door salesmen as a subscription, often by Civil War veterans with limbs missing.

[43] I later learned that the call to prayer used to be performed by a gifted caller, or muezzin, but because of Cairo's roaring, nonstop 21st century traffic, the *summon* has since been replaced by a tape recording over a loudspeaker, which is what I probably heard.

Gamal used his pass giving us free access through the guide's door, bypassing a turnstile with a lone guard reading a newspaper. If he had looked up he would have caught sight of an amusing pair, the slight, dark, bespectacled Egyptian in strict three-piece suit without a wrinkle, walking next to the lumbering American in T-shirt, sunburned neck, cowboy boots and jeans.

"We will save the Royal Mummies for later," Gamal explained. "They're in a room all to themselves."

"The only mummies I've seen so far have been on the paper money."

"You know, back in the 1800's, when Egypt's railroads were being built, mummies were in such supply that the engineers would use them as kindling for the boilers…feeding them in like coal."

I couldn't suppress a smile because I knew it already, then told Gamal about Mark Twain once claiming that, while in Egypt he overheard an engineer yell out to one of his assistants 'damn these plebeians, they don't burn worth a cent, pass out a king.'[xli] I couldn't tell if Gamal got the joke or not because his expression hadn't changed. He did strike me as the serious one, even more so than the previous evening. Something was bothering him and I couldn't help wondering if it had to do with Tahany asking him this favor.

"Didn't they grind up mummies for medicine?" I asked in an effort to keep the conversation going. Freshman history was coming back to me in dribs and drabs. Museums are good at reminding me of things I thought I'd forgotten all about and as we entered I caught a whiff of the musty smell they all carry. Fungi and other hidden decomposers were busy spewing out gaseous waste products.

"Yes. Europeans believed that mummies could have magical, restorative powers…because they are so old. Even Queen Victoria took a bit of ground up mummy with her afternoon tea…or so I have heard.[44] Maybe it is not so surprising in an age when patent medicines might be laced with opium, that people would turn to mummies for cures." Peering up over the black rims of his glasses Gamal added, "Because the Tut exhibit will be crowded this morning, we need to begin our tour on the first floor. These rooms are all arranged in chronological order, which means we will be visiting 42 containing thousands of years of history beginning with the earliest relics from pre-dynastic times – some over 5,000 years old…older than writing itself – and ending in the more recent Greco-Roman period."

I thought about how most great museums in Europe, if they were lucky, only *begin* with the Greco-Roman period. This museum actually ended in it, as if ancient Rome were somehow modern.

---

[44] She also was a beer-lover.

"You know," Gamal explained, "civilized people have always enjoyed museums...and libraries. If not for the Library at Alexandria, we would never have known of a poet named Homer, or the playwright Sophocles. In Egypt is where the tradition got started...of copying manuscripts for posterity."[45] I nodded while recalling that exactly one week ago I was grading papers and so was happy to be anywhere else. Just walking inside the atrium was worth the price of admission...if I'd paid one.

"Did you stop to consider why we are on the east side? And your hotel too? It's no coincidence, you know." I hadn't, and admitted as much. "It is because the east is where the sun appears each morning. The east symbolizes rebirth...the side of the Nile most Egyptians went about their daily routines on. It is also why the pyramids are on the opposite side, to the west. The west is where the sun went to go...to die each evening, return to the underworld...so the west has always represented death. You are visiting the pyramids tomorrow I understand?"

"Yes," I answered, slowly becoming aware that Gamal had given this talk before. He seemed comfortable with it, and yet I still couldn't shake the notion that something was weighing on him. Then again, maybe it was just his nature. Perhaps he was moody, I reasoned. I sometimes am in the mornings for reasons I can't always explain. And besides, when you've only known someone for a couple days, it's kind of hard to tell.

"Tomorrow evening you will be traveling to Luxor, Tahany says?"

"By train. Is that the best way? I've got nine days more, then it's back to assigning homework."

"You should be fine," he answered curtly.

We were at the beginning of a long atrium extending before us. Within it were various monuments, some reaching almost to the ceiling, most made of stone, including sphinxes easily recognizable even to a casual observer. Others were smaller and in the shape of people, pharaohs I assumed, some assorted arches and pillars were also present, even a small obelisk helping to echo back the 21st century voices. I was excited and we hadn't entered a single room yet.

"Zoomorphic," Gamal said as he pointed in the direction of the sphinxes, apparently noticing my being enamored by them.

"I didn't know you had sphinxes...with different animals on them I mean...other heads on the same lion's body. Falcons and hawks, there's even one with a ram's head over there. Look. It's as if they're interchangeable."

Gamal then explained how they represented local deities and how these sphinxes once had important jobs: to guard the temples. It was my first

---

[45] Only Egypt with its dry, hot climate has preserved any of the works by Homer directly from antiquity, all of which are on papyri.

surprise of what would morph into many that morning...finding out that not all sphinxes were alike.[46]

Gamal pointed to an inscription from 2000 years before Christ, hieroglyphs chiseled into stone, and with one finger running across he translated it, "The mouth of a perfectly contented man is filled with beer. My ancestors had a sense of humor. They made up words that sounded like those the object made. For instance, they called wine erp, like a slurp when you drink wine. Cats they called miw. With donkeys they said eee-aww. Tahany informs me you used to work with anthrax. That is caused by a bacterium, is it not?"

"Yes," I answered, somewhat hesitantly because it was the first time I'd thought about the anthrax grant since waking up. More unexpected progress. Venturing to an entirely different continent will do that for you apparently.

"My ancestors had a hieroglyph to represent anthrax. As you probably know, it was one of the plagues described in the Old Testament. Anthrax attacked Egypt's cattle...and then our people." [xlii]

"Didn't King Tut have some wine in his tomb?" I asked in an effort to divert the conversation. Gamal didn't know about my recent setback with the grant and I didn't feel like explaining it.

"Yes. Carter found 26 amphorae in the boy-king's tomb after unsealing it. We even know how some of Tut's wines must have tasted. Four were labeled sweet...some others had been imported...all the way from Syria in fact. [47] The wine's vintages were stamped on the jars and provided the name of the winemaker. Ancient advertisement of a sort I suppose. We can only guess at what the colors would have been. Egypt being in a desert means the wine evaporated shortly after the king was laid to rest, but the writings of later Greek travelers inform us that Egyptian wine was mostly red." [xliii]

We continued walking the perimeter of the atrium, Gamal explaining in his formal way how scientists can be so certain ancient people made wines. It reminded me of the way researchers studying grizzly bears at Yellowstone National Park will try to learn about them indirectly if possible, by studying the tracks they leave, including hair follicles, when they brush up against the rough bark of a tree.[xliv] It's a lot easier, not to mention safer, to collect specimens this way both for the bears and the people, than it is tranquilizing them. Likewise, when looking for signs of ancient wine[48] it is the chemical traces left behind in the jar, its molecular fingerprint so to speak, that one

---

[46] It's likely that Queen Cleopatra never laid eyes on the Great Sphinx since it was already covered by sand when she began ruling over Egypt in the first century BC.

[47] Wine is perhaps the most complex food, flavor wise, thanks to the yeast's unique chemical contributions.

[48] Ancient beer can be identified on pottery thanks to *beerstone* – a calcium oxalate mineral left behind by precipitation from the beer – in the Zagros Mountains of the late 4th millennium.

looks for rather than the wine itself. It turns out the DNA of yeast has been recovered and sequenced from the bottom of wine jars from ancient Egypt, stored in a database of a computer at NIH no doubt.

"Clay jars and amphorae are good at holding onto organic molecules. For thousands of years, even. There is a clay[49] winemakers still add to their wines to remove impurities like proteins…so the wine does not turn cloudy. It works because clay is negative and the proteins in the acidic wines are positive in charge. Does this sound familiar?"

"Chromatography," I answered confidently. It was all too obvious where he was going. [xlv] As most folks with even a passing interest in science know, opposite charges attract each other. Negative particles attract positive ones and vice versa. It's the same force that holds electrons and protons together inside of atoms. Separation chemistry really can be that simple sometimes.[xlvi] Often, to separate one chemical from another in a modern lab – different proteins from a bacterium or a yeast for example – we use ground up clay, highly purified and loaded inside a glass chromatography column (tube) because the clay particles bind selectively and, just as importantly, reversibly, to specific proteins as the proteins all pass through the clay from top to bottom. The implication is that ancient amphorae from the Middle East behaved like separation columns do today, a strange but fortunate coincidence for molecular archeologists. For the Egyptians, clay had more practical uses, though. The Egyptians coated their beer jars with clay during storage and it probably even helped clarify the beer, freeing it from impurities, as well as preserving it.

"Tartic acid is a good chemical marker for grapes," Gamal pointed out, "One pigment in grapes is a small molecule called syringic acid. It also binds to clay. So tell me, to a bacteriologist, is the yeast beginning to sound more interesting?"

"I suppose," I answered, "But I'm still not sure if I have the enthusiasm it takes."

"Well you have been involved with much simpler organisms, ones that are much better understood than the yeast. But what if I told you that, if not for the yeast, we would not be where we are today?"

"How's that?"

"If not for fermentation, civilization wouldn't have begun when and where it did. What I am saying is that the first farmers settled down and raised crops not so much to eat as to drink fermented beverages, which are rich in calories…made from wheat or grapes and fermented into beer and wine by microbes. [xlvii] Imagine a hunter-gatherer experiencing alcohol for the first

---

[49] The clay is *bentonite*, named by a geologist in 1898. He found a large deposit of it near Fort Benton, Montana. During the letter attacks in 2001, it was incorrectly claimed by the media that anthrax spores in the letters had been prepared (i.e. weaponized) using bentonite.

time quite by accident.   A difficult existence full of uncertainty is all of a sudden transformed into a relaxing, mind-altering one, a drink that can be shared with others, with a unique taste and a belly full of calories as a bonus."

LIQUID BREAD

I told Gamal about how, as a boy back on the ranch, I used to watch cattle in the winter dig down to get the grain at the bottom of a pile, down where the oxygen was scarcer and the microbes more plentiful.   These microbes were fermenting the grain into alcohol...creating their own beer...as if tiny workers in a microbrewery.

"Even animals enjoy getting tipsy. It is quite possible primitive people watched and learned from the animals all around, cultivating plants they knew would turn to alcohol more easily. Birds prefer eating the rotting berries more than the sugar-filled ones. Have you ever seen a drunken blue jay? It is quite a site."

"Animals are smarter sometimes.   Which is probably why I've never seen a horse placing a bet on a person."   Even though he seemed to be loosening up more, Gamal still didn't get my attempt at a joke as we turned the corner of the giant hall and came upon a glass case containing some clay pots. Whatever it was bothering him may have been losing its grip as we ventured further into the atrium.

"Fermentation is a way of preserving food.   It is the culinary art of allowing food to rot...but in a controlled way.   If food decays only a little, this decay can actually be good because it helps preserve the food.   Strange how that works, but in an age before refrigeration, fermentation would have been one of the important ways people had of keeping calories safe."

He explained how alcohol produced by yeast is not only an aphrodisiac but also an antiseptic, and why most mouthwashes today contain 20% alcohol.   In fact, alcohol will kill the yeast too, eventually, which is the reason wines are seldom above 12%.   The brewer's yeast will succumb to its own waste.   True enough, waste can be deadly if allowed to build up.   One of my earliest memories was watching the Apollo 13 astronauts on TV on their way back from the moon after the accident happened and how they nearly died because of the buildup of their own carbon dioxide they exhaled, which became concentrated inside the spacecraft.   The yeast release carbon dioxide too when they break down their food.[50]

---

[50] Many of the victims at Pompeii may have been killed not by lava but by carbon dioxide gas released from the volcano ahead of its pyroclastic surge.

We walked some more until Gamal, ever the perfectionist, stopped suddenly in his tracks. "Oh, I nearly forgot," he exclaimed, "the slightly acidic nature of fermented beverages. The yeast also discourages bacteria because of the acid it makes, but perhaps you knew that already…about the inhibition of bacteria using acid."

"It's why vegetables like tomatoes are easier to store in cans. They're high in acid."

"So you see the wild brewer's yeast begat the first agricultural settlements, in places like Turkey and Syria, because they gave us the ability to store liquids for longer periods…years if necessary…very important when you're talking about people living closer together in large numbers for the first time and having to share food. As for the Nile valley, it was a lack of water that encouraged the people to settle initially. Before the drought, they lived scattered in what was then a much larger Delta, portions that have since become desert. Changes on the Earth caused these marshes to dry up, so the inhabitants had no choice but to move closer to the Nile, and thus closer to one another. Egypt came about by a sort of…um…not explosion. What is the word you have in English…the opposite of explosion…?"

"Implosion?" I ventured.

"Yes. Thank you. An implosion. People living so close together for the first time meant they needed to be cared for…become better organized. It is no coincidence that hieroglyphics were invented when people were so much more in contact. They needed to keep track of supplies. Beer was a food…for them liquid bread. A source of calories. Famine due to the fickleness of the Nile caused a great many problems my ancestors lived in fear of, and for good reason, as we shall see. The first written language was numbers, boring yes, but important nonetheless, for they needed to keep track of how much beer could be brewed from a given harvest. In fact, every step in brewing has been found depicted on the walls of a tomb in Egypt somewhere."

"How about wine?" I asked.

"Are you asking about evidence? If so, then traces of wine have been found using techniques like mass spectrometry,[xlviii] from clay vessels dating to about 5400 BC in Iran.[xlix] I believe that is the oldest confirmation of wine we have. Every civilization has taken advantage of the brewer's yeast, or a relative, to make a fermented beverage. When Christopher Columbus sailed to the New World, he found the Native Americans already fermenting agave, manioc, and maize. And further north, in your neck of the woods as you like to say in America, they were making alcohol out of birch bark."[l]

I told Gamal about a display I came across as a student on a field trip to a museum along the Oregon Trail and about how the early pioneers on their way west used the succulent growths at the tops of pine trees to brew beer when nothing else was available.   There are sugars in pine needles as in other kinds of leaves.   They also used what was left of their dried vegetables, purchased before leaving St. Louis, to brew beer if they had any left.   They would supplement this brew with fresh blueberries gathered along the trail too.[li]

"Oh, and I almost forgot," he said, "Beer is a source of B-vitamins, and minerals like magnesium and phosphorus.   Especially the darker beers. Bread and beer are the two most important staples wherever grains are grown." [lii]

AN UNEXPECTED DETOUR

I was curious how Gamal had ended up working with the yeast and finally saw an opportunity to ask.   And why wasn't he still at the university, at least in an official capacity?   He answered as we began exploring a room devoted to Egyptian furniture, which was all wooden.[51]

"My original work was with another species of yeast…one that causes a disease. Thrush. Have you heard of it?"

"Heard of it."

"It is also called candidiasis and if you had children you most certainly would have seen it in the form of diaper rash.   Thrush is a painful yeast infection in the lining of the mouth and tongue, especially for those with compromised immune systems.   AIDS patients for example.   We all have this yeast in our mouths, but our immune systems keep it in place.   Doctors in Japan have found patients with so much *Candida* yeast growing in their intestines, they can make their own alcohol…enough to be legally intoxicated. We all have a tug of war going on inside us between microbes and our immune systems all the time.   Because fungi like *Candida* are more like our own cells than to bacteria, it is correspondingly more difficult to rid them from the body without harming our own cells."

"Trying to kill fungi with antibiotics is about as much fun as trying to teach a cat walk backwards."

"Whose cat can walk backwards?"

"That's okay. I didn't mean to interrupt."

"The same antibiotics that work against the fungus damage us, which creates side effects.   I was looking for new and useful compounds, small molecules actually, that could kill yeast blooms while leaving human cells unharmed."

---

[51] The Egyptians also invented folded furniture. A folding chair was found in King Tut's tomb.

I knew it was true what Gamal said about the yeast being more like our own cells than to any bacterium it shares the soil with, in fact this was the very reason I had gotten the grant in the first place: to make human proteins inside the brewer's yeast...proteins that are more similar to our own than to those made by a bacterium like *E. coli*. A protein being more human means less likely to cause side effects when given as a medication. Gamal had been doing what's called a *drug screen*.

He was silent for a moment, as if trying to make up his mind on whether to let me in on something or not. Finally, he must have decided.

"What happened is that I lost my lab. I tried to hang on...for some time, actually. Even Tahany tried to help out but gradually...what with no money coming in...which of course means no bench space...it is the same for you, isn't it? But at least I was able to turn my attention towards the brewer's yeast. I am doing quite well now thanks to *Saccharomyces*. But if you had told me just three years ago I would be working on making a better tasting beer I probably wouldn't have believed you." I didn't get the feeling he wanted to change the subject so I didn't say anything. "You know, in today's world, your thoughts are still on your career. But if you had lived in ancient Egypt and were in your 30's you would have been thinking about what your tomb should look like. Making arrangements for the afterlife was common even for someone your age in ancient times."

I was only half-listening for I couldn't shake the notion that, due to my losing out on the big anthrax grant, Gamal and I were traveling a more similar road than I had first imagined. At least he was resourceful enough to have landed on his feet...having found a place in industry. A lot of us in the states were turning to private industry too, biotech companies for example, as the government left many young scientists withering on the vine. Even though I was trying hard to stay in academia I was still being forced to switch fields, from a bacterium to the yeast, which may not sound like much but is, in fact, a major step biologically. I'd just as soon *not* have changed fields to be honest. Most microbiologists would definitely consider what I was doing – moving from a prokaryote like bacteria to a eukaryote like the yeast – not to be a very good career move on my part, but I needed to keep the money flowing in. The brewer's yeast was my only chance now. But worrying about finances is hardly new. As Gamal would mention later in the day, even scribes in ancient Egypt had to make extra income sometimes by taking in students and teaching them to read and write hieroglyphs.

Gamal stopped at what was for him a waist-high model of an Egyptian house from the Old Kingdom period, made out of mud bricks; mud being another gift of the Nile.[52]   It's where he picked up his lecture on the brewer's yeast.

"If I have given you the impression that all I have told you is proven, I must apologize.   In fact, no one knows with any certainty the world's first fermented beverage."   He stared off for a moment with a blank expression on his face and I almost regretted asking him about his research.   Something was clearly weighing on him and I didn't want to make it worse, whatever it was.   I had so many questions, though.

"Did you know my ancestors used to keep bees?"

"I hadn't.   There's a lot they don't teach us about your part of the world. Maybe they think we'll learn it all in church."

"It seems honey is a good source of the brewer's yeast."

We moved on with what would become a familiar pattern throughout the rest of the morning and into the afternoon, Gamal talking, I listening, as we wove our way in and out the museum's many rooms.   At one point I imagined myself as a thread trailing behind Gamal, the probing needle intent on stitching the museum together like some kind of quilt.   I listened and watched because it was becoming clearer just how much he knew about the yeast and besides, you're never supposed to stop a horse from galloping in the right direction just to give him sugar.   Still, I couldn't help wondering if the loss of his laboratory hadn't taken some kind of toll on him.   Once doing independent research is in your blood... let's just say other things can seem mighty tame by comparison.   When not showing visitors around the museum or working with Tahany in her lab, Gamal was probably taking orders from some technician half his age on how much contamination was acceptable in his beer vats; a far cry from being on the front lines fighting a disease.

Another of our unscheduled stops, this one brought about by the soreness in my back, led us to the nearest wall.   I leaned against it while Gamal continued describing the first fermented beverages and how they might not have been either beer or wine, but honey.   In the Middle East, high caloric drinks like mead materialized, often without any help, discovered after rainwater diluted a honeycomb inside the hollowed out trunk of a tree.   The ancient Egyptians were accomplished beekeepers and, as Gamal pointed out, pure honey is high in sugar and therefore a good source of wild brewer's yeast.[53]   It seems the yeast lies dormant, trapped within the viscosity of the honey... that is until the inhibitory sugars get diluted by rainwater, whereupon these same sugars reach a lower concentration and so the yeast springs back to metabolic life, fermenting them into alcohol.

---

[52] Dividing up 3,000 years of Egyptian history into the "3 *Kingdoms*" we still use today was the notion of the Greek historian Manetho, who lived during the 3rd century BC.
[53] The Egyptians had over 900 remedies requiring honey as an ingredient.

For the Egyptians, honey might have been what for us is the equivalent to freeze-dried yeast on a supermarket shelf. Whenever they needed its power to start a new batch of beer or wine fermenting, they may simply have added a cup or two of honey. Honey would have been what we today refer to as a "starter culture", not so different from how I sometimes transfer E. coli from an old Petri dish onto the end of a sterile toothpick to inoculate a fresh batch of broth just by tossing the toothpick in. Sugar is sugar, chemically. The yeast doesn't care where this carbohydrate comes from.

"Yes. It is pretty amazing, isn't it?" Gamal had noticed my increased interest. "My people were the first microbiologists and yet they didn't know about microbes. They would have passed fermentation off as the will of the gods, not so different than the rise and fall of the Nile I suppose...what for them was their recurring miracle. At one time there was no oxygen in Earth's atmosphere and so most cells had to make due by fermenting sugar to get all their energy, like yeast do today.[liii] In some ways, that makes the yeast a relic...more ancient than anything in here, that's for certain."

"Maybe we could start a museum for microbes," I added. "And we could arrange them in different rooms depending on how ancient their metabolisms are."

"It would have to be a big museum, though. Yeast and human cells are separated from a common ancestor by about a billion years." He laughed and I made a mental note to change the subject more often as it seemed to cheer him. I looked around, noticing for the first time that the museum had filled to capacity. Still sore from the previous day, I began looking for a place to sit, except that all the benches were taken by elderly tourists and young mothers caring for small children.

"Ancient Egyptians would have been well aware of changes...of what constitutes a normal fermentation...and felt a certain reassurance whenever it took place," Gamal continued. "Their lives could have depended on it. Ancient people lived in constant fear of famine. They kept careful watch over the changes fermentation brings as closely as they watched the height of the Nile each season, monitored their food's colors and smells, its textures, the roiling of carbon dioxide bubbles as the sugars turned into alcohol and carbon dioxide.[54] If all went well, the food would have changed predictably, from a sweet one to a slightly sour one. Next the distinctive aroma from the alcohol would have begun mingling in the air just above the pot as they whiffed the liquid with their cupped hands." Gamal closed his eyes as he described the scene, "Whenever they saw fermentation taking place, it would have filled their hearts with a warmth I suppose. Fermentation was the first process ever taken from nature and reproduced in a controlled fashion. Here was an

---

[54] If conditions are optimum, sometimes this roiling can be so rapid as to be audible, and is described in ancient texts.

event that could be found in the natural world, as unpredictable a place as we can imagine, and yet they made it predictable given enough skill. They improved upon it and eventually made fermentation into an industry. It may have been a source of empowerment and pride for ancient people, allowing them to feel in control of a very mysterious world."

A SHEPHERD'S LIFE

Gamal then explained how the brewer's yeast makes no powerful toxins, yet lives in perhaps the most competitive place on Earth...its soil. It is no coincidence soil was where scientists first obtained antibiotics – penicillin and streptomycin – in the 20th century. There's so much rotting food and competition for it down there, microbes need all the help they can get to survive in places like this, and they have to be good at making use of nutrients because everyone's waging war. But rather than produce a toxin, or an antibiotic, the yeast simply avoids this and instead relies upon fermentation...and patience. Like other fungi, the yeast doesn't have a problem waiting things out. Unlike plants, fungi don't need the sun's rays to make their food.

We walked out into the atrium and then strolled into another musty-smelling room where Gamal picked up his talk. "Strictly speaking, the alcohol is a poison since it is good at keeping unwelcome bacteria out. Unlike most bacteria, the yeast doesn't require oxygen to extract energy from food. It lives at the bottom of compost heaps and so bores it way through rotting vegetation, exploiting and digesting the plant's sugars as it goes, making alcohol while keeping oxygen out. Reducing the competition. There's not much oxygen inside decaying things anyway."[55]

"Fungi are nature's hidden kingdom," I added.

"Yes. And as for bread dough, the yeast was enlisted only later. The very first breads were flat, unleavened breads. The earliest evidence for leavening dough is from 4,000 BC. People in your part of the world would say meat and potatoes. In Egypt we say bread and beer...but they mean the same. The Mexican cuisine is based on not having a fork," Gamal continued, "Flatbreads like the tortilla are an ancient cooked gruel. Wrapping food in flatbread is a convenient way to keep your hands clean. Think of a bean burrito. The very first breads were unleavened and started out mostly as water. Then they cooked and hardened these liquids onto the surfaces of hot

---

[55] Some species of fungus like *Aspergillus flavus* make a poison called aflatoxin so powerful that a single molecule of it can mutate DNA. Indeed, the most toxic substance on Earth is not made by a nuclear reactor, but by a tiny bacteria cell: botulinum toxin (the active ingredient of Botox).

stones, producing a bread with very little moisture, one that could be stored and eaten anywhere. It is why we so often think of the nomadic shepherd's life as consisting of unleavened bread. And why Moses' people were in such a hurry once settling to begin leavening dough again. The smell of bread baking in a brick oven or inside a clay pot undoubtedly reminded them of civilization.[liv]

"The world's first comfort food?"

Again, Gamal didn't seem to notice my joke. "Letting bread dough rise is a way to enhance not only the bread's texture but also its flavors... to increase the variety of molecules responsible for taste,[lv] enhancing its digestibility. Here is where the Jews learned of it. This isn't even the oldest story in recorded history. That distinction goes to the Epic of Gilgamesh," and with that Gamal explained how one of its main characters – Enkidu the wild man – became civilized, made more human if you will, only after drinking beer and eating bread for his first time.[56]

"Beer and bread have always been closely related, which is why, when we find an ancient brewery, it is always next to a bakery here in Egypt."

"Because both are made of grain?" I ventured.

"And the yeast. These first beers were accidents, intended as liquid breads, but someone left it out too long... long enough for the yeast to settle in from the air, or perhaps from the Nile River's water, to colonize the gruel and begin fermentation."

He explained how the first doughs were made after chewing wheat kernels in the mouth. And how enzymes in our saliva would have begun the process of chopping up the plant starches into sugars required by the yeast. It's the reason when you chew a piece of bread or raw potato long enough in your mouth they begin to taste sweet. The enzymes in your saliva are cleaving and liberating smaller sugar fragments from the ends of long starch chains, getting these simpler sugars ready for absorption into the blood.

The first cities came about in Mesopotamia, and increased crowding meant more opportunity for spread of disease, and therefore an increased demand for fermented beverages. Beer and wine have always been safer than water to drink, wine because of its higher alcohol content and beer for its having been boiled during brewing.

"Did you know the Bible never mentions water for drinking? So it's not surprising that half the barley grown in Mesopotamia would have been earmarked for brewing. The first laws in recorded history were chiseled into stone at the behest of King Hammurabi and included regulations on how beer should be sold.[57] He used stone to emphasize the permanence of these

---

[56] ~3000 BC Sumerian tablets. Gilgamesh is also the first story in literature to feature a woman character – a Mesopotamian tavern owner who serves Enkidu his first beer.
[57] 1760 BC.

laws.   Not only was the price of beer fixed by the king in Babylon, but the punishment for owning a beer house and watering down beer for your customers was death by drowning."

We passed a display of ancient, yet familiar-looking, plowing tools while Gamal added with what sounded like pride that the Mesopotamian bread would have been more like flatbread, not leavened like bread in Egypt was.

"What about wine in Mesopotamia?   Didn't they have wine too?"

"Yes, but grapes did not grow as well as grasses like wheat and barely. Grapevines require more attention in difficult climates and therefore would have been tended to by priests in temples.   So wine was a luxury, reserved as a holy drink, or used in sacrifices.[lvi]   I am going to mention wine and religion more in a moment, but for now I will only call your attention to how similar religious ceremonies are in so many cultures...and to a priest in ancient Mesopotamia tending his grapevines.   When you care for a grapevine, you bring it an offering of water, don't you?   Then you kneel before it and tenderly sprinkle the water onto the ground...like a libation. Taking care of a grapevine is not so different than kneeling and praying before a god at an altar.   In both cases you're making a personal sacrifice.   When the Greeks – members of what most assuredly was a wine culture – visited Mesopotamia thousands of years ago, they claimed their wine god Dionysus fled from it as the people only wanted beer."

"But how about the yeast?" I asked, "Didn't they have some idea of the yeast?"

"They would have known of it indirectly.   The yeast is why they used straws to sip their beer.   Straws helped them avoid chunks of yeast and other impurities floating on the surface.   It was all a mystery to them, brewing, leavening and winemaking.   It is why there were so many deities for brewing...Ninkasi is the goddess most often mentioned...but there were others...many others...too many to mention here."[lvii]

Gamal continued without taking his eyes off the display, "Personally, I believe Egyptian civilization will turn out to be every bit as old as Mesopotamian, however it is true that the first written evidence for wine and beer come from clay tablets in Iraq.   The world's oldest recipe is for brewing beer."[58]

STARTER CULTURE

There is something comforting about history, its timelessness, the way it joins us with the past.   All roads lead to Rome as they say and Gamal seemed to enjoy immersing himself in history as much as I did...taking refuge

---

[58] Wine, 2750 BC.

in it almost.   And he was smiling a bit more as we ventured further into the museum's interior, passing twin barges with sails unfurled. Gamal was hitting his stride, confidently walking with his hands firmly clasped behind his back, staring intently at the floor, concentrating on his mental notes I supposed.

"Ancient people didn't know about microbes, but like brewers and winemakers today, they did know that the sweeter the starting material, the stronger the beverage would be, which is why the Egyptians and Hittites sometimes used raisins instead of grapes for making wine.   There is more sugar to a raisin than to a grape, more starting material for the formation of alcohol.   One can get the alcohol content up to 16% by starting with raisins. The Hittites were fond of saying that the 'raisin holds the wine in its heart'.   It seems wine could be so important that the first thing an invading army did was to destroy all their enemy's vineyards.   It was a way of fermenting unrest among the people, of encouraging resentment for their king's lack of authority. If the king couldn't protect his grapevines, then how could he possibly protect his own people?   My ancestors also made wine out of dates...in the oasis communities.   They carried this date wine in amphorae all the way to the banks of the Nile on the backs of donkeys, where it was offloaded and shipped around the Mediterranean.   But do not be fooled. Egypt was always a beer culture.   It never became a true wine culture like the Greeks or Romans later on.   Still, my ancestors did appreciate variety in their lives, just as people do today."[lviii]

Arranged before us inside a long glass case were shelves with toys on them...wooden ones that Egyptian children were given as gifts over a thousand years before Julius Caesar came here, some with the original paint still on them. I tried to picture a young Nefertiti tugging on one as a child.

"They have strings to pull their moving parts.   See the crocodile over here?" Gamal indicated by tapping the glass.   "Its mouth opens.   And this one," he moved to the other side to get a closer look, "It is three dwarves standing in a line.   Can you tell?   It was carved during the 12th dynasty. Pull the string by way of tiny pulleys and they still dance."

Another statue had been carved into the shape of a person bent at the waist, and it was obvious even after dozens of centuries that it was someone engaged in toil.   The figurine turned out to be the main reason we were in the room.

"This person is kneading dough.   While the rest of the ancient world was subsisting on flatbread, we Egyptians were taking leavening to new heights. Leavened bread also has more protein because of the yeast.   The yeast helped build the pyramids, doing their part by providing nutrition and sanitary drinking water in the form of bread and beer."[59]   Inside another glass case was what appeared to be a chess table, one that was on runners, looking somewhat like my grandmother's rocking chair.

"It is called Senet.   And all Egyptians played it, from the king on down to the workers in the field." [60] Gamal then went on to explain how more variety was obtained by making bread into different shapes, some as crescents, others more pyramidal, or perhaps in the form of a woman.   They also added honey and anise...even cumin.   Bread was so important that they gave up eating it for periods of time during mourning, just as many Christians do today for lent...as a sacrifice.   Bread may also have been toasted and used as stuffing inside fowl caught along the banks of the Nile.   Scenes in Memphis dating back 5,000 years portray every important aspect of baking.[lix]   I was suddenly reminded of the elusive papyrus I had yet to locate.   Maybe I'd try the museum gift shop, or perhaps Luxor, I resolved.

"They added the yeast by using dough from a previous batch of bread as a starter culture.   They may also have used bread to start beer brewing.   As long as the bread was not heated too thoroughly, the yeast would survive in toast, still able to begin fermentation of the next batch of beer wort given a chance.   Bavaria is a fine example of a beer culture.   Tahany says you have a stop over in Germany on your way back home?"

"I have to change planes.   A layover of about ten hours I think.   Why?"

He apparently didn't hear me as I followed him through the open door and back out into the giant atrium again.

As we walked along, the thought that perhaps Gamal had picked up on the fact that I was one of Tahany's admirers became a more likely explanation for his strange mood.   Maybe that's why he seemed so awkward, unlike the previous night.   For all I knew, maybe she had even mentioned something about my flirting with her.   And he was older than Tahany, while she and I were about the same age.   The thought left me wondering if perhaps Gamal wasn't the quiet, jealous type.   After two days of knowing someone, it's kind of hard to know these things.

"Language is an amazing thing," Gamal began, "From written records, we know that Ramesses III distributed something like seven million loaves of bread to his temples for distribution, and that the average worker got three loaves and two pitchers of beer a day along with some vegetables like onions."   We stood with our backs against the wall as a dozen school children began parading by, their teachers reminding me of shepherds, with one bringing up the rear, apparently keeping an eye out for stragglers.

---

[59] The ancient historian/tourist Herodotus compared Greek bread with Egyptian bread, marveling that "We all worry about food fermenting; but the Egyptians deliberately make dough so that it does ferment."
[60] Senet was a favorite pastime of King Tut's. He even had a portable version to take along on his travels.

"The temples devoted to Ra, the sun god, had their own vineyards. We have always stomped grapes in Egypt to make our wine, you can tell by the wall paintings. They show men holding onto short pieces of rope above their heads so they can steady themselves while stomping. They knew pressing grapes with stones could break open the seeds, infusing the wine with an unwelcome, bitter flavor. But the wine was not what you would call fizzy today, nothing like Champagne. In fact they went to great lengths to avoid carbon dioxide building up inside their wine. The Egyptians drilled holes in the clay plugs to let this gas escape. Then when the yeast had done its job, they sealed the plugs back up to keep the wine from spoiling due to oxygen in the air."[61]

After the last of the children and their teachers were past, we found our way into the next room.

"The yeast cell has always been a mystery and would remain so until the Enlightenment in the 1600's, so perhaps it is only fitting that the yeast was so influential in religion. In the Middle Ages its mysterious fermentative powers were referred to as 'godisgood'."

I had wondered about wine and bread's relationship to religion ever since taking communion in church as a child. None of my family could answer my questions. Everyone just performed the ritual. I realized that year that part of being religious meant not asking so many questions.

"Ancient people had many gods and some were caretakers. It is similar to the way you visit a supermarket today. If you want to ward off mosquitoes you buy an insect repellent. When you need to clean something, you purchase a detergent. Well with ancient people they chose a god for what they needed, took it off the shelf often quite literally, and then made arrangements, a kind of a tit-for-tat deal."

"You scratch my back I'll scratch yours I think you mean," I added gently.

"Yes. There is a difference, isn't there? Tit-for-tat implies something bad for something bad, doesn't it? Well for the Sumerians, it was Ninkasi who watched over the brewing, for us Osiris, for the Greeks Dionysus, the Romans of course had Bacchus to call upon.[62] The fear was always a *stuck fermentation* after all that work harvesting and preparing, of where instead of beer and wine one ended up with sour-tasting vinegar or far worse...putrefaction. Vinegar could at least be useful as a weak acid and as an additive to make food taste sour or for preserving drinking water. Homage

---

[61] Champagne was created by yeast while still fermenting in the bottle. The French were the first to do a *second fermentation* inside the bottle, making the first sparkling wine, or Champagne. The yeast is capable of producing pressures inside a wine bottle exceeding that of a car's tire (4 atmospheres of pressure) which is pretty small compared to *Magnaporthe grisea*, a pathogenic fungus that can produce up to 80 atmospheres of pressure in order to penetrate the tissue of its host plant.
[62] The Romans considered beer a barbarian drink, calling it a poor imitation of wine.

needed to be paid to the spiritual world, prayers said correctly, wine and beer libations sacrificed onto the ground in appropriate proportions. Yet they also knew how air could derail the fermentation process. One solution ancient people came across was to use long-necked vessels for storage. This helped keep the air out. Another solution was to fire-harden clay amphorae, making them more... how do you say... immovable? To the air, I mean. Some have even been discovered 8,000 years old."

I provided him with the word impermeable and then considered how familiar amphorae had been to me from a youth spent watching the Undersea World of Jacques Cousteau. All ancient shipwrecks on TV seem to have amphorae scattered about them, it's a requirement almost, containers shaped like a person's body, but without a head; just a curving trunk and slender neck. Some had a pointed end so they could be stood upright in the sand and handles at the top so two workers could carry them side-by-side. Different cultures had different styles it seems, yet the same basic shape remained unchanged for thousands of years, right up through the Roman Empire.

"The Greeks believed the god that gave humanity the gift of wine was Dionysus. Did you know his story is similar to the one you have for Jesus? Or that Egyptians were the first to believe in monotheism and resurrection over a thousand years before the Bible was even written?"

"No," I answered. Another thing I hadn't learned in church.

"Yes. Both Jesus and Dionysus performed miracles like turning water into wine, and both of their bodies would come to be represented by bread, their blood by wine... to be consumed during a ritual meal by their followers. Strange how that is, in fact it is probably why Christianity caught on so quickly in Egypt in the first century. And both Dionysus and Jesus rode donkeys and arose from the dead. Both were preoccupied with peace while on Earth. Both were viewed by their followers as someone who could liberate others from a world filled with pain. It is no coincidence the Greeks believed some of their gods were half human, like Jesus. Both came here as saviors, you see? When you stop to think about it, it is not so surprising how Dionysus could also be the god of wine. His father was the most important god of all, Zeus, his mother the mortal Semele. Dionysus ruled dreams and intoxication. When ancient people drank wine they felt more connected to gods like Dionysus I suppose."

In spite of the similarities between the two, I'd barely heard of Dionysus before.[ix]

Gamal continued, "I sometimes like to picture the yeast as an epic hero, sacrificing itself for the wine as the alcohol gains in strength. The invention of the theater came about as a result of the wine god too. There was one Greek who, in 540 BC got up and acted out the part of Dionysus during a festival. His name was Thespis, and his round stage was inspired by the shape of a threshing floor. Anyway, the debut of Thespis turned out to be the first play in history.[lxi] I sometimes wonder where Hollywood would be without Dionysus."

"Or the yeast," I added.[63]

Rather than spending our time talking shop, as one might imagine two microbiologists doing, we used the next hour or so to gaze at artifacts without saying much that I remember. On one level, we were like two schoolboys separated from their group as we took in objects that had been brought here, often at great sacrifice. Sometimes the temperature inside a tomb in Egypt can be high enough to melt candle wax. It's one of the reasons the government put electric lighting in Egypt's tombs as soon as it was available. I'm not sure how long we lingered because it's easy to lose track of time in a place like that, perhaps not so surprising Egypt being the closest thing to timelessness I've ever experienced other than sleep. Hours can slip by disguised as minutes when you're immersed in this degree of antiquity. Trying to keep track of ordinary increments like minutes and seconds seems almost laughable here.

There was a long stretch of silence before Gamal picked up again exactly where he left off by talking about religion and wine, apparently following some sort of an inner script he'd prepared.

"The first person to make wine in the Old Testament was also the first to get drunk on it. That would be Noah, of course. Even the practice of keeping new wine in old animal skins is cautioned against in the Old Testament, a reference to the buildup of carbon dioxide in the wine I suppose. The yeast can make this gas so concentrated it literally splits leather apart. The very first miracle performed by Jesus was turning water into wine if I recall. And as one who was fond of using stories to get his ideas across to people from all walks of life, Jesus likened himself to 'the true vine' while his Father he called the 'winemaker'. His followers were for him the branches of the vine.[64] Jesus also referred to himself as 'the bread of life' and his goal was to feed all of God's people. The Bible mentions wine no less than 200 times. I am not sure about bread."

In this way Gamal returned again to the topic of bread and the yeast. "It is still a mystery why God told the Israelites to eat only unleavened bread in the desert. Perhaps it had something to do with the unclean connotation leavening always carried. It is in our natures, I suppose, to fear that which we do not understand...things humans have little control over...and it must

---

[63] It's no coincidence that democracy and acting came about at the same time and place. In ancient Athens acting would have been a skill necessary to persuade an audience, just as it would have been to sway voters to win an election or prevail in court.

[64] "He who dwells in me, as I dwell in him, bears much fruit." John 15:5

have been rather disconcerting for ancient people to watch a once flat, inanimate piece of dough rise up all by itself... expand as if it were harboring a ghost.   To them, the flour had become contaminated... possessed is perhaps a better word... possessed by a spirit that could be carried along from the previous batch of dough into the next."

"The yeast made its initial appearance in bread as a spirit?   That's an interesting thought."

"Or perhaps they viewed the whole arrangement more like impregnation. I am often reminded of the baker in the ancient city of Herculaneum, the one who before the volcanic eruption had placed carved phalluses above his ovens to ensure his bread would rise and he would be prosperous."

"He was thinking of impregnation?"

"Perhaps.   What we today would refer to as inoculation of media.   You have seen it in your church, how during communion the bread used to represent Christ's body is always an unleavened flatbread.   A wafer perhaps? Pure and round, isn't it?   The apostle Paul compared leavening of bread to a sin, and warned that a small amount of fermented dough already raised could do the same to new dough.   Too much salt kills leavening, they knew this and today we know why.   Too much salt inhibits the yeast's metabolism and so slows the formation of gas.   It is why the New Testament instructs Christians to 'have salt in yourselves and be at peace with one another'.   And yet it is also says that 'a little leaven leaveneth the whole lump' so I suppose it depends upon which verse you read when it comes to the yeast. [65]

---

[65] Egyptians didn't have soap but used natron salt instead. They also made use of salt as a mouthwash, which would have killed bad breath-causing bacteria.

YEAST AS SAGE

"Wine played an important part in bringing about western philosophy too. You have been to a symposium before, haven't you Ben?"

"My first one was in graduate school," I didn't mention the topic was on bioterrorism.

"Well the Greeks invented them. Essentially they were drinking parties but often with a more profound purpose. The Greeks were the first to democratize wine too. At a typical symposium, Athenians would gather in someone's home, drink wine while enjoying various entertainments... talking about important issues of the day. Sometimes they would discuss subjects like 'what is the meaning of love', or 'what is logic'. [lxii] People simply showed up uninvited. The only requirement was to be a free citizen. Some who planned on doing a lot of drinking even brought their slaves with them... to carry them home afterwards."

"History's first designated drivers," I quipped, but Gamal still didn't seem to get my jokes and so continued on without changing his expression.

"Plato described a symposium that took place around 385 BC." [66]

"It's hard to imagine the scientific method if not for the Greeks," I added, "Einstein once said that he'd rather study the ancient Greeks than science."

"As Euripides said, Dionysus gave mankind the simple gift of wine, the gladness of the grape, to rich and poor alike."

"And all this time, no one suspected a living organism was behind the mystery of leavening and brewing?" I asked.

"No one saw the individual yeast if that is what you mean, except as clumps or foam on the surface of the vat. My ancestors knew that if they used the same vessels for brewing, they would get more reliable batches of beer. The clay containers with the most cracks and crevices worked best, no doubt because the yeast could hide out, hitching a ride into the next batch. They were chain culturing a microbe. The yeast co-evolved in the process, maybe at different times and places, but adapt the yeast did, as it is still doing today. Most yeast float toward the end of brewing. The term we use in industry is flocculation. There is an interesting quote from Pliny the Elder, the ancient Roman historian, about the Gaul's using this beer foam to make bread lighter, but I cannot remember the exact words, so I will not attempt it."

Gamal talking about the yeast foam took me back to my senior year in high school and our first fieldtrip, the one where we were to observe microbes in nature. Here I had been expecting to collect samples and take them back to the classroom and look at them under the microscope, but instead the

---

[66] It discusses the meaning of love. To the Greeks there were different kinds of love (e.g. "Platonic love"). Plato was a student of Socrates, who poetically once said of wine that it could moisten the soul and lull grief to sleep.

teacher merely pointed out microbes all over the place and we used our five senses to detect them, wherever we walked there were microbes that day. Our senses are enough to see, smell, touch, taste, and perhaps even hear them on occasion. You even can detect microorganisms by whiffing a handful of ordinary dirt. That day we observed circular orange mats that looked like layers of a giant onion, cut down the middle and alternating between different colors: red, yellow, green, and brown, all one inside the other within the pristine hot springs of Yellowstone. They are bacteria and algal mats. The different temperatures allow only certain species to thrive in them.[lxiii] In the center where the water first percolates out of the ground is the hottest and is where the colors were blue. Then they transitioned into yellow. We didn't take a thermometer along because just looking at the different colors gives an idea of the thermal energy the water possesses, depending on the species of microbe living in it.

But you don't have to travel all the way to Yellowstone to see microbes on any given day. Even the slime on the surface of your teeth you can detect with your tongue if you skip brushing for a day or two is a film of oral bacteria that continually springs back into existence given half a chance. If you have a bottle of wine that goes cloudy then you can blame bacteria...in fact the cloudiness you see are bacteria, millions of them in each drop. You can culture microbes with just a bottle of wine or a bowl of beef broth soup (Leeuwenhoek claimed the best source was in water laced with black pepper). You can even touch microbes with your fingers if you've ever left hotdogs in a freezer too long. That slime is a relative of brewer's yeast, a fungus that thrives in the cold.

On my way to a symposium I once flew over the Great Salt Lake in Utah at just the right time of day and while looking out the window caught sight of a purplish tint to the waters. This phenomenon was created by another bacterium, one that thrives in the salty waters of the lake, a microbe called halobacter. Even the immense white cliffs of Dover in England are made of the accumulated shells of single-celled algae that lived and died millions of years ago.[67] The black slime that accumulates on shower curtains or the sides of buildings wherever downspouts direct the runoff of rain can be places where bacteria will happily grow. You are experiencing them whenever you wince at the odor of a dumpster on a hot afternoon, and in fact over half the oxygen we breath is manufactured by photosynthetic life in the oceans too small to be seen without a microscope.

---

[67] Egypt was under the sea during the Cretaceous period. Over millions of years, single-celled creatures called *nummulites* lived, died, and sank to the bottom. In time their shells built up and were compressed into limestone rock, the oceans retreated and evaporated, and the creature's remains were cut into blocks and used as stones for the Great Pyramid.

Then there are the holes in Swiss cheese, created by gas given off from a microbe called *propionibacterium*, one that normally lives off the protein in our outer skin, and like the yeast, ferments sugar into carbon dioxide. Everyone's seen the orange color on the surfaces of Limburger and Muenster cheese; but fewer know that this pigment is from a *Brevibacterium*, the same species that causes unwelcome foot odor. The microbe consumes protein, whether it's protein in cheese, or someone's dead skin between their toes.[68] It's not picky; both are warm, moist places with plenty of nutrition. Turn over a compost heap sometime and you can feel the heat with just the palms of your hands. What you are detecting is the collective metabolism of billions of unseen bacteria, all furiously burning glucose, not so different than how our own bodies stay at a constant 98.6 degrees F due to our cell's collective metabolisms...the continuous burning of glucose by our ten trillion cells. [lxiv] On a molecular level, we're a lot more like microbes than it's comfortable to admit sometimes.

SAY THE MAGIC WORDS

We headed upstairs to the Royal Mummy Room, the soreness in my legs an unwelcome reminder of my overdoing it the previous day. I began to wonder how I was ever going to climb the Great Pyramid at Giza, as I had wanted to since reading Twain's description of it. Some cowboy I'd turned out to be. The only consolation was that my brothers couldn't see me coming up lame halfway up a staircase, one hand on the rail, the other tending to my sore knee. But something else bothered me even more. Tahany and Gamal had been so generous with their time, and yet I hadn't been able to reciprocate much. I made a mental note to find a way to show my appreciation before leaving for Luxor.

The Mummy room was darker and quieter than expected, heightening the suspense the way turning off the lights in a movie theater can do before the show starts, and my eyes took a few minutes to get adjusted. After a while I began registering details like the condition of the skin and the texture of the dried hair on Ramesses the Great. The ancient Egyptians seem to have been preoccupied with the afterlife and you can tell a lot about a society by what it values. With the Romans it was their baths, a reflection of the Roman work ethic and rich social life, the Greeks had their athletic events, theater,

---

[68] The blue color of certain aged cheeses is due to the fungus that makes penicillin: *Penicillin roqueforti.* / Pliny the Elder commented on blue cheese's flavor in the first century AD.

and other arts.[69]   While with the Egyptians it seems to have been mostly about pyramids and mummies.   If the king were granted eternal life after death, then his immortality would benefit everyone.   He became an eternal god.   More recently, Egyptologists have come to think of the pyramids as a joint building project that helped bring about not only the pharaohs' tombs but also Egypt itself as a nation.   The pyramid builders, priests, and embalmers were the ancient equivalent to NASA's space program in the 1960's.   And perhaps not coincidentally, both were concerned with placing someone up among the stars.

"Is it considered disrespectful to display a mummy?" I had to ask since this is the elephant in the room kind of question if you ever find yourself standing next to an Egyptian in a room full of mummies.

"In a way, yes…but in another way, not so much.   You see Ben…every time we say one of their names we are keeping them alive.   My ancestors believed words had magical, even restorative powers.   So in a way, we are doing them a favor by saying their names…of remembering them after all these centuries."

"If you look closely enough," Gamal continued, "you can see a family resemblance."

I glanced around at the dozen or so pharaohs lying in their climate-controlled cases, which is when I noticed others were also speaking in lowed tones.   "It's true," I whispered, "the shape of the nose is recognizable…with a little bit of imagination."

"Mummification was advanced enough by Ramesses the Fifth's time that we can still make out the smallpox scars on his skin, even the remnants of tuberculosis in the spine of this one here." [lxv]

I was familiar with TB, a disease caused by a bacterium that infects some of the same cells in the human body that anthrax bacteria go after, the macrophages, but with far different results for the patient.   Anthrax is a swift disease, TB a more gradual one. In the lab I'd worked with TB, a weakened strain of it, for a time in graduate school before joining the anthrax group. Gamal led me to another king lying in eternal state.

"Meet Seqenre the Second, he was a Theban who met a violent end at the hands of the Hyxos.   We can tell by the wounds on his body…five in all. You may count them if you wish.   We surmise he was either killed in battle or executed afterwards because the places on his body where he would have been wearing his armor have all been left unharmed."   Gamal explained how mummies have been found with hardening of the arteries, even though, surprisingly, they didn't eat much meat.   Ramesses the Second may also be

---

[69] The largest buildings in ancient Rome were the public baths, which were "social levelers" because rich & poor alike bathed together and the baths became a symbol of Rome itself. Most towns throughout the empire had at least one public bath.

known to us as Ramesses the Great, but this pharaoh suffered from arthritis and tooth decay. X-rays of his skull indicate he died of an abscessed tooth. In fact, he's been described as a dental cripple. [70]

"Look, he had red hair, but of a darker shade than yours," Gamal said pointing through the glass near his head.

As for me, I was too busy convincing myself I might actually be staring at someone mentioned in the Old Testament, someone who talked face to face with Moses. It seems dental problems were common, given all the sand in their bread. It came from grinding stones used to mill the flour. Eating bread brought with it a hidden price in ancient Egypt. Abrasive sand over time wears away tooth enamel, the hardest substance in the human body, harder than steel, eventually wearing teeth right down to the pulp.[lxvi] I guess they never made the connection. Tooth decay may have been another reason the Egyptians invented the breath mint. I started to feel sorry for him lying there; that is until I thought about his having fathered some 90 children while I hadn't a single one to my credit.

"Didn't the sand lead to lung problems?" I seemed to recall reading something about it as a graduate student.

"It is called pneumoconiosis. We see it in mummies all the time."

Oxygen in the air, it seems, combined with moisture is the unseen enemy if you're a mummy. It encourages bacteria and fungi to grow on you. In the 1970's it was realized that the royal mummies were decaying so steps were finally taken to address the problem.

Gamal somehow managed to find another connection with wine. I had come to realize by now he knew more about the yeast than I could have expected. In fact, I felt as though I could have asked him anything...the true sign of an expert.

"Ancient people realized air was the enemy and deliberately put additives in wines. Terebinth, a resin from pine trees, was added for its antimicrobial properties. It is one of the traces we can detect with mass spectroscopy, fingerprints we take as an identifier of the ancient wine. There is a wine in Greece still sold called Restina, which has resin added to it. It is an acquired taste, and it turns out I could not acquire it," Gamal made a sour face, "My ancestors knew resin was good for mummification too. It discourages bacteria on the human body for the same reason it does in the wine."

I got my fill of mummies quicker than expected – which is probably why I never wanted to go to medical school – and with my eyes fully adjusted began looking around; wondering what else was in the room.

---

[70] Current evidence suggests inflammation plays an important role in heart disease. / Within a year after their discovery by Röentgen, x-rays were being used to look inside Egyptian cat mummies in a museum (in 1898).

"Wine, in fact, is the world's oldest medicine still in use," Gamal continued, "Even today we see its health benefits appearing in epidemiological studies. It not only seems to be good for the heart, in moderation, but also may help prevent cancer when taken sparingly.[lxvii]   I do not suppose in Montana you have so many health problems?"

I instinctively looked into his eyes to see if he was really serious.

"Well according to my grandfather, back in the days when Thistle was being founded in the 1860's, they remembered to build everything except the cemetery.   But since everyone was so healthy, they had to go out and shoot somebody to get it started."

Gamal managed only the faint traces of a smile.   He should have laughed outright because that joke always worked.   I leaned up against a railing long enough to give my back muscles a brief respite before we headed into the next room.   Try as I might, I just couldn't nail down what might be bothering him.   It had little, if anything, to do with my admiring Tahany I had decided by them.   It was more like anxiety.   He was as nervous as a cat in a new house. It was as if Gamal was worried about some impending doom hanging over his head.   Or maybe it was just his nature.   I couldn't tell.

"I LOOKED UPON WONDERFUL THINGS"

"Look," Gamal pointed to an inscription in hieroglyphs on the wall of the next room that turned out to be a replica of a tomb from Saqaara.   "As a professor, you will no doubt appreciate this one.   It was written by a teacher who lived several thousand years ago and he is saying to his wayward pupil 'I am told you abandoned your studies and whirl around in pleasures, that you go from street to street and the place stinks of beer every time you leave.' And here is another. It is by a girl who describes a boy in the following way. 'He is like a date cake dipped in beer'.[71]   I imagine she was in love. Egyptian women had more rights than other women in the ancient world.   Did you know? We had taverns where men and women could socialize together. We know there were some women doctors in ancient Egypt but we do not know how many." [72]

Gamal noticed me looking around for a bench and so offered to sit while he continued.   I was thinking about the Citadel from the day before, concluding that was where I must have done most of the damage, which is what I get for trying to impress a pretty Belgium tourist.   A 50-year old man

---

[71] When a man asked a woman to marry him in ancient Egypt, he may have done so by offering her a sip of his beer. Egyptians probably married for love.
[72] Egypt had the first female monarch in recorded history.

with sad eyes and a bad case of nerves was outlasting me in a museum, and a pretty slender man at that.

"While we sit I will explain more about medicine before we go back down the stairs. The alcohol in beer and wine is high enough that it can dissolve some substances ordinary water cannot…valuable substances with intrinsic medicinal value. Water is an excellent solvent, but with alcohol, it is even better at dissolving organic things. So they used wine and beer as a base to dissolve drugs from plants you see?[lxviii] The yeast with its ability to make alcohol out of sugar juice…it…I am not quite sure how you pronounce it in English…but it keeps the medicines from forming the hard bit," [73] he held up his thumb and forefinger as if holding a small, invisible pebble.

"A precipitate?" I ventured.

"Yes. Thank you. A precipitate."

We sat while watching the other tourists pass by. It can be strangely comforting to see people from your own world with electronic devices like cell phones and iPods held up to their heads pulling you back to the present. It's easy to feel out of sorts, lost to a time warp that a museum can create. It can be a bit overwhelming after awhile. Gamal, however, was still thinking about wine's role as a medicine.

"The Romans had their laxative wines and as late as the 1800's many wines were still being used by doctors in the West to treat various ailments. When I was at Cambridge late one evening I came across an old book in the library written by a physician prescribing various wines, a port wine for women in the midst of childbirth or another useful as an antiseptic to dress wounds, a sherry for someone suffering typhoid fever might also be in order. When Alexander the Great brought his army through Egypt and then on to India, he added wine to their water to ward off disease. He founded Alexandria and as a Macedonian he would have preferred drinking his wine undiluted, something the Greeks in Athens would have been shocked at."

"It's one of the few things I remember from Ancient History 101," I quipped, but still no reaction from Gamal, perhaps the only Egyptian I'd seen so far with an aversion to smiling.

"His army even had a mobile city that followed them around, bakers baking bread for example. After Alexander led the way to Egypt, the Romans found their way here too."

We headed back down towards the Greco-Roman exhibit. Egypt was a culture that maintained its independence even after invasion by the Hyksos and later the Assyrians, the Persians, and lastly Alexander. In some ways Egypt was never conquered, at least not until the end. Some claim Egypt

---

[73] Egyptian papyri list over 100 medicines requiring beer as an ingredient (1600 BC). / The first physicians to specialize in areas of the body like the eyes, teeth, and the gastrointestinal tract, were in ancient Egypt.

changed her invaders more than the other way around. But with Rome things were different. The Romans were the only occupiers who turned Egypt into a vassal kingdom. The land the pharaohs once ruled over became Rome's breadbasket. To their credit the Romans did adopt Egyptian ways, and yet they never left a significant monument of their own here. Still, they did colonize her, and some of the most beautiful, well-preserved mummy coffins in Egypt were commissioned by the Romans. They're made of wood and the owner had his or her face painted on the lid. It's striking to see an intricate portrait of someone who lived two thousand years ago; exactly how they would have looked in life over top of their mummified remains.

"When the Greeks came to Egypt, they couldn't help noticing how only priests and the wealthy drank wine. This did not sit well on the Greek mind. Here was a people who gave wine even to their slaves because they believed it imparted strength and vitality."

"The Greeks did democratize wine then," I asked, echoing what Gamal had said earlier, just to make sure I'd heard him correctly.

"Yes, but the Romans would take this further. Everyone including the children drank wine in ancient Rome. The Romans had a saying – In Vino Veritas – in wine there is truth, because they knew of wine's ability to loosen the tongue...to act as a social lubricant. The Roman historian Justin declared winemaking to be one of the three greatest achievements of civilized people, the other two being a constitutional government and urban life."

We came upon a set of glass cups lying on their sides with pointed ends. Noticing my interest, Gamal explained how the pointed end was so the drinker couldn't set his glass down until all the contents had been emptied in a single gulp. Ingenious, and a typically Roman solution to a problem.

Gamal explained how the Romans were the first to experiment with glass wine bottles, but are mainly known today for using wooden barrels. Apparently they learned coopering from the Gauls.[74] Wooden barrels were not only more airtight than clay, but could hold more wine and, unlike amphorae, moved by a single person rolling it on the ground. Sometimes they aged wine for ten, twenty, even one hundred years or more if destined for the emperor's table. In fact, the Romans aged their wines so long that red wines could lose their color and take on an amber tint.[lxix]

It is tempting to believe that things were always purer in the past. But, as Gamal pointed out, the Romans put what we'd recognize today as additives in their wines, a surprising variety in fact. They diluted wine with seawater by as much as 50 percent and having a sweet tooth meant the Romans added honey whenever possible too, sometimes as much honey as the wine itself.[75] They even added ground up seashells, flour, herbs, and marble dust as these

---

[74] The Gauls sometimes traded a slave for an amphora of Greek wine.
[75] The different varieties of honey would have provided more choices of wine.

ingredients would have cut down on the acidity caused by organic acids, which create a sour note, allowing the sweetness of the grape's leftover sugars to predominate. There was still no pure sugar in ancient times so they made due with whatever they had.

Another thing they might do was to boil the fruit juices, "reducing them" as a chef would say today, and then add this concentrated sap to the wine. They did this boiling in lead containers. In fact, the Roman orator Cato preferred lead boilers because he said it imparted a sweetness to the wine. The natural acidity of the fruit combined with heat was enough to liberate lead from the boiler out into the wine. Chemical analysis of a victim at Herculaneum indicates he suffered lead poisoning, perhaps from sweetening his wine this way.[lxx]

Gamal took off his shoulder bag and laid it on the floor, then pulled out a hefty stack of papers. "We can learn a thing or two from Roman burials." After shuffling through the stack, he came across what he was looking for and began reading.

"The love of wine was even expressed by the deceased. On one Roman tombstone a citizen had carved the following epitaph: 'In the ground I lie, who was once known as Primus. I lived on Lucerine oysters, and often drank Flarenian wine. The pleasures of bathing, wine, and love aged with me over the years.' Here is another, also from a tombstone. 'Flavius Agricola was my name... friends who read this listen to my advice... mix wine, tie the garlands around your head, drink deep. And do not deny the pretty girls the sweets of love'."[76]

With the unsettling image of a lovesick toga-wearing Flavius in my mind, I continued walking behind Gamal, then my mind drifted onto another subject, this time Mark Twain and some of the experiences he wrote about on his trip to the Old World. One of their stops was the Roman town of Pompeii, in southern Italy and he had been surprised by the amount of excavation already done there. Pompeii had lain undisturbed, buried by a volcano for nearly two thousand years. It was starting to get late and the crowds were thinning as I noticed Gamal leading me out of the Greco-Roman room and back into the atrium.

"The Romans even added an incense called myrrh to their wines. Can you guess why?"

"Taste?" I suggested.

"Perhaps," Gamal answered courteously. "But more likely it was because of its pain-killing properties. Myrrh was an 'aspirin' if you like, and would

---

[76] Our use of tombstones goes back to ancient Egypt; indeed the oldest stone building in the world is in Saqqara. When an Egyptian wanted to say someone had died, they explained that he "went west". / Among ancient Roman cities in Italy, Pompeii seems to have had especially strong ties to Egypt. Scenes on the walls of homes there often include hippopotami and pygmies from along the Nile.

have made inferior wines taste better...perhaps by dulling the senses of the drinker."

The gift shop didn't have any aspirin or papyrus for that matter and the soreness in my back and legs returned full force. At least it was easy enough to locate an unoccupied bench so we sat and talked before going back upstairs to the Tut exhibit. Missing the last two rooms downstairs would be my punishment for overdoing it the previous day, and for eating so many doughnuts during final exams week.

"Pliny the Elder recorded how the Gauls used foam from their beers to make bread dough rise. In effect, they taught the Romans to recycle yeast, using leftover beer yeast for leavening their bread dough. Bakers have always had a long relationship with brewers. Bakers got their yeast from breweries as late as the 1800's...that is until they started cultivating their own, more bread-friendly strains.[77] The same way dogs can be selected into different breeds...the yeast was slowly undergoing its own conversion...genetically becoming more like the baker's yeast we know today. Most vineyards and breweries these days use special strains of yeast so they can control the quality of their finished product more carefully and reproducibly."

We sat for several minutes more without speaking. Then Gamal came up with an idea.

"I have a thought. Since you have a layover in Germany, perhaps you wouldn't mind taking a side trip for the day...down to Munich. They have wonderful trains in Germany and Munich is a fine example of a European city that emerged out of the Middle Ages because of brewing. I have a colleague there who might be able to help you. He is retired. And I should warn, a bit eccentric. But he might know a thing or two about your project."

"Yes. I'd like that, Gamal. I wasn't relishing the idea of spending ten hours in an airport bar anyway."

"Very good. I'll contact him then. And there is just one more thing," Gamal reached into his satchel, pulling out the same stack of papers he'd read from before, then lowered it between the both of us on the bench, "Tahany says you have a cousin in New York? And she is in the publishing business?"

I vaguely recalled having mentioned something about it in one of our e-mails.

"Yes," I answered hesitatingly, "but I only see her at Christmas...or whenever she decides to come home for a visit."

"Well, I was wondering if...if it wouldn't be too much of an imposition...you see I have been working on this for a very long time now." Gamal put his hand on top of the papers. "It is my manuscript...and has taken me all of four

---

[77] By 1887 the Pabst brewery was adding pure strains of yeast to brew their beer.

years to complete, but it is finished, at least I am hoping so. It is all about the brewer's yeast, Ben. I hope it is not too much of an inconvenience? If it is, please say so. You are the first person I have shown it to. Except for Tahany of course. It concerns the brewer's yeast and its place in industry and in history. I was wondering if perhaps you could find someone who might want to publish it. I would be most grateful. Maybe you could look it over too...for spelling errors and the like...if you can find the time," his hand was actually trembling. The stack of papers obviously meant a great deal to him.

"I'd be happy to, Gamal," I said, barely able to contain my surprise, yet also relieved that I now had a way to repay he and Tahany for their time. "I've never known anyone who's written a book before."

"Well it's not a book...not yet anyway," Gamal carefully guided the manuscript back inside its Manila envelope and then handed it over to me. I tucked it under my arm as we went upstairs to the King Tut exhibit, which was everything the guidebook promised and more if that's possible, but oddly enough we didn't talk very much after that. It was as if all that needed to had already been said, the dead having been disturbed enough for one day. So most of what was left of the visit was taken in silence, paying our respects to this civilization that had lasted so much longer than any other but, in the end like a candle's flame, it too flickered out.

Gamal and I were from different cultures and yet we were more alike than I had first allowed. Maybe I shouldn't have been so surprised, though. I often see bits of myself mirrored back at me in other scientists from time to time. It's kind of reassuring and yet a bit unsettling because it makes one wonder if there's enough room for someone like yourself sometimes. But that probably says more about me than anything.

As the turnstiles came into view, Gamal turned to me, looked up and asked, "So is there anything I did not get to?"

"Yes. There is one thing that's been on my mind since I got here. Why does everyone in Egypt smile so much?" I was collecting opinions on this matter, having already interrogated Tahany's brother and just about every other Egyptian I'd been able to strike up a conversation with.

"I don't know. Do we? I suppose when you live here all your life you are like the perch swimming in the Nile. The last thing you are in a position to comment on is the water. Or as Tahany says, the camel never sees his own hump."

I shrugged, which must have prompted Gamal to come up with some kind of a theory.

"Perhaps friendliness says something about desert communities in general. Perhaps you really miss the companionship of others when you do not have it for long periods out in the desert."

We shook hands and as we parted I felt the sense of relief that comes with the realization that you didn't do anything to offend someone like you thought you might have. Maybe I'd feel that strongly about something I'd write too someday.

I vowed to get back in touch with him when I returned from Luxor to finalize the meeting with the microbiologist in Munich. After reading the manuscript I would come to realize just how much he and Gamal had communicated. In fact, most of the quotes seem to have been attributed to this Wolfgang character. If I learned half as much from the self-proclaimed "Yeast Master of Bavaria" as I did from Gamal, I was definitely looking forward to our meeting however brief it was to be.

SECOND WIND

That night was my last in the Bedouin before beginning the second leg of the journey down to Luxor. The anticipation was welling up again just like it had the morning I left for Egypt; strange the way excitement interspersed with overwhelming fatigue can come and go on a long trip, kind of like the tide coming in and out. Except this time I was journeying farther into the heart of Egypt, several hundred miles up the longest river in the world, which would allow me the opportunity to compare different cities in Egypt for the first time. The guidebook provided some hints about what to expect, as well as displaying some nice snapshots, the kind of photos I'd like to take more often if I had the time…and the ability. On the east side of the Nile it promised Karnak temple, the largest religious complex ever built. In fact, its map took up two full pages. The book called it "the most expansive open-air museum in the world". Just to the west lay the Valley of the Kings, where tombs had been carved into hillsides thousands of years ago and could be toured by the still-living, among them the tomb of Tutankhamen himself.

But my trip wasn't just about Egypt, or Tut, but also the brewer's yeast and how Egypt came to appreciate its special gifts and how both changed one another, and as if our conversation had ended too abruptly back at the museum, I felt the urge to pick Gamal's manuscript back up and thumb through it one more time before turning off the lights and resting my eyes.

I stopped midway through his chapter on Rome to ruminate on what I had just read. It seems Greek bakers taken as slaves taught the Romans to leaven bread in 168 BC. Like the Egyptians, the Romans leavened bread by inoculating new dough with old dough from a previous batch. Bread apparently caught on so well that by the Emperor Augustus's reign there were more than 300 bakeries in Rome alone. One Roman baker even had his tomb built in the shape of an oven.   Leavened bread found its way into religion, too, in fact the Romans had a festival devoted to bread on their calendar, complete with decorated donkeys paraded through the streets sporting baked rolls fashioned into jewelry and worn around their necks.

I thought back to freshman biochemistry and the way the same simple chemical reaction that takes place inside bread dough is also what makes beer and wine so valuable.   With bread, it's the carbon dioxide one wants, for this creates the tiny air pockets that can, thanks to the heat of the oven, expand and force the flour apart, adding to the unique texture and mouth-feel of fresh-baked bread we all crave, while with beer and wine it's the other product of fermentation, the alcohol, that matters most.   The alcohol is present in the dough too, but only for short while.   Because of its low boiling point the alcohol evaporates away during baking, contributing to the complex aroma of fresh bread.   During leavening, the brewer's yeast is consuming the plant sugars and shredding them into carbon dioxide and alcohol, whether from dough or grape juice, it doesn't really matter to the yeast what the source is.   This extraction of the chemical energy locked within the bonds of a sugar molecule is how most cells on Earth get their energy to manufacture ATP molecules too, ours included.

I kept reading longer than expected for it seemed Gamal's infectious passion for the yeast had rubbed off on me.   According to Gamal, from the middle of the second century BC on, bread was within easy reach of most Romans.   Even spectators in the Colosseum watching gladiatorial games and wild animal fights could enjoy bread during a match.   They purchased it from stalls, baked in special molds, perhaps even engraved with the image of a gladiator who was in a fight that particular day.   The baker may have added his own name to the loaf too...for advertising.   And in addition to staging gladiatorial games for free, Roman emperors also gave out flour and bread to Rome's less fortunate, hence the phrase keeping the masses happy with "Bread and Circuses."[lxxi]

It's strange how nature herself – rather than humans – has been what best preserved ancient cities for centuries.   Angkor Wat and Machu Pichu were abandoned and forgotten about while being covered by jungle.   Luxor temple, where I was headed the next day, was buried by sand and even had a village on top as late as the 1800's, when it was moved and the site finally excavated.

In AD 79, a powerful volcanic eruption and the ensuing lava and mud buried two cities in southern Italy, Herculaneum and Pompeii, effectively allowing us a glimpse into what life was like 50 or so years after Jesus walked the Earth.   Of the two, Herculaneum was more completely buried, and in fact two thirds of this town still remain underneath the modern city of Ercolano. Naples and Rome on the other hand, have been continuously inhabited and therefore, like portions of the Roman Colloseum or the great pyramids at Giza, were slowly built over top of, or carted off and destroyed.

Pompeii it seems would have been a difficult place to go around thirsty in ancient times. Excavators have uncovered no less than 200 places a Pompeian could have purchased wine, including bars, taverns, brothels, and snack stalls, which were a first century version of a fast-food restaurant, in this town that once boasted 20,000 inhabitants.[lxxii] According to Gamal, during important gladiatorial matches, vendors would set up temporary stalls outside the Amphitheater to dispense wine to a thirsty crowd.[78] On just one block alone near the public baths of Pompeii there were eight bars selling wines, including one local vintage called Vesuvinum. Pliny the Elder, who would also perish in the eruption of AD 79, had categorized 91 different wines available in the Roman world. [lxxiii]

Back at the museum I had felt some pride when Gamal entrusted me with something that obviously meant a great deal to him, so I gently slid the manuscript back into its envelope, closed the metal clasp and stowed it away inside my bag before turning off the lights and trying to get some sleep. Tomorrow would be a long day of travel and there was yet another reason the bag would never leave my sight. But I still couldn't get to sleep until I turned the lights back on and dug out my other book, the one that meant so much to a 13-year old kid growing up in Thistle.

I remembered how Twain had described Pompeii and for some reason wanted to read from it one more time, so I flipped through the pages with the worn-down corners until I found it again. The *Quaker City* had anchored off the coast of Italy in 1867 while Twain, along with some shipboard companions, ascended Mt. Vesuvius riding on the backs of mules. Pompeii had been rediscovered not long before and Twain was surprised at how extensive the excavation had been.[79] Walking around Pompeii's abandoned streets the bachelor on the verge of becoming America's most important writer was moved by how devoid the place seemed, yet still recognizable after nearly two millennia. I savored his words one more time:

> "...I went through shop after shop and store after store, far down the long street of merchants, and called for the wares of Rome and the East, but the tradesmen were gone, the marts were silent, and nothing was left but the broken jars all set in cement of cinders and ashes; the wine and the oil that once had filled them were gone with their owners."

---

[78] Roman trackers near the Black Sea captured lions for gladiatorial games by getting them drunk on wine first. They poured the wine into a trough near the animal's favorite drinking hole.

[79] Some roads in Pompeii had crushed wine amphorae used as pavement called *opus signinum*.

Exploring still further Twain came across one of Pompeii's bakeries.[80]

> "In a bakeshop was a mill for grinding the grain and the furnaces for baking the bread; and they say that here, in the same furnaces, the exhumers of Pompeii found nice, well-baked loaves which the baker had not found time to remove from the ovens the last time he left his shop, because circumstances compelled him to leave in such a hurry." [81]

I turned off the lights and fell into a sleep so deep I don't even recall dreaming.   With no particular time to get up I was barely aware of the call to prayer going on outside and was rewarded with the first truly good rest I'd had since arriving in Egypt.

Feeling refreshed, I finished my complementary breakfast while glancing at the pictures in the newspaper Tahany's brother had left outside my door, loaded my backpack and then left the key inside the room, only to come across a handwritten sign posted on the elevator door that read "Elevator in Despair."   Now that's interesting, I thought.   Only in Egypt do the people go around smiling while the elevators are sad.   Somewhere on my way down the stairs it occurred to me that whoever wrote the sign probably meant to write the word "Disrepair".   Still hungry, I checked out of the Bedouin and spooned some fuul sprinkled with olive oil inside an empty piece of flatbread, confident enough now in my abilities to handle traffic in modern Cairo, so much so that without consulting the guidebook I caught a public bus down to the Giza plateau.   I was making progress; maybe not so over the hill as I'd begun to fear.   Maybe I still had some scraps of youth left in me yet.

Planning was essential to the builders of the pyramids and work was done while the Nile was at its height.   It was easier to float the giant stones to the site while the fields lay inundated and the construction project also gave the farmers something to do.   A pyramid worker was fed well and got the best medical attention when needed, which would likely have been often.[82]   The Pharaoh had a special city built for his workers, one that had a fish-processing plant, some bakeries, a beer factory and even its own cemetery.   Experts today believe it was the building of the great pyramids during the Old Kingdom that helped provide the first sense of social cohesion for Egypt.[83]   Giza would

---

[80] Pompeii's bakeries were on 4 main streets, yet the town had no street signs.   Residents of Pompeii would have navigated around the city by using landmarks or drawings on the walls of buildings depicting various trades that took place there.
[81] There were at least 30 bakeries in Pompeii and the one Twain visited was discovered with 81 round, leavened, and charred loaves still inside its oven.
[82] Now believed to be among about 10,000 to 20,000 other workers.
[83] Sometimes the Old Kingdom is referred to as the *Age of Pyramids*.

have been home to a diverse assortment of characters from all over the kingdom, both upper and lower. When not constructing their tombs, Egyptians shared ideas, styles, even taught one another songs, games, and how to dress. Another bread vendor crossed in front of me with a wide tray balanced atop his head, apparently unaware of the contrast he made with the modern world all around. Even today the Egyptian government subsidizes bread. In fact it was dirt-cheap. The guidebook claimed I could have found over 80 different varieties in Cairo if I'd taken the time to look. [lxxiv]

When Twain visited the pyramids, he was hustled to the very top, or "dragged" as he put it, but nowadays there's no climbing all the way up. In fact, I had gotten there so late in the day that the 300 tickets giving tourists access to Kufu's burial chamber on the inside had already been distributed. The entrance I would have walked through had I gotten there sooner was created by an 8th century caliph mentioned in *1001 Arabian Nights*, the same one who bashed a hole into the side mistakenly believing the pyramid would be filled with gold.

Fortunately, standing outside and looking up turned out to be enough for a first-timer and I made a mental note to get here earlier on the way back from Luxor. I watched in fascination as another tour bus pulled up and the postcard sellers closed in for the attack. When Twain's party passed by they too complained of being hassled for baksheesh.[84] Even in the 1840's, tourists complained of graffiti they found at the tops of the pyramids. I closed my eyes, trying to imagine Cleopatra giving Julius Caesar a personal tour here two thousand years ago in her chariot. Strange to think how she wouldn't even have known whom King Tut was, let alone have been able to locate his tomb. That's how good Tut's successors were at making sure his name all but disappeared from Egyptian history, as if he had never even existed.

When Twain was here, people were just coming around to believing microbes could cause infections. In the 1850's, during the height of the Crimean War, ten times more soldiers would die from unseen bacteria in drinking water than from bullets and bayonets of the enemy.[lxxv] Most, including a young nurse who ventured to the battlefront from England to care for the wounded, thought these diseases were spread by "bad air", the so-called Miasma theory of disease that was then popular. But in time this same nurse became convinced the Miasma theory was wrong. At her insistence,

---

[84] There were so many opportunities for horse and camel rides around the pyramids in the 1860's that some tourists complained Giza "smelled like a stable".

ventilation was improved and sewers installed in hospitals and the death rates dropped from 42% down to just 2%.[85]

My guidebook had a list of some of the well-known tourists throughout history that stood at the foot of the pyramids where I was now. [lxxvi]   One of them was that same nurse, the Lady with the Lamp, Florence Nightingale, which was what got me to thinking about her.   She wrote that Egypt's pyramids seemed as if they were boring holes in the sky, while the writer Gustave Flaubert claimed they were trying to come down and crush him. Always the opportunist, Napoleon saw the pyramids as a way to increase his chances for victory by reminding his soldiers just before doing battle with the Mamluks that 40 centuries of history were watching over them.   And there's little doubt his men would have been impressed.   The Great Pyramid was over twice as tall as any building they would have seen during their lifetimes. In fact, the Great Pyramid still weighs more than 17 Empire State Buildings combined in spite of Saladin's requisitioning efforts.

Wondering how they were built has been a question tourists have asked for thousands of years and I turned out to be like the others.   It's been calculated that workers managed to lay a two-ton rock on average every three minutes...for 23 straight years.[86]   These weren't slaves as had been assumed since antiquity, either.   As for me, I came to believe they did it mostly out of pride.   The guidebook claimed the average lifespan of a pyramid worker was 32 years and like so many who've visited, I too became reflective, thinking not only about the pyramids, but also of time and biology and how short life can be and how diverse it is too, and in so many ways, more than a lifetime's worth of study would ever allow a biologist to uncover.

There are bristlecone pine trees still growing in California that were mere seedlings when the pharaoh and his scribes set about directing the layout of the Great Pyramid, while at the other extreme are insects I swat at all summer long that will live out their entire lifetimes in less than a week even if I don't manage to get one of them.[lxxvii]   Bacteria like anthrax – which simply divides in two over and over – can be thought of as immortal, while there are other bacteria that may reproduce only once every century, if that.   It was the first time I'd thought about the anthrax grant since the previous day.

From this vantage point, I looked out as far as I could into the distance and caught sight of a train winding along the Nile almost like a toy.   It was mingling with the dusty air and headed in the same direction I would be going

---

[85] She later published her methods in a book in 1860 called *Notes on Nursing*.   The US government during the Civil War sought out her advice on improving sanitation in field hospitals.
[86] Even the smaller stones weighed 2 tons and workers were sometimes divided into teams that competed with one another during construction. / Napoleon brought a mathematician to Egypt who calculated that there were enough stones in the pyramids to build a wall between Germany and France 3 meters (10 feet) high and 1 meter thick.

later in the evening. Another wave of excitement rolled through me and I had the sudden urge to begin on my trip down to Luxor then and there, but instead settled for walking around the pyramids and chancing the Sphinx's gaze one more time.

Gamal had mentioned how wheat, beer, and bread were so important to Egypt's stability that the pharaohs ordered their soldiers to help out in the fields during harvest time. [87]   Egypt was the granary for the ancient world, in fact Egypt's wheat supplies provided a third of Rome's needs and helped keep Constantinople Christian for centuries after the Roman Empire fell in the west.[lxxviii]   Gamal said something else I'd been thinking about, too, about how it wasn't unusual for boys to marry at 13 in ancient Egypt and here I was pushing 40 still without a wife.

After my walk I had a few hours more to kill, recalling that I'd placed the copy of *Innocents Abroad* on top of my clothes in case I wanted to read from it.   I carefully thumbed through the worn, tissue paper-like pages until I came across Twain's description of the pyramids as he found them in 1867:

> "At the distance of a few miles the pyramids rising above the palms looked very clean-cut, very grand and imposing, and very soft and filmy as well. They swam in a rich haze that took from them all the suggestions of unfeeling stone and made them seem only the airy nothings of a dream – structures which might blossom into tiers of vague arches or ornate colonnades maybe, and change and change again into all graceful forms of architecture, while we looked, and then melt deliciously away and blend with the tremendous atmosphere...A laborious walk in the flaming sun brought us to the foot of the great pyramid of Cheops. It was a fairy vision no longer. It was a corrugated, unsightly mountain of stone....Each of its monstrous sides was a wide stair way which rose upward, step above step, narrowing as it went, till it tapered to a point far aloft in the air. Insect men and women – pilgrims from the Quaker City – were creeping about its dizzy perches, and one little black swarm were waving postage stamps from the airy summit – handkerchiefs will be understood."

I looked up and tried to imagine how small a person would seem from the very top.   Twain climbed to the summit after paying some "draggers" to take him there.   He wrote of it:

---

[87] Egyptian soldiers were paid 10 loaves of bread a day and carried flour to make bread inside mud ovens they'd build on campaigns.

"Each step being full as high as a dinner table; there being very, very many of the steps; an Arab having hold of each of our arms and springing upward from step to step and snatching us with them, forcing us to lift our feet as high as our breasts every time, and do it rapidly and keep it up till we were ready to faint…who shall say it is not lively, exhilarating, lacerating, muscle-straining, bone-wrenching, and perfectly excruciating and exhausting pastime climbing the pyramids?…I iterated, reiterated, even swore to them that I did not wish to beat anybody to the top; did all I could to convince them that if I got there the last of all, I would feel blessed above men and grateful to them forever."

It's just as well tourists can't get to the top anymore these days, I concluded. Twain had already done a better job of describing it than I could have.

I closed the book that helped bring me here, put my glasses back on, and looked for another train but saw only the spray from irrigation sprinklers and a flock of birds providing any movement. I marked off another experience from my list and was about to begin a new one. The ancient capital of Thebes (modern day Luxor) lay ahead to the south. Every culture has its creation stories and it's interesting, I thought, how the ancient Egyptians believed that in the beginning there was only water…that is until the Nile receded, providing everyone with land.

I got myself to Ramesses Train Station too early and found I had a couple more hours to kill after locating my bunk. While watching the porters making up the beds and with what was left of the sun filtering in through the coach's window, I read some more of Gamal's manuscript, drawn to the past, but was then seized by anticipation of what lay ahead and replaced it with the guidebook.

According to a sidebar, the former Baptist minister and temperance advocate Thomas Cook didn't begin bringing tourists all the way to Luxor until 1869,[lxxix] but Nile cruises quickly caught on; becoming popular with Europeans and even Americans by the late 1800's. A young Theodore Roosevelt would become a lifelong hunting enthusiast after shooting his first bird from the deck of a houseboat on the Nile while traveling with his family upriver in 1872.

My stay in Upper Egypt was relaxing overall and one of my most enduring memories will probably always be of sitting on the balcony of my guesthouse watching the sun set over the river each evening, the tall shadows thrown down by the ancient arches of Karnak Temple seemingly intent on swallowing everything up within reach. One of the best things about Luxor was being able to stay right within the ruins. There were times I thought of Tahany, which heightened my loneliness, but at least I had the consolation of knowing she had found someone good for her. Science has its share of couples. When you do research, you usually spend so much time in a confined lab it's not that surprising relationships develop.

Luxor was more tourist-oriented than I expected but less humid[88] and the backpacker guesthouses and Internet cafes stood juxtaposed within the ruins. On my first day I rented a rickety bicycle and took it with me on the ferry across the river to the west side. Waking up in the east and traveling west could have been a daily ritual for an Egyptian artisan, a stonemason or a painter perhaps, carrying the tools he paid for himself to work each day. He likely lived during the New Kingdom period, when Thebes was rising to prominence as a religious capital and he no doubt worked for the king as a member of a team. His pay would have been in beer and food, and if he was exceptional, perhaps some precious metals could be counted on as well. If he were young and talented but not very experienced, he would have been put to work on the less important tombs, only later having been allowed to work on the royal ones.

I used my book's insert map while peddling and sweating my way around the west bank throughout the Valley of the Kings in increasing increments each day. King Tut's tomb was along the route and turned out to be smaller than I expected. His mummy has recently gone on display there, too. I also learned he'd been married at 12, not 13 (my first possible correction for Gamal).[89]

The second night back at the guesthouse I became aware of a change taking place. For one thing, I felt only the vague remnants of where the sharp muscle pains had been in Cairo, replaced now by a dull ache, a good kind of pain – if there is such a thing as good pain – and I even caught sight of myself one evening in a full length mirror, noticing that my pants were fitting better, which made me look like I was standing more upright. I was becoming a well-oiled machine…yet another unexpected gift of the Nile. I tried to make some headway on Gamal's manuscript but most nights I wasn't able to concentrate enough to give it the attention it deserved. It felt almost sacrilegious coming all the way to Thebes and then not learning more about Thebes, so I read all I could in Luxor while peddling around town – tourist pamphlets, local guidebooks left lying around the guesthouse, anything about

---

[88] Less rain in Upper Egypt is why – during rare downpours – tomb raiders were able to locate ancient tombs. They knew where to dig by noticing where the water disappeared down into the ground.
[89] Tut's wife, also his half-sister, was just 13.

the area I could find from travel agents and museums, even pamphlets describing the local geology and soil put out by the government.

Several days passed this way, during which time I had built up my own library of material and a comfortable routine, exploring what I hadn't seen of the ruins the previous day by bicycle, then returning to the guesthouse with just enough time to watch the sun set over the river and enjoy a cold bottle of beer in front of me.   Then, for some reason I suddenly up and decided to return to Cairo early, to rediscover its fuul stands and other surprises, its colorful hats and scarves, its children playing jump rope in the cemeteries, the sense of cohesion that can make a place special and that Luxor seemed to lack somehow.   I missed the human activity apparently, which was strange for me, as I've never really seen myself as that much of a people person.

So I arrived back in Cairo by train the next morning and spent my last few days in Egypt either sightseeing or helping Tahany in her lab in the evenings before returning to the Bedouin to sleep.   One afternoon I returned early because I'd left the guidebook in my room only to find Tahany's brother had bug-bombed all three floors, which meant I had a decision to make.   I could either go without my book for the rest of the day or run upstairs with my shirt covering my nose, find my room through a thick blanket of fog, grab the guidebook, and then head back down all in one breath, which turned out to be the toughest thing I did the rest of my time in Egypt, doubtless something I would have been unable to accomplish the night I bought the ticket back in Montana.   And I was barely out of breath afterwards.   One of the most consistent findings in medical research is how good regular exercise can be for the body.   Exercise is one thing that never seems to get "recalled" by some other study later on.   My last evening Mustafa gave me a lift to the airport and within a couple hours I was headed towards Germany the way the crow flies.

LET THEM EAT CAKE

As the plane gained in elevation, then veered off towards the northwest, finally leveling out above the clouds, I reclined further into my seat, trying to sleep but soon realized that wasn't going to be possible, so I pulled Gamal's manuscript from my pack.   Even unbound it was considerable and I wondered if I'd ever feel passionate enough to write that much about anything, including anthrax, or Montana, or Lewis & Clark, and so I began reading where I had left off down in Luxor, with the yeast having gained a culinary foothold in Europe.

Even though on separate continents, Cairo and Munich share some interesting similarities. Both were at a crossroads, geographically – Cairo where East met West along the Silk Road, Bavaria in the heart of Europe, in close enough contact with the northern parts of the continent while at the same time still near the trading routes of the Mediterranean – providing for a constant influx of new travelers with new ideas from different directions.

The Roman Empire's fall in the west marked the beginning of the Dark Ages in Europe, a period which was to last about a thousand years.[lxxx] Gamal had a theory that the invasion of Rome by Germanic tribes from the north was due at least in part to the climate getting cooler, not to mention a desire for Roman wine. Northern climates don't allow grapes enough time to make all the sugars necessary to impart a sweetness to wine so northern civilizations tended to become beer cultures, as is obvious today. Who likes sour wine?[90] I knew I didn't. Grains have a better tolerance for the cold and so it's no coincidence we have three microbreweries in my hometown but not a single winery, and Montana being the second largest producer of barley in the United States helps out too, I imagine. Northern grapes tend to be sourer; perhaps the climate got colder and the barbarians simply had a sweet tooth is what Gamal was implying. What is certain is that, rather than wither on the vine, viticulture became more widespread after the Roman Empire declined, when their technology for winemaking would have spread further afield.[lxxxi]

Religious authority stepped in to fill the void left by the collapse of Roman order during the Dark Ages and monasteries needed to keep a constant supply of wine and bread around for the Eucharist.[91] In Germany, beer was one of the foods monks were allowed to take in for extra calories while still adhering to fasts. Like the roiling of carbon dioxide bubbles produced during a good fermentation, for many in the Dark Ages life remained in a constant state of flux too and ordinary folks just didn't stay put for very long. It's one of the reasons we have so few records of them today. [lxxxii] [92]

I had selected a new guidebook – this one on Bavaria – while in a gift shop in Cairo and to be reminded of Bavaria's connection to monastic brewing I needed look no further than the full-page ads placed by brewers like Augistiner-Brau and Paulaner. Indeed, even the name Munich comes from the German word for monk. [lxxxiii]

But brewing was hardly new to Europe following Rome's fall, in fact the Celts had been making their own version of a beer called curmi on Britain long before. They also made mead by mixing honey and hard cider.[lxxxiv] Brewing

---

[90] In fact, the French still sometimes add sugar to their grape juice to allow the yeast to make extra alcohol.
[91] Some nunneries also brewed beer.
[92] During the Dark Ages, there were no fewer than 500 monasteries brewing beer in Germany, the most prodigious of which were the Benedictines. The daily ration of beer per monk was one gallon.

in England was one of the enterprises the king didn't tax and so each medieval village had its own ale-taster,[93] an official charged with keeping up local standards. As had been the case with Egypt, brewing and baking was most often entrusted into the hands of housewives, in fact being a good alewife could be reason enough for a man to ask for a woman's hand in marriage.[lxxxv] England doesn't have the tradition of monastic brewing like Bavaria because King Henry VIII put brewing in private hands.[94]

And it turns out even the layout of the typical European city today was influenced by monasteries, making these religious orders the equivalent of "embryonic cities".[lxxxvi] Within the confines of a Middle Ages monastery one would have found much of what you'd expect to find in a modern town today, including a brewery, bakery, place of worship, and boarding houses for pilgrims, workshops with seller's stalls, even a small hospital perhaps. And it was in this relatively stable environment that monks began experimenting with different varieties of grape, improving growing methods by taking seriously for the first time what seems obvious to any rancher worth his salt today...the quality of the soil and the amount of sun and water their crops received. Even the clarity of wines was improved upon in monasteries by adding egg whites to remove impurities. Monks examined the connection between climate and vintage and experimented with grafting more productive shoots onto sturdier but less productive rootstocks in an effort to increase grape yields. It was within the confines of the brewing monasteries of Germany that hops would be added to beer for the first time in the 11th century, which not only imparted a distinct bitterness to the beer, but also increased its shelf life by discouraging spoilage-causing bacteria.[lxxxvii] This natural antimicrobial property in hops allowed beers with less alcohol to be stored for longer periods.

The monk Dom Perignon has been credited with inventing champagne but according to Gamal this misconception began with an advertising campaign in the 1800's. In reality, the monk spent most of his time trying to eliminate bubbles in wine, which he himself found excessive. Perignon did introduce innovations that would someday make champagnes more practical, though, like thicker glass bottles along with a rope to hold the cork in. Interestingly, according to Gamal these French sparkling wines came about by accident after a cold spell one fall, which caused the yeast in the developing wine to become dormant and stop fermenting. This was then followed by an early spring, which allowed the hibernating yeast to awaken and rapidly resume fermenting full force all the leftover sugars still in the grape juice, producing a

---

[93] William Shakespeare's father was an ale taster. / The Roman emperor Julian wrote a poem 1,600 years ago in which he described Celtic beer as smelling "like a billy goat".
[94] Ale was sold in bottles at the Globe Theater during Shakespeare's plays / Drinking beer would become so popular that water removed from wells by brewers would cause London to sink three inches beginning in the 1860's.

sudden burst of carbon dioxide, shattering the bottles and creating the first Champagnes in the process.

Looking out my oval window I could just make out the jagged tops of the Alps towards the west, and I wondered if I'd been wise enough to place my passport near the top of my backpack like an experienced traveler would have. When ancient Rome fell it fragmented, eventually giving rise to many of the countries in Europe today.[95] I skipped the airline's breakfast and went back to Gamal's manuscript.

Incredibly, it was the monastic thirst for improving winemaking and brewing during the Dark Ages that helped lay the groundwork for the scientific revolution that lay just ahead. It was this need to spread information about crops and winemaking and other manufacturing processes that led to some of Europe's first guilds, scientific societies, and journals. The typical tavern owner in England was also a vintner and likely benefited from this spread of information obtained by monks and the winemaking guilds.[lxxxviii] In the Middle Ages wine was added to drinking water during times of siege to help ward off disease and King Edward II of England provided 4,000 barrels of wine to his army during one particular siege.[96]

I flipped forward through Gamal's manuscript until I came across another footnote, this one on bread, surprised to find that baking bread wasn't all that popular in Paris until the 1700's. Northern Europe has a much longer tradition of eating meat than baking bread. The practice of leavening dough did catch on when its time came and the role of the yeast changed as it spread throughout the Mediterranean region, radiating outward from Greece. In the Dark Ages, people didn't use plates but instead ate off of thick slices of bread called trenchers. These soaked up the grease and were then donated to the poor or thrown to hungry dogs after a meal. Baking was slowly becoming an industry in medieval Europe; however bakers still got most of their yeast in the form of skimmings (leftover yeast) from breweries. It wasn't until the explosive growth of cities during the Industrial Revolution that bakers would be forced to turn to a more reliable source of yeast.[97]

Bread's scarcity, its high price and poor quality, became a direct cause of the French Revolution and Gamal claimed that the hungry mob storming the Bastille prison in Paris was as intent on finding flour to make bread with as they were obtaining gunpowder. The French king made the unfortunate

---

[95] A more recent version of what occurred when the Roman Empire fell was the fragmentation of the Soviet Union in the 1990's, doubling the number of Eastern European countries in less than a decade.

[96] When Mark Twain toured a castle in Heidelberg, Germany he described a wine cask that held 221,726 liters of wine (the Heidelberg Tun) as being as big as a cottage and usable as a German dance floor.

[97] In 1871, a factory in Austria was the first to produce yeast just for bakers. Bakers today use strains of yeast more efficient in carbon dioxide production and with more *invertase activity* to cleave the sugars found naturally in the flour.

mistake of deregulating the sale of flour after several poor grain harvests so there was hoarding, which all combined like the perfect storm to create a massive bread shortage. Incredibly, a single loaf could cost as much as a month's wages, so it's not surprising that bakers in Paris accused of hoarding flour, or of making bread with spoiled flour, could be seen hanging in the streets following food riots.[lxxxix] Louis XVI's coronation took place at the height of a bread shortage, which is when Marie Antoinette was supposed to have said, "If they have no bread then let them eat cake," though there's no contemporary evidence for it.[98]

My plane landed on time but I still missed seeing Frankfurt itself, blaming it on the efficiency of the German train system because it turns out there's a station in the basement of the airport, which is where I bought my ticket from a vending machine. Soon my backpack and I were being whisked through the rolling hills of southern Germany on towards the Alps of Bavaria with only a quick change of trains in Mainz first. I measured our progress by the gradually increasing slope of the landscape as it transitioned from flood plain into alpine forest.

Equally as important as the monasteries were to brewing in Bavaria was another act of divine intervention...the local geography. The weather patterns and soil profiles in Bavaria are nearly ideal for growing beer's ingredients, barley and wheat, not to mention hops. An area larger than either Ireland or Portugal and situated near the Alps, Bavaria has the advantage of being fed by rivers of melting snow with little limestone in the soil as this mineral has the nasty tendency of turning good beer cloudy.

Perhaps I shouldn't have been surprised to find that my Bavarian guidebook had devoted an entire chapter to brewing, explaining how it was the natural ice caves of the Alps (and later the Bavarian's attempts to imitate the year-round coolness of these caves by digging beer cellars near breweries) that would bring about the most significant change in brewing in thousands of years...the birth of the world's first lager beers.[99]

I held the colorful pictures of Munich's beer halls up to my nose, taking in a whiff of fresh ink from the pages, the sharp stinging helping ward off my fatigue for a moment before I continued on. Unlike much of Europe, Bavaria remained stable, politically, for a very long time. The kingdom was ruled over by the same family for more than 700 years and has the world's oldest food purity law to show for it.[100]

---

[98] What we call cake today is an enriched bread containing milk, eggs, and sugar.

[99] The coolness of the ice caves increased not only the shelf life of beers but also their clarity, producing the amber color which makes most ales look like used motor oil from my pickup truck by comparison.

[100] Reinhettsgebot Law – first proclaimed in 1516 by Bavarian Duke Wilhelm IV – said that only hops, barley and water could be used to make beer.

As my train rounded another bend nearing the outskirts of Munich, I couldn't help wondering if Wolfgang (or Wolf, as Gamal referred to him) had any intentions of honoring our agreement or whether he would leave me sitting in the Hofbrauhaus. The retired microbiologist apparently had no mobile phone or any other way to get in contact with him, which made me wonder how Gamal had been able to keep track of him. In fact Wolf, as far as I could tell, didn't even use a computer, which is virtually unheard of even for a retired professor these days.

"Somewhat of a character," was how the diplomatically inclined Gamal had described him.[101] There was even a list of things I wasn't supposed to bring up to him. At the very top of it was his forced retirement. In Germany, professors faced mandatory retirement and apparently Wolf felt he'd gotten the shorter end of the stick. The thinking is that scientists do their most productive work by age 45; then one should make room for the younger scientists coming up. It certainly seems true enough, sometimes. Einstein did his most significant work as a young man even before he found his way back into academia, as a newly minted Ph.D. just out of college while working at a government patent office. [xc]

I propped my glasses on my forehead so I could continue reading but still glancing out the window once in a while to gauge my progress. After skimming through the section on breads, my stomach began a low but audible rumble. It seems that at first monasteries were making beer for their own needs, but King Ludwig eventually changed the law so lager beer could be made available to the public.

Ludwig realized he could tax beer and so it wasn't long before Munich's outdoor beer gardens began to dominate the city and Munich's poorer residents found they could bring their own food and eat under the shade of chestnut trees (the trees did double duty by keeping the underground cellars dry, absorbing the rainwater with their roots). This picnicking resulted in a crisis when brewers realized how many profits they were losing from the sale of food, so the King stepped in and hammered out a new agreement and as a result the poor would continue to bring their own food to the beer gardens, but could eat only at uncovered tables, which can still be seen today as the self-serve tables of Munich's beer halls.

I caught the unmistakable whiff of charred meat from a vendor's wagon through an open window while flipping pages in an effort to find out where Hofbrauhaus was exactly. That's where I was supposed to meet Wolfgang and the only information I could find was a warning about a "touristy feel", but

---

[101] Gamal had mentioned that Wolf made a name for himself as a successful "yeast breeder" earlier in his career and that he came up with the first strain of yeast capable of surviving in up to 25% alcohol, but apparently failed to cash in on the recent 'extreme beer' craze, microbreweries, and home brewing.

the upside was that the beer hall was guaranteed to be like Oktoberfest any day you drop by. Hofbrauhaus became a brewery in 1589, and then like so many beer halls in Germany, it became an institution. Entire families made a day of visiting beer halls here and have for generations, still bringing picnic lunches to the gardens. This is, after all, a city with a college that doubles as a brewery for teaching Europe's next generation of brewmasters the ancient art.[xci]

THE YEAST MASTER

I left the train station on foot barely aware of the heavy load over my shoulder and so it didn't take me long to find the beer hall. The train had dropped me off in the heart of the city – as is true with most trains in Europe – and I simply walked the rest of the way to Hofbrauhaus. The first thing I noticed near the hall was a costumed blonde waitress about college-age walking out the door carrying a tray, charged with the task of delivering two pitchers of beer through a crowd of tourists. To my surprise, the pitchers would turn out to be drinking glasses and after entering and locating a seat near the stage, giving my eyes a chance to adjust, I was able to see that the place was filled mostly with tourists hoisting beers while someone with a camera or cell phone always seemed to be angling for a photo. On the stage closest me were two Alpine singers who could have been cast members from *The Sound of Music* – a man dressed in black lederhosen chopping wood with an actual axe while next to him stood a woman, unconcerned and adorned in traditional dirndl dress and broad red apron.

German beer has a reputation for being strong, apparently because it's purer, often having been brewed on the premises and so without any preservatives, which means it has a better taste. And served in larger portions means many first-timers like myself simply drink too much. The guidebook had neglected to mention this, plus the fact that I was drinking on an empty stomach meant a hangover was in my future. The table next to me had a plate of fresh bretz'n – soft pretzels – heaped on top of one another, but when the waitress came by it slipped my mind to ask for any even though I was hungrier than a woodpecker with a sore beak. It seems I was still on autopilot, nervously searching for the elusive Wolf. In fact, I had begun to suspect he didn't really exist, or more likely that I'd missed him and he'd stormed off in a huff and was already phoning Gamal at that moment. Giving up, I began scanning the hall for the nearest restroom, the image having popped into my mind of the great Danish astronomer Tycho Brahe, the one who had helped Kepler with his calculations and how in 1601 had met his unfortunate end at the hands of a burst bladder while drinking too much beer at a festival in Prague.

In spite of the hall's enormity, I was the only one who seemed to be alone at a table and so with an oversized beer as my companion I pulled the guidebook on Bavaria back out, sipped the foam off the top of my glass, and picked up where I had left off, reading once more about Munich's beer gardens.

The book claimed that some beer gardens could hold 7,000 drinkers, about twice the summer population of my hometown (not including the university students who always arrive in the fall). [xcii] Some of the more established ones have chestnut trees so full that their tops can keep customers dry even during a downpour. But it's not just about drinking since beer gardens are a part of everyday life in Bavaria. Some even have petting zoos and playgrounds for children while others may boast a small lake with a cycling path.

I had made it halfway through my liter ration of beer when a cold sensation came over me. Looking around, it felt as if everyone in the hall had dropped what they were doing and were aware of only me.

But before I could turn completely around I heard in a thick German accent: "Zo you are the vun who thinks he can learn the vays of the yeast in only two veeks."

I swiveled around in my seat and there towering over me like a shade tree was a very large man. My first thought was that he must be listed in the Guinness Book of Records, or at least was in the running. He had pale eyes surrounded by sunburned skin and the sturdy frame of a much bigger man than Gamal had led me to expect. His hair was straight, long, and thinning at the temples but gathered into a ponytail neatly behind his head. The beer hall was dark and he was still too far away from me without my glasses on to make out the tiny blotches of broken capillaries that would come to remind me of spider's webs on his cheeks; the remnants of too much sun on trips to the south of Spain as a youth perhaps. He was wearing faded blue jeans dotted with small holes, and a t-shirt beneath a blue-jean jacket, the jacket matching his pants in its degree of fadedness. All in all, not exactly what one expects to encounter in a retired microbiology professor.

"Wolf?" I asked, gingerly extending out my arm, forgetting to ask if I could call him by his nickname, "Gamal has told me about you."

He reached out and suspiciously pumped my hand while I wondered why Gamal hadn't mentioned his size. Most people would have mentioned someone this enormous when describing them.

"Gamal says you are eager to learn all zee vays of zee yeast. But zee real question is…do you believe it dock-torr?" With one hand he guided the wooden chair opposite me out from beneath the table and slowly sat down.

There would be no need for further formalities. He motioned for the waitress to return, then leaned back and stretched his legs out in front of him, crossing his ankles. I could see he had tennis shoes on but not bothered with any socks. Leaning further back still, he placed his arms behind his head, interlocking his fingers to use as a headrest. He did this all the while keeping his eyes trained on me. My initial impression was of a man who couldn't be bothered with what my impression of him might be; it would only be his that mattered. I wondered how Gamal had managed to find him, let alone gotten him to sit for all those interviews. He said Wolf had been a brilliant scientist at one time, before...whatever it was I wasn't supposed to ask him about.

"Yes," I wanted to reply, "I'm just starting out with the yeast. Not really a eukaryotic scientist, and definitely not a fungi expert by any stretch. Gram-positive bacteria are more my speed," but what actually came out was, "Um...well...I did work with *Bacillus anthracis* in graduate school," as if I were somehow ashamed of it.

"That causes anthrax," Wolf replied and an elderly couple, a man and a woman, turned around in their seats to look at us as soon as they heard the word anthrax. It's a strange sensation even for a scientist to hear someone talking about science in a beer hall with an Oompah band in the background.

Trying to regroup, I added, "You come highly recommended. Gamal says that if anyone can bring me up to speed, you certainly can."

"Ah, little Gamal. Vell I hope you are up to zee task. He seems to think so, but I'm afraid I already have my doubts." The seriousness in his tone made his words feel more like a diagnosis and the microbiologist then began strumming his fingers on the table. He was staring at me.

"Zee yeast is not so easy. You are finding dis out. And time is short, yah? You are not so young anymore, are you dock-torr?" He emphasized the "dock" half of the word, giving it what I felt was an unnecessary air of sarcasm. Playing possum, I decided to take the role of listener, assuming that is probably what Gamal had done much of the time he sat across from him. Just lay low and keep an open mind is what the little voice inside me kept saying. Besides, they say that in spite of appearances, Germans actually enjoy their pessimism and that no foreigner has a right to deprive them of it.

"I worked with anthrax in grad school, that is until," I listened to the sound of my own voice fade into insignificance, easily replaced by the band, then leaned my face down towards the top of my mug, feeling what was left of my confidence drain into it. 'Was he really implying I couldn't handle the brewer's yeast', I wondered? Then I finished the rest of my beer.

"Yes, I know. You used to verk with bacteria," his face crinkled as if saying the word left a bad taste in his mouth. "Gamal has told me."

I immediately thought back to my conversations with Gamal in Cairo, trying to recall if I'd ever mentioned anything about losing the anthrax grant. No, I was pretty sure I hadn't. But I did mention it to Tahany in one of my e-mails right after it happened, I was pretty sure. Yep, sure enough. Wolf knew too. I tried to take another sip from my empty glass as the waitress returned with two new ones. At least someone was showing good timing.

Cradling his beer, Wolf, his elbows planted firmly in the middle of the table, advanced his upper body towards me, his voice directed into my ear closest to him, speaking as if he were letting me in on a secret.

"The brewer's yeast may not mean much to you. She is not so pretty to look at. She does not move around by herself. Why? She has no means. Yet she harbors more secrets, possesses more talents, than you nor I nor anyone else vill ever know, dock-torr. She has everything to teach us if we vill only listen to her...take the time to examine her...ask all the right questions of her. She is not an *E. coli* and you don't simply dab a toothpick into some sterile broth and feed her some sugars. You must remember dis if you vant anything verth publishing. You don't streak her onto a Petri dish, lock her away in some incubator and then forget all about her while expecting she vill reveal everything in a single afternoon in between teaching science to children. Dis isn't a game dock-torr. Dis is a dialogue...a dialogue between you and one of nature's greatest creations."

He spoke as if the yeast were someone he knew personally and felt the need to give some sort of a character reference to. Pretty strange, I thought, but then again, if I hadn't learned anything at all from all those years of college, it was that microbiologists can be an eccentric lot. So far, this one had them all beat, though. Wolf went back to leaning in his chair, apparently content to study me from a distance again, hands interlocked behind head and using questions to probe me the way scientists often do an experiment, gauging my reaction after each perturbation of me.

I tried to sip some foam off the top of my glass, the foam itself having collapsed and was now trying to retreat down the side of my mug, my chair becoming larger and larger, the same chair that just a moment ago seemed to fit me fine. I decided to give the conversation another chance to develop in a meaningful direction. My tenure might just depend on it.

MULE OF THE ICE CAVES

Only when the music lowered enough and the yeast master felt sure he had my full attention did he continue. "Vell, I vill begin by explaining that it is YOU dock-torr who is the student and the yeast who is the teacher. Is that clear?" It wasn't really a question so I didn't try to answer.

"Dis..." he proclaimed while holding up his mug, "...is a lager. The pretty yellow color you see here...the fresh taste...it would all not have been possible if not for dis humble fungus. A microbe that gets no respect. Hundreds of years ago, not far from where you and I are now, dis tiny microbe vent its own vay...changed and the verld would never be the same. It changed much more rapidly than any bacteria ever vill. Have you ever seen a mule dock-torr?"

Of course I'd seen a mule before. I grew up on a ranch. Still, I just shook my head no. It seemed the safest thing to do.

"Vell the mule is a hybrid...vhat happens ven a horse and a donkey mate. The same thing happened vith the brewer's yeast inside an ice cave high up on the mountain," he gestured towards the Alps, "It came in contact vith dis other yeast, a different kind, and it didn't know vhat to do so it vent ahead and did its own little experiment. It mated vith it." The audience let out a loud gasp, then a spontaneous roar of laughter as I craned my neck towards the stage just in time to catch sight of a fully-grown man in lederhosen trying to plant a kiss firmly on the cheek of a rather plump woman with red cheeks. He failed and looked on forlornly as she ran off the stage giggling uncontrollably. The microbiologist waited for my attention before resuming.

"The yeast accomplished its task, and yet it failed at the same time. Have you ever had an experiment give you more than one answer depending on how you looked at the results dock-torr?"

Once again, I didn't answer other than to just nod my head yes this time.

"Dis produced a hybrid...something we would call a lager yeast.[xciii] Dis new hybrid yeast, dis mule of the ice caves, she came about by happenstance, and yet she could ferment beer in the cold at the bottom of the vats in the ice caves better than any yeast before. And it's all because zee yeast can have sex. Can your bacteria claim that? No, I don't think so. Certainly it was not the haphazard random kind bacteria might boast of once in a while when they shed a small piece of DNA and give it to another, but the true kind when two cells merge to become one, not so different than how human cells come together at conception you see?"

"I worked with anthrax, not *E. coli*," I added, looking for signs he even heard me above the band.

"Vell this new yeast, this mule of the ice caves, it produced dis vunderbar new beer with a clearer, crisper flavor, not the complex profiles of those dusty ales, zee ones the British seem so fond of.   Mine got I vill never understand dis," his accent seemed to wax and wane and I soon found I could use it as a gage to his emotions, "You know vhen ve first began adding hops to beer in the 11th century, the British on their island vay up there, zey actually thought ve vere trying to poison them?   I sometimes dink it is the reason dose limeys still drink ales…and they drink zem warm, too," he made a sour face, "They still think ve vant to poison them, I suppose." [102]

I found it amusing how he seemed so obsessed with my having worked with bacteria before.   To him a yeast cell on its worst day was much more clever and interesting than any bacterial cell having the audacity to lack a nucleus on its best day. [xciv]

"You know, because of our lager beers, people have always come from all over Europe to study brewing in Munich.   Zee man who started Carlsberg in Denmark…Jacobsen vas his name…vell he came here too, and he studied beer like all the others.   And it vas in Munich dat refrigeration vas used for making beer first. [103]   You probably didn't know that either.   No need for ice caves anymore," Wolf's voice became calmer the more he talked of Munich and its unique contributions to brewing.   He then pointed to a clear drop slowly dribbling its way down the side of his mug.

"You see dis dock-torr?   Each drop it takes a million yeast to make.   Talk about your slave labor.   They are so amazing, though," he paused while admiring the droplet rolling down his glass and I half expected to see a tear form in his eye.   "They do fermentation not vith DNA but vith proteins.   If you dry out a yeast cell almost two thirds vill be protein," then he erupted "So, vy do you think there are so many breweries in Wisconsin? Certainly you must have some answer.   It is vere you are from after all."

"German immigrants, I suppose?   Actually, I'm from Montana, Wolf.   It's a different state."

"Yes, of course…because Germans vent there, but vaat else?"

I thought for a moment, then gave up and only managed a puzzled look.

"Zee lakes…vith their ice of course.   To make goot lager beer like in Germany, dey needed ice…to recreate the coldness of the brewing caves of Bavaria…but in Wisconsin…or is it St. Louis?"

As if a consolation prize, the waitress returned and placed a plate of soft, salted pretzels in the center of the table, still steaming, and it occurred to me I was halfway through my second liter of beer without having eaten anything since Egypt.   Wolf picked up one of the pretzels and began studying it in his hands like a bug.   I was relieved whenever his attention was drawn away from me, however momentarily, but I braced myself because I sensed he was about to become sentimental again.

---

[102] The English began using hops in their beer around 1700.
[103] Spaten brewery, 1873. It allowed for the first year-round brewing of lager beer.

"Soft bretz'n. Bread made vith handles," he cooed fondly over the golden-brown, leavened pretzel while turning it over in his hands, caressing it almost, his hands so large that the pretzel almost disappeared, the bread perhaps reminding him of his own Schwabish childhood. "Pretzels like dis could be hung on hooks, or exchanged between knights on horseback...vith just the tips of their spears. They were invented when a monk used them as bribes for the kinder, so they would memorize their prayers.[104] See how dis one it looks like a child praying...with the arms crossed?" He held it up to the light and I scooted my chair in closer to the table, having trouble hearing him now that the Oompah band had picked up additional members. "You see deese holes? They represent the Trinity."

Sometimes Wolf spoke so calmly and quietly he seemed almost normal and yet his demeanor could change out of the clear blue, so I didn't want to miss anything if he should say something important about the yeast. I sensed I could learn a thing or two, if I could just get past his exterior. There was no sense in asking questions, though. I knew that. It's always best to ride a horse in the direction he's already going. A little braver I decided to test the waters of this latest mood.

"You know, bacteria are pretty complicated too despite their small size. It was, after all, someone from your own country, Robert Koch, who worked out anthrax's lifecycle...of it alternating between germ cell and spore."

"Yes," Wolf replied wistfully, still fondling the pretzel, "he was a very detailed, careful thinker our doctor Koch," then he looked straight at me and erupted, "Not like dese dumkoffs today, dese gene jockeys vith their enriched media. Do you know zey actually bragged zey vill make paleontology obsolete someday? Mine Got! The entire field of paleontology! Obsolete! Just like that," he snapped his fingers, "vith only DNA! More publications, more and more, all the time, dat's all dey dink about, Christ vy es dis, and dey don't care how zey get zem, either," he pounded his fist on the table without harming the pretzel any, "And bringing in grant money, traveling to symposia...dat's another thing, deze are de ones who are de bane of goot science. Deese dummkopfs wit der machtpolitik."

I clearly had touched a nerve and backed off, hoping he had gotten it all out of his system, considering myself lucky not to have been the object of his wrath. Mark Twain once claimed that whoever invented the German language must have done it after staying up all night with a toothache.[105] After regaining his composure, Wolf went back to describing how proteins from cells like the brewer's yeast, and human cells as well, have been relegated to what he called the "stiefkinder", or stepchildren, of the life sciences, while accusing DNA of grabbing all the headlines, of being a "dead

---

[104] AD 600's.
[105] Twain also claimed that German words were so long, they had a perspective.

molecule in disguise of a living one" as he put it. [xcv]   I couldn't argue with him too much because on one level I actually agreed.   The DNA chromosome is mostly a storage place – a repository if you like – for information, when compared to proteins.   In spite of its length, DNA is a relatively modest molecule, both physically and chemically compared to proteins, which is why DNA can be recovered from mummies and Neanderthal bones even after thousands of years and still retain useful information.   It's simpler.   Proteins on the other hand are the most "alive" molecules in our bodies.   They are the closest things we have to robots working away deep inside us.   It takes proteins, for example, to make a chromosome divulge all its information to the cell, or to make another copy of a chromosome during mitosis, or move about if you're a muscle cell, or to turn sugar into alcohol if you're a yeast.   But at least I'd made my point and that was enough…for now.   I silently chalked one up for the bacteria. [xcvi]

Cooling somewhat he then added, "As a prokaryotic person you are no doubt familiar vith the Dutchman...van Leeuwenhoek then, eh?"

"The first person to lay eyes on bacteria," I answered confidently.

"And the yeast.   Do not forget about the yeast.   Leeuwenhoek's family on his mother's side dey were all brewers.   Many people do not know dis.   He vas interested in brewing because brewing vas in his blood."

"I'm not too surprised.   Wasn't even Fredrick the Great a brewer?"

"Yes, but right now vee are discussing the Dutchman," he looked away as if disappointed in me and I silently resolved not to change the subject anymore if I could help it.

After a slight pause, Wolf leaned in closer and began again, "He vas the wine-gauger of Delft even though Delft had a reputation as a vunderbar place for a beer.   It was Leeuwenhoek's job to go around and visit all the merchants and measure the wine…for taxation purposes.   So anyways, dis Dutchman, he makes deese perfect tiny lenses in his spare time until one night he gets the idea to have a look at his beer, and not the finished beer, either.   Oh no.   He is a true scientist.   He vants to know vhat is happening vile his beer is still being made…during fermentation.   So he puts a drop of sweet barley wort on his microscope and he fumbles around vith the light from his candle.   He has hundreds of handmade microscopes by the vay.   He made a different one for each microbe he wanted to look at.   He was a very self-reliant man, dis Dutchman, not like deese spoiled kids today.   Even a music student can pop open a catalogue and order a microscope dese days.   It does not take that much…but to be a goot scientist…ah that dock-torr…is another matter.   So Leeuwenhoek he sees the yeast, but he mistakes them for pieces of starch from the barley.   Mine got, as brilliant as this man vas...he simply missed them you see?   Yeast do not swim around.   They do not have to.   So they just sat there in front of him and he vas fooled into thinking they vere nothing.

And as a result, the yeast would keep its secrets for 150 years more. By the way, why do Canadians wear those flags? The ones vith the red leaf on them. They sew them onto everything, their backpacks, their shirts, even their luggage. Mine got, I even saw this Canadian once who had one on the seat of his pants."

"I don't know. Why are you asking me?" I asked, dumbfounded.

"I just thought maybe since you are from America, and Canada is right next to America, vell maybe you might know."

"Pride?" I suggested, "It's just a guess."

His question gave me a sense of vindication for who was changing the subject now? Wolf picked up his beer, wiping his mouth from one corner to the other with a sweep of his sleeve and then began looking again into the golden liquid inhabiting his glass, clearly taking delight in it before downing another large gulp, apparently unconcerned that I was watching and capable of forming my own opinions.

"He was a very curious man, dis Dutchman, and he vanted to know what caused all the bubbles in his beer, that is all. Curiosity. Why is just finding out the answers not enough anymore? He didn't do it for the fame, although he did become famous. And he didn't do it for the money, either. No. He only vanted to know if any of his little animiclules could live without air. Yes, he found out, they could. But the Dutchman, he missed the fermentation part because...vell because he didn't see the yeast as alive. Mine got he simply didn't see it even though they vere right in front of him,"

Wolf slammed his mug down on the wooden table and it made a hollow ringing sound. The same couple as before turned around and stared at us again while I scrunched down in my seat a little further. His accent increased in thickness but then thinned out as he was capable of calming just as quickly.

"But he did scrape zome bacteria from his teeth and he found he could kill zem vith heat and vinegar. So Leeuwenhoek he becomes not only the first person to see bacteria, but also the first to kill zem. You should know."

I didn't reply even though I knew what he was saying was true. Wolf wasn't the only microbiologist with an appreciation for history. They didn't need government money to run their labs in those days, which was why I currently had an affinity towards scientists like Leeuwenhoek. Most scientists back then were wealthy, or had sponsors, and governments didn't seem to feel the need to support research much in those days.

"Yes," Wolf said wistfully, leaning back in his chair, "He invented a mouthwash too...just a mixture of vinegar and wine. Dat's all it vas. And dis Dutchman he had over 400 microscopes he made vith his own hands. Did you know?" Wolf was working on his third beer while I was still nursing the foam off my second.

"Yes. You said something about that, Wolf."

We sat without talking for a while longer, perhaps ten minutes more, my attention focused mostly on the band but not really listening to them.   It seemed the right thing to do as they were putting an awfully lot of effort into whatever it was they were doing.   Feeling higher than a Colorado pine tree, I had little idea what Wolf was thinking and didn't try to guess anymore.   A few minutes later and the yeast master picked up where he left off and as was becoming my habit I took up my position, leaning forward towards the middle of the table, my cheeks cradled in the palms of my hands in an effort to hear him as the band had begun playing a merry polka.

"The yeast kept its secrets so long because it likes to vait, you see?   Now that is patience.   A single yeast can ferment its own weight in sugar every hour if you let it.   You must remember dis.   Imagine yourself eating 70 kilograms of sugar in one hour.   Could you do that?   I don't think so.   Nor could I, nor could anyone in your department back in Wisconsin."

"No," I agreed, wondering if the pretzels were beginning to mop up the beer in my stomach, "I'm from Montana."

"Vell the yeast can do the equivalent.   Eat it's own weight every hour. You cannot imagine and barely can I.   Vell ven you can, you vill begin to appreciate all the vonders this little creature holds in store.   So it was left up to a Frenchman to take up the challenge…of proving it was indeed the yeast behind perhaps the greatest miracle of all time…of fermentation.   It had baffled so many all the vay back to ancient Egypt.   Dis is what I could never get Gamal to understand, the importance of asking questions just for the sake of asking.   Vell, de next person vas Louis Pasteur and he vas a chemist by training, but he wasn't a coward and he wasn't afraid to change fields. Maybe a bit like you that vay."

My first grudging bit of respect from him and I wondered if he was warming again.   The fact is, I'd known about Pasteur, and about how, as a young professor, he had found himself in a remote French village teaching chemistry in the mid 1800's, which is also where he began his historic work on fermentation.   Leading chemists of his day thought fermentation was strictly a chemical process so they assumed there was no need for a living organism. I identified with Pasteur too…as he was under some of the same pressures I was now, to prove that scientific work can be meaningful in a practical sense. That's been the case for a while and I expect it will continue.   Wolf was right in a way.   Just finding out the answers isn't enough anymore.   We need practical reasons for doing our work now.   There's even a place on your grant application where it's a good idea to try and answer this if you want your grant approved.

"Anyway, Pasteur he turns into a biologist and decides to use some of the same skills he learned as a chemist in Paris to determine once and for all why it vas that grape juice had this tendency to turn into wine.   You must understand, the French are always more concerned vith their wine...more so than their beer. They are French, after all. They cannot help it. It is in their nature. Vell as a lecturer in Lille, he became quite vell known. Lille was an industrial town so one day a boy whose father owns one of the sugar beet factories for making distilled alcohol, he asks Pasteur to find out why his beet juice keeps going sour. It was costing the family a fortune each time a new batch of beet juice spoiled and had to be thrown away.[106] Pasteur vas a vinophile so he knew about the same problem in wine, in fact he sometimes returned to his family's farm to relax.[107] They had a vineyard there and Pasteur he walked around it in the evenings, dinking about fermentation, I suppose. He even brought his microscope home vith him, perhaps much to the displeasure of Mrs. Pasteur, I don't know. So he decides to help the boy's father which means taking his microscope into the factory to examine the vats of beet juice up close. It does not take him long to realize dat in the vats dat keep going sour, he finds tiny rod-shaped bacteria...millions of them. But in the goot vats dat produced the alcohol, he finds only the yeast. And not only dis, but in the bad juice the Frenchman he sees that the yeast cells are becoming deformed, changing into strange oblong shapes, in the sour juice, the ones with the bacteria...as if the yeast are becoming sick or something. He was quite observant for a Frenchman. And of course being French he had an imagination to go along with it. What Pasteur saw vas a kind of *war* taking place vithin the beet juice...it was a battlefield between good versus evil. Whoever won got to make their product."

"To the victor belong the spoils," I added, confident the pretzels were doing their job and so I hoisted my mug and took another dent out of my beer.

"If the bacteria won, then the product would be vinegar. If the yeast won it would be alcohol. It's how Pasteur became convinced of germ theory," I added with a certain triumph, "The notion that certain diseases, infections in the body for instance, were caused by bacteria. No one had any good evidence for this before."

"Yes. And the brewer's yeast led him to it, you see...that bacteria can cause diseases the same way they ruin good wine. Pasteur astonished the people of Lille because he got so he could even predict ahead of time, like a fortuneteller, which vats would go bad, and which would not, just by looking at a single drop of juice under his microscope. Mine got he was a true scientist."

"He eventually found out what Leeuwenhoek knew before, to use heat to get rid of bacteria," I added.

---

[106] The French were using sugar beets rather than sugarcane because of an embargo by the British in the West Indies.

[107] So did Pliny the Elder, who wrote about diseases of wine almost 2,000 years earlier.

"And dis is why the first use of pasteurization was to preserve wine, not milk.[108]   It became so widespread that the emperor Napoleon, the third one, what's his name, he wanted Pasteur to use this new process to solve the problem of sailors deserting vhen the wine vent bad on ships, which, of course Pasteur did." [xcvii]

Wolf continued, his voice now steadier and more even than before.   It probably helped some that the band had gone on break.

"Pasteur's insights were not lost on the English surgeon Joseph Lister, who takes Pasteur's knowledge and he applies it to fixing compound fractures, the kind that always got infected and ended in amputation.   So Lister he keeps the germs out of the wound using carbolic acid, one of the first modern antiseptics applied to the skin.   And it works, so he confirms germ theory and gives Pasteur the credit.   How is dat for finding a practical application for your work, eh dock-torr?"[109]

"Pasteur was a great man," I lamented.

"But there vas at least one thing he failed to do."

"What is that?" I wasn't too surprised there had to be a catch as far as Wolf was concerned.   Pasteur had the misfortune of, after all, being French.

"He never could prove how the yeast accomplished fermentation.   He tried and he tried...for more than half a century he tried to break open the yeast cells...to isolate, chemically, whatever it vas inside the yeast that accomplished dis miracle.   But he vas never able to do it.   No, it vas left up to another chemist...a Bavarian dis time...and his name vas Eduard Buchner.[110]   After much effort, Buchner saw a way to extract the yeast juice, its cytoplasm, from inside the yeast, and along vith dis spilled out the enzymes that could accomplish fermentation.   Enzymes are of course proteins.   It was like squeezing oranges to get the juice out.   He used sand to grind open the yeast cells, scraping them dis way and dat, and then he pushes the broken cells through a filter using high pressure and so the clear juice that flows out it goes into the beaker, leaving behind the broken parts while capturing only the cell's interiors.   When he adds some pure sugar granules from a jar on his shelf to this pressed yeast juice, he sees bubbles form, the same bubbles people had seen for so many thousands of years while making beer and wine.   Only he vas at dis moment the only person who could make alcohol but without any cells.   The very first.   He did fermentation vith just the chemicals...the proteins from inside the yeast, some water, and a bit of sugar."

---

[108] 1886.

[109] Lister's father was a wine merchant & amateur scientist. *Listerine* mouthwash is named in Lister's honor and was the first over the counter mouthwash available in America, in 1914. The ancient Egyptians made due with salt as a mouthwash.

[110] Before beginning study at the University of Munich to become a chemist, he worked in a canning factory in Munich.

"Buchner became the first biochemist by doing that one experiment," [111] I added.

"Yes. You do know your history, don't you dock-torr? But it only seems simple now. Buchner did dis in 1897 because he tried something new. He did vhat we would see as the very essence of biochemistry today," Wolf was patting his open hand large enough to be a catcher's mitt on the table rather than pounding it as a fist like before, to more gently drive home his point. Then he began pointing his index finger straight at me and exclaimed, "He recreated a process that took place only inside cells before, but he does dis in a test tube. No bacteria. No yeast. No living cells of any kind. Just the proteins along vith some sugar water. He vas the first to do cell-free fermentation. And it changed forever the vay people would look at life, not just the yeast, but all of life by this one experiment." [112]

I had already known about Buchner and his brother, who was also a chemist, having taken a course in the history of biology as an undergraduate.

"It's also true," I added, "that this experiment helped end the debate that had gone on for years, the notion that life was too special to understand using chemistry or physics, that living things had a divine spark that could never be reproduced in any test tube."

"Yes," Wolf was becoming more excited so I braced myself. But it turned out to be for nothing, as remained calm, "You are the cleaver one, aren't you dock-torr? You know your history... for an American that is. This idea of vitalism, an idea that went clear back to the Greeks and Aristotle, that all living things possessed special forces that made them impervious to the same things ordinary matter is subject to. Made them impossible to do experiments on."

We both sat for a while, I for one contemplating all the implications of Buchner's work. How if it wasn't for the yeast and Buchner, how biochemistry wouldn't be where it is, how we might not have molecular biology and my new yeast project, which grew out of Buchner's choice of the brewer's yeast as a research subject.

Wolf was smiling now, no doubt savoring the contribution Bavaria and his beloved yeast made to the history of something he clearly cared about. I also knew from researching the grant that if my lab was fortunate enough to discover any new yeast proteins, we'd be depositing their sequences of amino acids right here, electronically, in a database called the Munich Information Center for Protein Sequencing.

---

[111] He published his results in an 1897 paper entitled "Alcoholic Fermentation Without Yeast".
[112] Buchner wasn't the first to work with proteins. At least 1000 years earlier shepherds realized they could cut open the stomachs of goats and use seawater to extract a solution (the protein is called *chymosin*) to make cheese from goat's milk.

"You know, some time after Buchner accepted the Nobel Prize in 1907, you know what he said don't you?"

"No," I replied, "I have no idea."

"He told everyone that the reason he had succeeded where the Frenchman had failed for so long was that he, dis humble Bavarian, had made use of the leftover yeast from the breweries of Munich... the same hardy lager yeast that preferred the ice caves up in the Alps. Do you remember? Back in France, Pasteur had been using yeast that made only ale... at room temperature. Kind of wimpy compared to what makes dis," he hoisted his mug high enough to signal to the waitress for another beer. Class apparently wasn't over for the day.

"Sometimes the devil is in the details when it comes to experiments," I added.

"Pasteur died before they gave out Nobel Prizes, or he would have gotten one... for his work vith the yeast," Wolf acquiesced.

"Or anthrax," I added, knowing that what I was saying was true, yet it was also true that I was thinking more about a plate of steaming hot bratwurst topped with sauerkraut that had just traveled by at eye level – more so than I was arguing with an eccentric German who could be a few pickles shy of a barrel, one who had obviously tossed back a few before he even sat down.[xcviii]

THE GLOCKENSPIEL

After the waitress returned, he downed his beer in one gulp, looked at his watch, and then demanded we take a walk to see the town's Glockenspiel. The room had been getting more crowded and the thought crossed my mind that it might be easier to catch a bite to eat from a street vendor than wait for the waitress. Before I knew it, we were out the door and a cool breeze sweeping down from the Alps was helping wake me out of my lightheadedness. Perhaps the pretzels were doing their job as well. Things were looking up as we headed towards the large public square called the Marianplatz. The plaza has been the center of trade in Munich since medieval times.

Along the way Wolf began helping me with some of the more practical aspects of the project while probing me with questions regarding my plans when I got back. He offered up a cornucopia of advice both practical and theoretical, sprinkled with the kind of details you won't get from any journal, including which supply houses to order our chemicals and yeast from, which houses to stay away from, as well as whom to trust and whom not to when it came time to share any preliminary results we might get. He apparently knew all the players in the field by their area of expertise including which university or brewery they happened to be at, even which proteins they were interested in, however he didn't always seem to remember their names, which at the time seemed odd, but maybe not so much given his penchant for beer.

With darkness descending and the lights being switched on inside the houses, the neighborhood felt like one of those Grimm's Fairy Tale villages and at one point we passed a gingerbread-looking building on the opposite side of the street that turned out to have special significance.

Wolf noticed my attention drawn to it and explained, "By 1750 there were 4,000 breweries here in Bavaria. Beer was safer to drink than river water. It still is.[113] Even in England in the 17th century, the kinder in boarding schools would take bottles of beer to school vith dem. You see that building across the street? It is the Bavarian National Theater. It caught fire in the winter of 1823...yet it was saved by beer not once, but twice"

"Twice? How's that possible?"

"The first time was to put the fire out. It was so cold in Munich that the water had frozen over, so they had to use beer from a brewery to douse the flames."

"And the second time?"

"To rebuild it, they raised the beer tax." There was silence for much of the time as we walked the rest of the way to the square.

"Yes," Wolf began again finally, his mind apparently more receptive in the open air. "We had peace for over 700 years here...and the first streetlights in Germany that ran on electricity, too. How strange to think that Munich would someday become the birthplace of Germany's greatest sorrow."

"Greatest sorrow?"

"I am speaking of course of Nazism. You see, after the First World War, Munich along with the rest of Bavaria had fallen into political instability. It vas terrible and it affected everything. You probably don't appreciate it. Inflation was so terrible the bankers weighed out the money instead of counting it. German farmers in beer halls ordered two beers in case the price went up before they got done finishing the first one. There vas starvation in the streets and the people sometimes they ate rats just to stay alive. Most of the intellectuals had fled to Berlin, leaving Munich to the reactionaries. Even Vladimir Lenin lived here for a while. For a time, Munich was a Soviet Republic. Vell, as you can imagine, dis was a good nest for radicals, and one in particular who set his sights on overthrowing the Weimar Republic. In 1923, hoping to spark a revolution that would spread all the way to Berlin in the north, with his small army Adolph Hitler he walks through the doors of a beer hall here in Munich."

"I've never understood that, really. Why a beer hall? And why here?"

---

[113] If not for its unclean drinking water 2 centuries ago, New York City wouldn't have grown into such an important brewing center in North America.

113

"I vas just coming to that. You see, like most politicians, Hitler had developed his speaking skills inside Munich's beer halls. Beer halls have always been important, not just as places to drink, but to socialize," he emphasized the word socialize by saying each of its syllables more slowly than the other words, "To live life, to discuss issues of the day. These enormous halls were natural gathering places, as you just saw. They draw people like moths to a flame, and if you are a politician, where better to find a receptive audience? In America you prefer your soapboxes in public parks... vell in Bavaria we have a fondness for our beer halls. It's that simple. The famous Bierhallenputch was planned by Hitler inside the very hall you and I were just in.[114] Hitler and about 600 of his followers stopped a rally already in progress at the Burgerbraukeller, a beer hall not far from here. Like some animal, Hitler is wild with excitement, so he leaps up onto the table and fires his pistol into the ceiling, announcing to the crowd that the national revolution has already begun. In another hall just across town, at the Lowenbrau Keller, his accomplice Ernst Rohm is waiting with members of his own Fricorp, hoping to widen the coup. Like I said, their idea was for all of dis to spread to Berlin. But when Hitler fails to gain support from the leaders in his own beer hall, Rohm's men near the other one are easily defeated."[115]

"I guess it's hard to believe it all actually happened. Everyone's been so friendly here. Countries change I guess."

"Yes. And Hitler vas wounded. They capture him and put him on trial where he undergoes this transformation. He vas put in prison but the judge liked him so he serves just nine months. Nine months. And unfortunately for Germany, he writes a book and became a vell known figure after dis.[116] So you see...a beer hall launched Hitler on his rise to power. It's not so strange, the yeast being involved in social change. Take your own history. You probably knew that the American Revolution was planned in taverns in and around Boston and New York didn't you? The Boston Tea Party? Or that John Adams and George Washington met for the first time in a tavern? Of course you knew. And how about Jefferson? Didn't Jefferson begin writing the Declaration of Independence inside a tavern? Of course he did. In less than ten years that document would change everything."

---

[114] In *Mein Kamph*, Hitler described the first time he spoke to a large crowd (in the Hofbrauhaus; February, 1921). He used a beer table as a podium.

[115] Hitler took Bavaria's 3 highest officials at gunpoint into a back room to negotiate his terms while 3000 patrons in the beer hall grew increasingly restless. Fellow Nazi Hermann Goering tried calming the crowd by yelling "Shut up. You've got your beer, haven't you?"

[116] When Hitler ran for President of Germany in 1932, he promised the voters "brot und arbeit" (bread and work). / Berlin still has an "underground city" built in the 19th century by brewers to store beer since refrigeration wouldn't be invented until the end of that century. Hitler would later turn one brewing cave underneath Berlin into a weapons factory.

When we arrived at the Marinplatz I looked up at the stage, which is when it dawned on me I'd forgotten my eyeglasses back at the Hofbrauhaus. It was already dark and I was left with the light of the streetlamps to go by. I've needed glasses to see distance ever since doing too much reading back in graduate school. The scientific literature is a vast resource, or depending on how you look at it a torture chamber, of small print and graphs all competing for the same space.

So as a result, when the bells of the Carolinian on the Glockenspiel began to chime and the wooden figures started circling high above our heads, all I could make out were some brightly painted, out-of-focus, blobs. I chalked losing my glasses up as punishment for drinking two liters of beer on an empty stomach while trying to keep up with a native Bavarian. I looked up at Wolf, who seemed transfixed by the display. It was as if he were seeing it for the first time. I took my pack off and placed it next to my leg, then dug into it looking for the guidebook so I could get an idea of what I was missing. The book had a photograph reminding me of a puppet show I once saw. The figures in the photo were all wearing traditional Bavarian costumes of the Middle Ages.

After about twenty minutes, the music was over and everyone had disappeared back into their cloisters. Wolf didn't bother looking away during the entire show. I don't think he even knew I was there. And besides, it was too late to go back to the beer hall. My train was due to leave in just twenty minutes. On top of this, I still hadn't eaten or slept in two full days.

After the crowd dispersed, Wolf finally tilted his head down to look at me, then said softly, "You probably know by now I am a mycologist... a dedicated yeast person. I am... I mean I vas a scientist and I don't think much of dese gene jockeys dese days."

"Yes, I think I figured that out Wolf."

"Bacteria and viruses with their DNA, they get all the headlines... and the grant money... and what's left for the rest of us? But that's how it goes, I suppose. The young take the place of the old. As it should be. Just the same, without fungi where would we be?"

"Up to our necks in undigested leaves?"

Wolf didn't smile, but instead continued speaking softly, but deliberately. "You know, I cannot deny bacteria have their place. Did you understand the story?"

"Story?"

"The one the Glockenspiel just told us."

"No. I had no idea it was a story. Something about dancing?" I thought about my eyeglasses lying next to the pretzels but didn't mention it as I didn't want to give him yet another reason to look down on me.

"It was about the Black Death."

"But the figures on the stage…they were all dancing. Weren't they? The guide book said they were."[117]

"Yes…they were dancing because as coopers of the town, the barrel-makers, they knew the plague was over. The year was 1517. Their job before was to make barrels for Munich's brewers, so they could store beer, and now the plague was gone and business was going to be goot again, if only the people would come out and start living, you see? But the coopers, they were the only ones who realized dis. They promised the people that if they would come out of hiding, then they would do the dance for them in the Marienplatz. The people came out and the coopers kept their promise. In fact, you can see it. They still do dis dance every seven years…right here."

"So the people wouldn't even come out of their homes?"

"Some sealed themselves inside with bricks and mortar…they laid them in their doorways and windows. The wave of Black Death 150 years before, in the fourteenth century, had already wiped out maybe a third of Europe…almost 40 million. Well anyway it kept returning, the rats and plague. No one saw the connection. So the people were afraid. It is quite natural. You are the expert when it comes to bacteria, perhaps you can enlighten me for a change." I couldn't tell whether he meant it as a compliment or something else. Then he cleared the air by adding, "You seem to know your history. Not many do these days. Except for Gamal, maybe."

"Well, the plague was caused by the bacterium called *Yersinia pestis*," I began, still wanting to impress him for some unsettling reason, "The disease had been known since antiquity, but not the microbe that caused it. It's even possible King Tut ascended the throne at such a young age because of plague."[xcix] I threw in some more ancient history as we walked because it seemed to keep him calm.

"It was bubonic plague that set the stage for the final events of the Roman Empire.[c] It lasted 6 decades in Rome, killing millions. It had to have hastened Rome's downfall. Plague was also the final push that got the French out of Egypt in 1801. In the 1300's, ships from Asia carried black rats to Italy. Well when the rats died, the fleas feeding off them weren't too particular about where they got their next blood meal from, so of course they jumped onto the people, biting them and in the process transferring the microbes into their blood," I paused for a moment to make sure he was still listening.[118]

---

[117] The dance is the schafflertanz.

[118] Robert Hooke published a detailed drawing of a flea he made for his book *Micrographia* in 1665.

"Yes," he added calmly, still staring up at a full moon, which was just visible above the rooftops.   He was clearly expecting me to continue.   "Go on."

"An interesting thing about the plague is that it isn't even a human disease.   Bubonic plague is a disease rodents get," I continued.   "It's one of those zoonotic illnesses that happens when we get caught in the crossfire.   In fact, about 75% of emerging diseases today jumped from animals...like the plague did...and still does.   Maybe that's what's going on with Ebola now. The Ebola virus is becoming more humanized as it jumps from animals to us. Hopefully, Ebola and Hanta virus never evolves the tricks smallpox or *Yersinia* did...to cross over the human lung barrier and get spread through the air."

"Yes.   I've heard many diseases came from animals.   Because of agriculture."

"Lyme disease, measles, HIV," I rattled off a long list seared into my memory from cumulative exams back in graduate school, "smallpox, brucellosis, all came from animals.   These are domesticated diseases that somehow crossed over.   The flu virus is another...one that is still evolving and jumping, evolving and jumping, becoming a little more human each time. We can't eradicate it, and probably never will.   The virus remains safe inside birds and pigs.   It rearranges its genes in pigs as if they were mixing vessels, and reemerges as a new strain of flu every year." [119]

"It was a big mistake killing all the rats," Wolf replied, steering the conversation back to something he clearly cared about, "People make so many mistakes.   They thought the Great Pestilence was punishment from God.   They practiced self-flagellation, whipping themselves on their backs. All they did was bleed so much that they gave the microbes to others.   Their devotion made things worse, not better.   People can be so silly," [ci] Wolf said wistfully.

"But at least there were some positive outcomes.   Along with the horse collar, the coming of plague helped bring about an end to the feudal system in Europe, didn't it? [cii]   And because the people felt let down by the church they began doubting more.   They had prayed but it didn't help.   Ironically, it was because of the shortage of clergy due to plague that the colleges of Oxford and Cambridge (where the structure of DNA would someday be discovered, but I didn't mention that to Wolf) were founded, to train more priests.   The loss of security in religion helped lay the groundwork for skepticism during the Renaissance.   Modern science is based on skepticism, so in a way plague

---

[119] The only animals the Native Americans deliberately bred were horses, which is why Lewis & Clark encountered so many Appaloosa after crossing the Divide and while living with the Nez Perce.   The tribe routinely gelded inferior male horses.   That day Lewis & Clark witnessed the largest herd of horses on the North American continent, tens of thousands while west of the Divide.

played an important part in bringing about the scientific revolution."[120]   I started to wonder if taking me to the Glockenspiel wasn't Wolf's way of apologizing for his harsh treatment of bacteriologists earlier on.

Wolf continued with my thought, "So the people found themselves in demand and were rewarded vith a living wage for the first time.   German farmers had disposable income.   And do you know what they bought vith their newfound wealth?   You saw some earlier…in the beer hall we were just in.   Mostly used as decorations these days.   People can be so silly the way they relegate things to shelves."

It must have been obvious by my expression I didn't know.

"Would you like a hint?   They were the first luxury items in Germany."   I still drew a blank.

Embarrassed, I added, "No.   I can't think of anything," wondering if he was disappointed in me.   Strange how I still wanted to impress him for some reason.

"Beer steins, of course.   Dey became more than just vessels to drink beer from.   In time, beer steins found their way into the Bavarian hearts as works of art.   Dey had scenes from the Bible or someone's family shield carved into them.   A cottage industry emerged making pewter and earthenware beer steins, wooden and glass ones too.   They found dat, fired in a kiln with high temperatures, clay fuses like glass.   It became less porous, stronger, and could even be shipped long distances.   Dis new, harder surface was also more aseptic because it had less places for bacteria to hide out.[121]   Germans thought drinking beer out of steins should be an event, a feast for the eyes as well as the stomach.   But a strange thing happened along the way, you know?   One of those little ironies of history we both seem to appreciate. Germany's new lager beers invented here in Bavaria were too pretty. Everyone wanted to look at their new beer…but not the container anymore. No, they only wanted to see through them," his voice became quieter, almost reverent, "And dis is why most people today drink beer out of a clear glass. It's a shame, really.   No one cares about beer steins anymore.   We changed the ale yeast into the lager yeast and look what happened.   We Bavarians are good at shooting ourselves in the foot sometimes."

"But lager beers are more popular than ales.   That's at least something, isn't it?   When I think of a beer, I always picture a lager beer," I added in an effort to cheer him up for he seemed to be lapsing into melancholy again.

---

[120] Skepticism only *seems* new but there were actually ancient Egyptians that questioned (and even mocked) the belief of resurrection and an afterlife.   Aristotle claimed that the mark of an educated mind was being able to entertain a thought without necessarily accepting it.
[121] This is also the reason your toilet and sink are made of ceramic.   One of the reasons so little is known about ancient Greek houses today is that they were made of unbaked clay, which didn't stand up to the elements like fire-baked clay.

"And what about the lids on beer steins?   What's up with that?   I've always wondered about the lids."

There was more silence and for a brief moment I thought that maybe he was ignoring me.   Finally, he replied.

"Yes, the lids.   Well they have a thumb lift and these also had to do with plague.   The people were so afraid of plague returning.   It was a constant fear for hundreds and hundreds of years.   Dis fear brought about the first hygiene laws for food.   You see, in the late 1400's, while Columbus was discovering your part of the world, Germany was visited by swarms of flies, and since a connection had been noticed between filth and the plague, by the 1500's a law was put in place to have a lid on the beer steins... to keep the flies out." [122] [ciii]

It was too bad he had to retire, I lamented.   It was becoming clear that behind all the bitterness still lay a competent, perhaps even brilliant, mind. He was one of those people you sometimes come across in science, a maverick that you know belongs in research, and can also make a newcomer like me feel a bit inferior.   Taking me to the Glockenspiel must have been Wolf's way of mending fences, I decided as we walked.   I was slowly coming to realize this.   And I'm not sure how much of it had to do with the beer, but I was definitely warming towards him.   I decided as the train station came into view that he wasn't so strange after all.   Just misunderstood more than anything.   A genius, like Gamal had described him, but in his own peculiar way.   I might feel the same way if I'd been forced out of academia the way Wolf had.

The next thing I knew, out of the clear blue, Wolf had me by my collar and with enormous strength began reeling me in like a fish tethered to a line.

"So tell me something before you go, dock-torr," he insisted as I arrived to within an inch of his face, "There is something I've been vundering about for a very long time now... about your country.   Are there any Schwartz's in your department?"

"I'm sorry?" was all I could manage, close enough now that I could smell the beer on his breath and even see the gray whisker stubble on his Adam's apple, which seemed to be bulging to the point that it actually looked painful. I instinctively looked up into his eyes hoping for some signal he was only joking.   Instead he merely cocked his head to one side, seeming more curious than anything, as if he hadn't done or said anything out of the ordinary.

---

[122] Superstition was also why people killed the cats, which, it turns out, were partially immune to plague.   A smaller cat population meant more rats and therefore more plague. / Biological warfare was waged in the Middle Ages when the bodies of plague victims were catapulted into walled cities under siege.

"Schwartz's...you know...black people," he reiterated and then paused, clearly waiting for an answer.

"Well...not in my department exactly," I squirmed, still firmly within his grasp, "but there is an African-American in another department." Before I could finish he was already nodding, as if he had all the information he required.

"Dis is vat I thought. Dare are no Schwartzes in Montana. Dey all prefer the big cities for some reason. You, on zee other hand, you like zee mountains. Vhy is dis?"

I didn't know how to answer him. It was true enough there were less black people in Montana than most states for some reason. I never knew the reason. I never asked.

I craned my neck to look over my shoulder, realizing that the train was my best hope for escape.

"Well, I guess I'd better get off...if I want to find a seat. Haven't had a thing to eat all day," I added and began reaching towards the ground for my backpack.

He let go of me as suddenly as he had grabbed ahold, then hunched his upper body over into a ball as if contemplating something while I slung my bag over my shoulder. Diesel exhaust was already making its way into my lungs, replacing what was left of my appetite with a sickish, bloated feeling.

"It looks kind of crowded. Thanks again for everything."

"Yes. We should keep in touch, shouldn't we," he announced even though he'd never given me a way to contact him. Maybe he meant through Gamal.

YEAST MEETS WEST

After locating my seat, I relaxed and my appetite eventually returned. I was finally able to eat and even sleep for a couple hours. When I awoke, we were pulling into the airport at Frankfurt and a mad dash ensued thanks to my inability to read any of the signs without my eyeglasses on. With boarding pass in hand and only my backpack to worry about, I saved time and soon found myself on the right plane, doing my part to get ready for take off. Looking out the window at the baggage handlers placing luggage onto the conveyor belts, I couldn't help thinking about how much I'd learned. Just two short weeks ago I had little knowledge of the brewer's yeast beyond what it took to apply for a small protein expression grant. Now a fungus was taking the place formerly occupied since graduate school by a bacterium Robert Koch called *Bacillus anthracis*.

I also couldn't help thinking about Wolf, and how I'd known a couple professors back in graduate school like him. They were used to doing microbiology back when it was what other scientists condescendingly referred to as "stamp collecting".[123] What microbiologists did mostly for the first 100 years or so was go around finding new microbes, bringing them back to the lab and growing them on dishes and then characterizing them with various tests to determine what kind of food they required, whether they used oxygen to respire with or not, where they lived and so on, then they classified them by their appearance or ability to hold a gram stain. Things tended to fit neatly into nice little boxes in those days.

Then, something unexpected happened in the closing decades of the 20th century. DNA sequencing got cheaper and easier so that, by the turn of the new millennium, not only did we have the entire 3 billion base sequence of the human genome on an NIH computer, available to any person in the world with access to the Internet, but there were 20 complete genomes of bacteria available too, base for base.[civ] And so because of DNA sequencing, things weren't fitting so neatly into little boxes anymore when it came to bacteria. It was becoming clear that bacteria swapped a lot more genes with each other, even with viruses sometimes, than few had previously allowed themselves to imagine. The insides of a cow turned out to resemble a trading bazaar not so unlike the Khan al-Kalili. Swapping DNA is why bacteria can pick up antibiotic resistance genes so easily in hospitals.[124] They get them from other microbes in our gut.

Instead of a "tree" with a trunk at the bottom and branches radiating outward towards the top, bacteria when they are grouped together on paper nowadays to show their interrelationships look more like an intricate cobweb, more like one of those food webs in biology textbooks used to show which animals eat which in ecosystems. All in all, it can be a little disconcerting to see things you thought were once set in stone all of a sudden appearing to go their own way for someone used to a career based on absolutism. It even calls into question the very definition of what a species is these days. There can be so many shades of gray that some scientists think we shouldn't even be using the word species anymore when it comes to describing bacteria. Things don't fit so neatly into boxes. Some things are, for better or worse, a continuum the closer you look. It's one of the reasons many of us in the life

---

[123] Ernest Rutherford – discoverer of the atomic nucleus – famously said that all science is either physics or stamp collecting.

[124] Bacteria mutate much more rapidly than our own cells. In fact, it was mutant strains of anthrax that eventually led the FBI to their prime suspect in the letter attacks. Dr. Bruce Ivins (the Ph.D. kind) did something unusual even among anthrax researchers. Ivins had mixed mutant strains along with normal ("wild type") strains within the same flask of bacteria. This provided a unique signature (uncovered by directly sequencing the bacteria's DNA), allowing the anthrax used in the terrorist attacks to be traced back to his flask. He committed suicide with an overdose of Tylenol in 2008.

sciences still suffer from physics envy. Yes, the days of "one gene, one Ph.D." were definitely over, I lamented as the earphones were being passed out and the trays placed into their upright positions. The days of the amateur scientist like Leeuwenhoek working alone in his home making important discoveries were no doubt over as well.

Aviation science, while half a century younger than microbiology, is based on far simpler principles, and so within a few minutes we were safely off the ground and headed towards the Atlantic. By the time the plane leveled out, the moon had slipped beneath the horizon and it was dark outside. Everyone seemed intent on sleep but oddly enough I was getting my second wind and so I turned my overhead light on and took out the manuscript Gamal had entrusted me with, flipping through it, noticing that I was now about halfway through.

After the German doctor Robert Koch figured out a way to grow bacteria separately – like the ones that cause anthrax and cholera, on dishes with nutritious growth media in the 1800's – it wasn't long before commercial brewers all over Europe and North America were using these same techniques to grow brewer's yeast in their own laboratories. Being good businessmen, brewers simply wanted more reliable batches of beer. It was more economical to control as many of the variables as possible. Now, thanks to Leeuwenhoek, Pasteur, and the others, the yeast was one of those variables that could be controlled too.

Breweries like Carlsberg in Copenhagen, begun by the young Jacobsen, the one who had studied brewing in Munich, became pioneers of yeast research. By the late 1880's not only were there methods for growing yeast as pure cultures, but brewery scientists were also able to keep track of the amount of sugar in beer wort using sacchrometers, and they could control the pH too, in fact the pH scale so familiar to us today was developed by chemists working at the Carlsberg Brewery. The old hit-or-miss way of pitching yeast from previous batch into new one was, after thousands of years, finally on its way out. It was too easy to pick up bacterial contamination that way.[125] The time had come for a change.

For quite a while it seems, biologists classified the yeast as a plant, simply because it forms buds when it reproduces, looking the way a plant does when

---

[125] The Danish fermentation physiologist Emil Hansen at Carlsberg in 1883 became the first to work with brewer's yeast in pure culture. He began his culture with a single yeast cell he isolated, the first time a pure culture was obtained after beginning with a single microorganism. Similar to the way Pasteur distinguished bacteria from yeast in the sugar beet juice, Hansen was able to distinguish brewer's yeast from other species of yeast in nature. While many industrialists would have closely guarded their secrets for culturing yeast if they had discovered it, Jacobsen insisted Hansen always share his methods with other researchers by publishing in scientific journals.

it forms a new bud on a stem.   In fact, it was this same budding process that finally proved Leeuwenhoek wrong and showed that the yeast was in fact a living creature when Cagniard-Latour took his turn examining it in 1835.   But it was Schwann who would come to realize that the single-celled brewer's yeast was – in truth – a fungus and so more like a mushroom, biologically, than any plant.   While working at the Carlsberg brewery in the 20th century, Dr. Ojvind Winge (the Ph.D. kind) looked closely enough to see that the yeast not only formed tiny buds, but could also undergo sexual reproduction with another yeast of the opposite type.[cv]

Two different brewer's yeast cells can, when conditions are right, actually fuse together to become a single cell so their chromosomes can mix. Genetic exchange occurs among their chromosomes to produce new combinations of genes, like shuffling a deck of cards to produce new hands during a poker game.   Biologically, the yeast would at this point be the equivalent to a fertilized human egg cell at conception.   But rather than developing into an embryo like we do, the yeast can live this way as a normal yeast cell, except that now it has the ability whenever it needs to, of forming 4 hardy spores, or gametes, almost like going to seed. [126]   The closer brewing scientists looked at this curious phenomenon, the more they discovered.   Not only does the yeast cell have a choice of reproducing either non-sexually by budding, or by undergoing sexual reproduction to produce more diversity in its offspring, but it can even change its mating type, which would be like us going from a male to a female or vice versa whenever one needs to in order to increase one's chances of finding a mate. [cvi][127]

Having sex is nature's most efficient way of mixing up the DNA, of reshuffling the deck of cards that constitutes an organism's genome.   And since the Earth is an ever-changing place, life needs to change too in order to keep up.   This is what natural selection and evolution are all about.   Shuffling your DNA by creating new offspring is the best way to introduce more variety into your lineage.   Each new individual is, in effect, a new experiment nature is performing, and those best adapted to the changing conditions will survive and reproduce the best; the rest won't and so their genes will not get passed on, it's that simple.   This is, of course, the very definition of natural selection. But something interesting happened in the yeast's association with humans.

Along the way, through our hand in the process, the brewer's yeast became hobbled, sexually.   In fact, it's now recognized that most yeast used in brewing, baking, and winemaking these days can no longer have sex at all. After hundreds of years, brewers and bakers inadvertently and without even

---

[126] For example, in nature if the yeast found itself in a situation where the juice of a berry was drying up and it needed to form spores to survive an impending drought.
[127] Scientists believe the brewer's yeast may have once been a multi-celled fungus, similar to the mushroom, but for unknown reasons reverted back to life as a single cell.

seeing the cells directly, selected for yeast strains that made the most consistent, reliable batches of beer, bread and wine. So it was to their advantage to choose yeast strains that *didn't* change very much. Therefore, most yeast used in industry today can only reproduce by budding off identical pieces of themselves (i.e. clones). They can no longer have sex and reshuffle their genomes anymore. They lost that ability long ago thanks to mutations and our preference for them.[cvii]

Winemakers have even learned since Pasteur's time that the yeast contributes much more than just alcohol and carbon dioxide to wine. In fact, it produces some 150 volatile chemicals – many aromatic – and the yeast even provides a valuable contribution to the "mouth-feel" and "cheesy flavor" of wines, which is why wineries will often leave their developing white wines in contact with dead yeast cells for several months after fermentation; during the aging process.

Wolf wasn't quite right about DNA being a dead molecule, though. One of the more useful aspects of DNA is that it can change, or mutate, its sequence. An important mechanism in evolution is, it turns out, dependent upon our cells making mistakes. When populations of cities began to increase during the Industrial Revolution, bakers could no longer keep up with the increased demand for bread by using the leftover skimmings from breweries. There just wasn't enough yeast to go around anymore. As a result, bakers began growing their own strains. Today, because of this, the baker's yeast has taken a turn and gone down a different path, evolutionarily, from its brewing cousins. Baker's yeast DNA these days have mutated in such a way as to allow it to become hardier, physically, so they can better withstand the rigors of modern manufacturing inherent in baking bread.[128] They are also able to produce carbon dioxide more rapidly (desirable for rising dough) and enzymes that are better adapted to consuming the sugars found naturally in bread dough. [cviii]

Today breweries like Carlsberg not only keep their pure yeast cultures separated in "yeast banks", but they may even use DNA fingerprinting, similar to what Tahany was doing to identify schistosome worms, but to keep better track of the changes in the yeast's DNA within breweries.[cix]

I flipped through the manuscript in my hands, cheating by looking towards the back to see how far I had to go. So far, except for losing my eyeglasses, the trip had been characterized by good timing. By the time the plane was over the Atlantic I had moved on to one of Gamal's later chapters entitled "Yeast and the New World". I looked back through what I'd read so far. Just a few circles I'd made, highlighting a typo here and there and some minor

---

[128] Centrifugation, vacuuming, pressing into cakes, and freeze-drying of yeast. Before 1825 bakers used yeast from breweries in liquid form, or as a paste. In 1825 someone figured out a method to press yeast into a "cake", making it more easily measured out and sold.

grammatical errors, that and a couple of suggestions in the margins about including some things I thought might be interesting, nothing to take away from the flow of it. Gamal had done his research...there was no doubt of that. Maybe a bit too much I was afraid. It tended to be dry in places – like the interviews with various mycologists I had never heard of – but all in all not bad. If it ended where I was now, I'd be glad send it off to my cousin in New York as soon as I got back.

For some reason, Gamal began his next chapter with a discussion on distillation, for it seems that the only source of alcohol during much of western history was wine, mead, or beer, that is until around AD 800 when something new happened...or maybe it wasn't so new after all. Aristotle described the way seawater could be made into fresh drinkable water as far back as the 300's BC using distillation. But compared to distilling water, alcohol is a good deal trickier to obtain in purified form this way, which is why it took so long to attain pure, 100% alcohol. During the Middle Ages, Arab alchemists like Jabir had been making improvements in the still – an important one being the addition of what's called an *alembic*. This glass bulb extended the still's physical range, meaning it expanded the temperature gradient inside the still so that, when the alcohol became a gas and evaporated away from the heated wine down below, it could get farther away from the flame and so the vaporous alcohol could re-condense at the top of this cooler glass dome and fall like tiny raindrops, eventually coalescing and dripping down the sides of a tube leading outside the still, keeping the liquid separate from the heated wine inside, to be deposited as pure alcohol.

They called this new, combustible substance *aqua vitae*, or water of life, and it was so loaded with calories that Henry Ford would one day run his first automobile on the same stuff. The Moslem alchemists believed they had captured the spirit of the wine, but since the Koran forbids alcohol consumption, in the Moslem world pure alcohol remained mostly a curiosity, useful as a base for dissolving oils from plants to make perfumes, or for perhaps suspending various medicines in.[cx] Brandy, for example, is distilled alcohol obtained from wine, and the word itself comes from the Dutch meaning "burnt wine". And so it was up to non-Moslems like Arnold Bochmove in the late 1200's, to popularize distillation methods for making more diverse alcoholic beverages. His recipes caught on quickly in the west. [cxi] Being more concentrated, pure alcohol obtained by distillation was also easier to transport, not a small consideration in an age before steam travel was possible. Brandy, for example, is 8 times more concentrated than wine when it comes to alcohol.

I leaned my head back into the groove of my headrest and closed my eyes, shielding them from the dry cabin and began thinking back to my American history class, the place I had first come across the importance of whiskey to the early settlers. It was something I knew Gamal could use in his book, if only I could remember all of it.

In a time before highways over the Appalachian Mountains, it would have been easier for farmers to transport their crop to market as a liquid, concentrated as corn whiskey. In fact, the only major revolt against the US government between the Revolutionary War and the Civil War was the Whiskey Rebellion of 1791, which began when the federal government tried to test its new authority by taxing whiskey to help pay down the Revolutionary War debt. The farmers of course thought the tax was unfair; being the very thing the colonists had fought against during the War for Independence.[cxii] Meriwether Lewis took part as a volunteer. Even the president who ordered the uprising ended, George Washington, owned the most productive whiskey distillery in America, producing some 10,000 gallons of alcohol from grain in one year alone using 5 stills.[cxiii]

According to Gamal, within a hundred years after Columbus's last voyage to the New World, there were already so many vineyards in Mexico and South America that authorities back in Spain had asked King Phillip II to issue a royal edict discouraging the planting of more grapevines in the colonies. South Carolina had a vineyard as early as 1568 and Benjamin Franklin outlined winemaking principles in one of his pamphlets in the 1740's and within 13 years after the signing of the Declaration of Independence, there were perhaps 2,000 whiskey distilleries in the US. Washington and Jefferson were both interested in viticulture and improved on it whenever possible. I already knew that Jefferson considered himself a scientist above all his other trades, including politician, but was surprised to learn from Gamal that, back in 1620, the Pilgrims had stopped short of their intended destination, settling on Plymouth Rock mostly because the Mayflower's beer supply was running low.[cxiv]

From this point on, the chapter was a little difficult to discern and I wasn't sure why Gamal had arranged it the way he had, but I still managed to mine some interesting details before my eyestrain set in permanently the rest of the flight. I didn't realize, for instance, that it was the advent of faster clipper ships in the mid-1800's that allowed Bavarian brewmasters to bring their favorite lager yeast strains to America while the yeast were still alive, eventually expanding these yeast cultures in cities like Philadelphia, Milwaukee, and St Louis, or that the first paved streets in America were in New York City so the beer wagons from the nearby breweries didn't shake the kegs excessively, which could ruin the beer.[cxv]

I added a note in the margin recommending to Gamal he add a section on a subject I'm partial to…Lewis & Clark. Napoleon once said that an army should be supplied with just enough beer and wine to make it far enough so that no one would desert when the supplies ran out.[129]   During the War for Independence, George Washington wrote letters to the Continental Congress when his soldier's alcohol ran low, so when Lewis & Clark set out to explore the newly acquired Louisiana Territory in 1804 looking for an all water route to the Pacific, it comes as little surprise they carried with them 120 gallons of whiskey.[130]   They also held some in reserve to trade with the Native Americans.   The party of 45 men and one dog set out from St Louis on May 14 and ran dry over a year later the following summer, near present-day Great Falls, Montana.   And as Napoleon could have predicted, not a single soldier deserted when they ran dry.[131]   The expedition did manage to obtain more whiskey again before they got all the way back to civilization, on their way down the Missouri River, by trading information with settlers who were traveling the other way.

Gamal had somehow ferreted out some other details I'd never heard before, like how Captain Cook learned from Native Americans how to prevent scurvy – by using spruce boughs – which Cook took along and fermented into a kind of beer on his voyages, and that alcohol was used in the first ships compasses to float magnetic needles in, and that the yeast's metabolism can also help reduce the natural toxins present in barley, which makes for a better tasting beer.[132]

Some of Harvard University's first students paid their tuition as malt grown on their family farms, and the benefactor of one of the earliest women's colleges in America was also a respected New York brewer named Matthew Vassar. He would later use half his fortune to start a women's college of the same name and brewers like Vassar relied on leftover yeast skimmings to help pay their bills. [133]

Gamal even had a paragraph on Prohibition for it seems that before Pasteur discovered heat pasteurization, there was no easy way to prevent grapes from fermenting into wine.   Fermentation happens naturally because the yeast is always present on the grape so it gets pressed into service along

---

[129] He was also fond of bread sticks called grissini.

[130] Many of Custer's soldiers carried whiskey ("Dutch courage") in their canteens, in fact Custer's highest-ranking subordinate officer, Major Marcus Reno was, by several accounts, drunk on whiskey during the entire Battle of the Little Big Horn.

[131] The actual number in the expedition varied because some members joined while others left, which reminds me of a modern research laboratory in that way. / Early on, Lewis & Clark did court-martial two of their men for trying to steal whiskey – ordering 50 lashes on their bare backs. / When British soldiers occupied Monticello during the Revolutionary War, they helped themselves to Jefferson's wines.

[132] Two thirds of the Lewis & Clark Expedition suffered from scurvy, as meat became their main source of calories.   None died of it, though.   When the Pilgrims ran low and had to begin rationing beer on the Mayflower, scurvy struck shortly afterwards and two on board died (one was a sailor).

[133] Michigan's Lower Peninsula – which is shaped like a mitten – has a town in its *thumb region* named after Vassar. To this day the brewer's namesake is still straddling a portion of the Cass River that turned out to be an adequate site for a lumber mill in the 1840's.

with the juice.   In 1869 an American dentist came up with what he called Dr. Welch's Unfermented Wine using heat pasteurization, which stopped the yeast from fermenting in its tracks, thereby allowing temperance Protestants to take communion using grape juice without partaking in the alcohol.   His innovation also made it easier for Prohibition to be enacted in the 1920's.[cxvi]

Just before drifting off to sleep somewhere over the Atlantic, I thought about Wolf one last time.   I still didn't know what to make of him.   That was the problem.   Was he just being curious, or was there something more sinister to his question?   And if so, was he aware of recent mitochondrial DNA evidence showing that the entire human race is more closely related to one another than a small band of chimpanzees are to each other in the wild? Growing up in the West there were some things I did know, though.   I knew that just as Lewis & Clark had been the first Americans to see Montana, so too was Clark's slave York the first black man to see it.   History books hardly ever mention that, or that they were boyhood friends and Clark had grown up with York because he inherited him.   In fact, York's father once belonged to Clark's father.   Slavery in those days was all in the family it seems.

SHUTTLE VECTORS

Within two days after being back, I found myself on yet another plane, this time headed to Seattle, Washington.   It had completely slipped my mind – what with preparing for Egypt and switching to the new project – that I had signed up for a symposium about the yeast back when I was researching the grant proposal, which meant I was now making up for my lack of travel, all in the span of two weeks.[134]   Because classes were already starting for fall semester, I could afford only a couple days in Seattle, so I resolved to pack in as much as I could.   I wasn't thrilled with the prospect of taking another trip so soon, but if you want to know about horses, it's probably always best to go directly to the horse.[135]

While fumbling through the brochure at the convention center, one of the first things that caught my eye was a poster session.   It's a good thing the folks who arrange these get-togethers provide maps because the entire first floor was wide as a cornfield with row upon row of bulletin-boards, each with its own presenter standing next to their poster detailing some aspect of the yeast and eager to share his or her work with anyone showing the slightest interest.

The researchers who would draw the biggest crowds this day would likely be the ones who had published recently in journals.   Others would draw the curious simply because they were using the latest techniques someone might need.   These researchers had all the buzzwords in their titles, like shuttle vectors, microinjection, artificial chromosomes, and single-molecule detection. Poster sessions are a good chance to hear the nuts and bolts of research from the ones who are actually doing the experiments – on the front lines so-to-speak – often a graduate student, an undergraduate, or a postdoc with their mentor waiting silently in the wings to bail them out in case any of the questions get too technical (except for postdocs who often know more about the project than their mentors do).

I was still in the process of filling out my nametag when I got wind later in the afternoon of a Nobel Prize winner scheduled to talk on the brewer's yeast, the same strain I was planning to use, and interestingly enough, he was a cancer researcher.   Seattle's Pike Place would have to wait as I set aside all other plans for the afternoon.   Most researchers drop everything they're doing at a symposium whenever a Nobel Prize winner is scheduled to talk.

---

[134] Unlike the Greek symposium with its 30 or so guests, and where Athenians could expect to drop by uninvited, I had to sign up for mine ahead of time and pay a fee.

[135] It wasn't until Champollion traveled to Egypt and saw the plants & animals of the region firsthand that he knew for certain hieroglyphics had to have originated in Egypt.   The Egyptians used what they saw around them (such as the Ibis) as inspiration for their hieroglyphs.

Attending a memorable seminar can be a mileage marker we gauge our careers by.

But try as I might, I just couldn't figure out how the brewer's yeast could shed new light on a disease as complex as cancer. And, as I hope you know by now, the yeast is a single-celled fungus that spends its solitary existence in the soil or on vegetation and the like. Yet cancer is a disease multi-cellular organisms like we humans get, a disease that produces tumors and spreads throughout the body. Yeast don't get cancer because they can't form a tumor. They're single-celled.

I looked at the map stapled to the inside of my booklet. Whoever arranged the session had tried to group all the projects with similar themes together so as to avoid randomly zipping about the room. Yeast projects in health and medicine were more towards the center, while projects having to do with industrial uses of the yeast were situated at either end of the hall. The farthest side was reserved for food scientists interested in making improvements in winemaking, brewing, and baking while at the end nearest me the yeast was being used to produce electricity in batteries, ethanol for cars, even bioremediation to help clean up toxic oil spills. If I spent even a couple hours and made it halfway around, it should be time well spent.

So with this in mind I began nibbling away at the edge nearest me as if it were a giant pizza, strolling up first one aisle then down the next, ingesting, absorbing, assimilating, and then moving on to the next keyword or friendly face to catch my eye. After attending a number of symposia, I tend to gravitate towards the less popular posters, where it's not as crowded and the students tend to look a bit unnerved, like a plucked chicken. It's more relaxing and it makes them feel better, plus I get their undivided attention in return. My first year as a graduate student I gave a poster presentation at this huge symposium and the only person who stopped by my poster the entire two hours was a competitor.

I had made the mistake of assuming my work would speak for itself, but scientists are people too and pay more attention to nice visuals just like anybody else would. My research was still unpublished and I also had the misfortune of being sandwiched between two other students on either side, both from Ivy League universities and had obviously spent a great deal of time and effort, not to mention money, on their posters. Both veterans of previous campaigns they had the foresight to have them printed up professionally in one nice glossy sheet of laminated plastic in amazing computer-generated Technicolor I hadn't even seen on a poster before. Just the simple experiments they'd done looked impressive, and maybe even more so was the fact that they needed only four tacks, one on each corner, to hold whole thing up, while I had been struggling with a whole box of tacks to display my meager shreds. As the morning wore on I began to feel more and more like a mud puddle the others had to figure out a way to jump across without looking too obvious in order to get on to the next beautiful poster. I thought back to what my grandfather liked to say about how life is always a lot easier when you plow around the stumps, except that when he said it, I had never counted on myself actually being one of the stumps. I couldn't wait for the whole episode to end so I could hide out among the crowd for the duration of the conference, like the insignificant researcher I felt certain I was to become.

I was slowly working my way around the perimeter when I came across a young man standing beside his poster and, judging by a lingering case of acne, was just now leaving behind the last vestiges of adolescence. You never know whom you'll run into at a poster session. It could be a Nobel Prize winner, a salesman trying to interest you in the latest piece of equipment you don't need, or it might be a gifted high school student taking extra classes at a nearby college. I noticed this one had braces on when he began explaining how his mentor had been interested in using the brewer's yeast to generate electricity in fuel cells by feeding it organic waste from a landfill. He turned out to be a gifted high school student in an honors program, as I'd guessed.[136] His self-consciousness reminded me of myself at that age. He also had a good grasp of the history of the field, and began by admitting how the idea of using microbes to generate electricity was nothing new, really. In fact, the same bacterium that lives inside all our intestines – *E. coli* – has been used to generate electricity in experimental batteries for decades, lighting dim flashlight bulbs in the corners of chemical engineering labs while being fed a continuous stream of glucose as an energy source, producing light in return.[137]

THE YEAST AGE

Standing next to a poster a bit further down was the boy's opposite – a snappily dressed dark-haired city woman, probably an undergraduate with an eye towards medical school. She explained using her laser pen how her mentor had been putting their yeast to work detecting poisonous gas, presumably from a terrorist attack. Their goal was to float the brewer's yeast in clear plastic boxes in subways, not unlike smoke detectors, using the yeast

---

[136] It might surprise a lot of folks, but doing basic manipulation of DNA – like splicing a human gene into a bacterium – can be quite straightforward, in fact it's been compared to baking a cake. I've seen video of elementary students adding human genes to *E. coli* using the *heat-shock method*. The same *E. coli* that lives inside the human gut was the first organism genetically engineered by scientists in the 1970's.
[137] Glucose can let go of its electrons without the help of enzymes too. Ordinary chemicals will steal them, and in fact this is the basis for some readout devices diabetics use to monitor glucose levels in their blood.

basically as a "canary in the coal mine". The way it worked, she explained as her laser pen light me through all the intricacies of a rather impressive poster, was by making the yeast change color in the presence of the gas. She and her professor had genetically altered the yeast's DNA to produce a jellyfish protein that glows green whenever a poisonous molecule triggered a change in the output (i.e. "expression") of the "green gene".

Next down the line were two researchers engineering into their yeast some foreign genes from a grapevine. Their goal was to produce some of the same chemicals found in red wine that seems to have an effect on preventing cancer and heart disease, but their project beer instead. Next up was an assistant professor who had engineered his "designer yeast" to produce aromatic chemicals, pleasing ones that made research labs smell better he promised, like bananas or strawberries; basically a yeast air-freshener fed a steady diet of sugar as its raw material.[138]

Another professor a bit further down was trying his hand at turning the brewer's yeast into a tiny vessel for making diesel fuel. Right across from him one had taken a gene from a fungus responsible for breaking down indigestible sugars like xylose (found as fiber in herbivore feces like the elephant's), and inserted this xylose-digesting enzyme's gene into his yeast in hopes the yeast could break down many of the unusable sugars in plant bark and make more ethanol from it.[139] Another two researchers were planning on using brewer's yeast to detect the breakdown products of TNT basically by turning it into a bomb-sniffer. The senior of the two had gotten the detection gene from the nose cell of a rat and placed the rodent's receptor gene inside the yeast creating what they called a "biosensor". Not having gotten any new ideas, I soon had enough of industrial uses and moved on to the medical posters.

The pharmaceutical companies were all grouped into several aisles, some with a goal of making morphine in yeast, others focused on producing malaria drugs or fragments of monoclonal antibodies designed to fight cancer.[140] Another had harnessed her yeast into producing toothpaste additives guaranteed to prevent plaque, and her colleague from the same company was after the Holy Grail of all yeast research…trying to engineer a strain that produces a chemical for his "Frankenwine" to counteract hangovers. Finally, I came to the end of a long row where a student and her mentor were each taking turns in front of a large crowd like a tag team putting on a show, eager

---

[138] One of the advantages of using brewer's yeast in the lab is that they tend to smell better than bacteria, reminiscent more of bread baking. Bacteria often produce more of a "rotting smell".

[139] Xylose is a type of glucose molecule chain that normally goes to waste when making gasohol.

[140] Malaria was well known in the Missouri River region in the 1800's, which is why Lewis & Clark took along 15 lbs of quinine-containing Peruvian bark as a powder.

to explain how they'd gotten brewer's yeast to produce an enzyme that mopped up malic acid, the compound produced by grapes that makes wine taste sour. What they were trying to do was a more sophisticated version of what the ancient Romans had already done: counteracting the sour taste in wine using marble dust, seawater, or ground up seashells, all of which are alkaline and so able to neutralize the grape's acidic (sour) taste.[141]

My confidence on the upswing I ventured on towards the heart of the room where the intensity was turned up a notch and I would find researchers humanizing the yeast to produce proteins that cause Huntington's and other devastating diseases. This allowed them to screen for potential life-saving drugs in hopes of "curing" the malfunctioning protein, the idea being that if you can correct a particular damaged human protein inside a yeast cell, then the drug has a better chance of working inside the person later on. Another wanted to use his yeast to display molecules (proteins) on its outside belonging to the HIV virus that causes AIDS, with an eye towards using his yeast as a vaccine vehicle. Whether he planned on putting the vaccine-coated yeast in bread or wine, I forgot to ask. Another had inserted a gene suspected of causing Lou Gehrig's disease, or ALS, into the yeast and was testing chemicals (another drug screen project) to see if she could break up the clumps of aggregated protein associated with this disease. Yet another had a yeast strain that produced the protein that causes mad cow's disease.

Eventually I found myself in front of a distinguished, older-looking gentleman with gray muttonchops as his only facial hair and an unlit pipe that had been extensively nibbled on. Standing next to him was his student, clearly a first-timer who merely listened as the older man, an expert on progeria, explained in a serious tone how his lab was exploiting the fact that the yeast has the same protein that causes premature aging in children, a genetic disease called Werner's syndrome. When the mutated gene produces its faulty version of the protein, this damaged protein can no longer unwind the chromosome like it's supposed to during replication. Copying DNA is always a critical event that takes place just before a cell divides in two, for example when renewing tissue that has been damaged or simply gone past its prime.

A researcher right next to him was using the yeast as a window into normal human aging. It seems that when the brewer's yeast cell buds to form a new cell, it can do so only about 30 times before it begins to show signs of slowing down. This "elderly" yeast cell even becomes erratic, senile if you will, producing mutant yeast progeny with damaged DNA in the process.

---

[141] She got the gene to digest malic acid from a cousin of the brewer's yeast, S. pombe. The ancient Romans had still another way to cut down on the acidity of wines: by tossing a piece of burnt bread into the wine jug, which is where we get the saying "to propose a toast" in English.

Understanding all the details of how this "mother cell" ages could yield clues for understanding human aging as well he promised. I saw on his poster his grant had come from a pharmaceutical company so I questioned some of his data more thoroughly. One of the ways to increase the value of a drug is to find more patients who need the drug and so there is always this unsettling relationship between academia and industry, and sometimes it's a bit of a mystery trying to figure out what is real and what isn't. I became convinced after listening to him that he was honest.[142]

Some of the other memorable projects I came across that morning involved researchers genetically modifying the yeast so it could use synthetic amino acids not found in nature. The idea was to construct entirely new proteins with never before seen characteristics. Another planned on using her yeast to produce proteins from an arctic fish, natural antifreeze proteins which could then be purified and added to ice cream to preserve its smooth texture, while another was trying to create a yeast that could capture and convert carbon dioxide from the air into rock in the hopes of someday reducing global warming.

In fact, I became so engrossed in the last of these posters I forgot all about the scheduled talk and as a result got there just as the first speaker had reached the wooden podium and was arranging his notes. My punishment was a seat in the very back of a theater-sized, echo chamber of an auditorium.

NOBEL YEAST

As I sat waiting for the main speaker, something I had been hearing all morning kept coming back to me. When you hear the same thing over and over it begins to stick more. Everyone, it seemed, especially the ones just starting out like me, were having trouble getting research grants lately. Things were really tightening up over at NIH. Maybe I needed to be more resourceful, I reasoned. After all, before finding King Tut's tomb, Howard Carter had to sell watercolor paintings he drew for tourists just to make his ends meet. Even the ancient Egyptians built the pyramids without the benefit

---

[142] The average lifespan of a yeast cell is about a week. / Another scientist was working on a joint project to sequence 1000 yeast genomes. They'd found so far that, rather than being domesticated just once, the brewer's yeast has done so multiple times throughout history in different places, and their chromosomes are mosaics that reflect this (when compared to brewer's yeast in the wild). It turns out there is considerable variation between different strains used in brewing today. Scientists have found that any two brewer's yeast strains, when compared to one another, can differ by 4 times as much as the DNA between a human and a chimpanzee differ.

of a pulley since they didn't use wheels.   But then again, who needs wheels when you live in the desert?[143]

The first speaker was a colleague of Hartwell's who, I could tell by the affection in his voice, must have known him for some time, and he set the tone for what was to follow, beginning by explaining how early biologists in the late 1800's had recognized cell division, or mitosis, when it occurred.   They could watch this process single cells use to reproduce themselves unfold in surprising detail on account of how optics and lenses had improved considerably throughout the Industrial Revolution.   In fact, by the 1880's their light microscopes were almost as powerful as the ones we use today are.[cxvii]   Another advance, this one brought to the table by organic chemists, was the availability of newly synthesized dyes able to stain cells and their specific parts like the nucleus, parts that would otherwise remain invisible.   Most of a cell is water and therefore transparent to light.   Almost all parts of a cell need to be dyed to see them; which is how the word chromosome came about. Chromosome literally means "painted body" in Greek.

These early biologists, many of them German, knew that certain things had to happen when a cell divided in two, whether that cell belonged to an animal, a plant, or a microbe.   Eukaryotic cells share many of the same processes in nature and the long thread-like chromosomes appeared under the microscope to these early investigators as if they were taking part in some kind of carefully arranged dance.   A researcher in 1879 could have researched more than 200 scientific papers scattered throughout various journals describing chromosomes and the predictable motions they undertook inside cells as these cells divided into two.   It was noticed, for example, that a cell's chromosomes always became visible and brightly stainable only at certain times throughout the process, and they tended to meet in the middle of the nucleus.[144]   They could see the chromosomes actually bending, as if being tugged at by some unseen force, pulled this way and that within the cell. The membrane barrier delineating the nucleus eventually disappeared, leaving the chromosomes in the middle of a large parent cell.

Then, like partners in the middle of a square-dance floor, the chromosome pairs each separated and went their own way, exactly halving in number, with a member of each pair moving away from the other to the opposite ends of

---

[143] The Romans invented concrete because – unlike the Greeks – they didn't have easy access to marble.   An ingredient of concrete is calcium carbonate (limestone) and the Mars lander Phoenix analyzed Martian soil in 2008, finding it to be about 5% calcium carbonate. At least one astrobiologist has claimed you could grow asparagus in it on Mars.
[144] Brown named the *nucleus* in a paper he published in 1833.   The nucleus had been observed by others before, going all the way back to Leeuwenhoek, but Brown recognized it existed in all plant cells and he called it the "organ of reproduction".   He is the same scientist who discovered *Brownian motion* while examining pollen grains in water using a microscope in 1827.   Einstein would later use Brownian motion in 1905 to prove mathematically that atoms exist.

the cell.   The final refrain came as the cytoplasm of the cell would simply pinch into two equal halves.[145]   Each new "daughter cell" ended up with exactly half the chromosomes of the "parent cell", no more, no less.   It was amazing and they didn't have any idea how it happened.   Like Newton's laws that had been so successful in predicting the regular motions of the planets, or Galileo's when he discovered that the pendulum followed certain rules after noticing a chandelier swaying back and forth in a cathedral, it was as if the cell had laws too, laws that were governed by an internal clock it could use to calibrate itself and keep track of what was happening during mitosis.

In the late 1800's, Dr. Theodor Boveri (the Ph.D. kind) had realized there were exceptions and that on occasion things didn't go so smoothly.   Boveri had been observing sea urchin embryos and worms when he noticed that, once in a while, one of the two daughter cells ended up with more chromosomes than it was supposed to have.   He even speculated that having this unusual number of chromosomes could explain diseases like cancer.   Boveri believed mitosis wasn't always perfect and that tumors could arise from these mistakes, errors that took place when one of the key actors got out of step during the "dance" that constitutes cell division.   Like the Earth's motion, cell division (mitosis) wasn't random.   But what was this unseen force, or glue, that held the whole thing together?[146] [cxviii]   No one seemed to know and few even bothered to guess for the next half a century.

Part of the problem was that animal models for learning about human diseases hadn't yielded much useful information.   In the 1930's a researcher could buy a strain of mice that would reliably develop cancer, and there were even certain chemicals – in coal tar for instance – that could be applied to the bare skin of rabbits and guinea pigs to speed the process of forming a tumor along, but the mystery of what actually controlled cell division, and therefore cancer, remained out of reach.   Even small laboratory animals like mice were too complex to get at these fundamental questions.   Scientists needed a way around animal studies.

The first speaker had done his job and our appetites had been whetted. The stage was now set for Hartwell, a distinguished-looking, tall professor with a full head of grey hair and the eyes of a young man.   He strode to the podium amid a polite applause.   While arranging his notes I thought about what it must have taken to put his faith in the yeast while looking for answers to cancer.   Here he was in the early 1960's, a newly minted Ph.D. interested in an organism so many had scoffed at when it came to complex animal

---

[145] They also knew different species had different numbers of chromosomes. / Even into the 1920's common belief held that cancer was caused by blunt trauma arising from a force strong enough to produce a bruise.
[146] The brewer's yeast has 16 chromosomes, but they're short and therefore couldn't be seen easily, even with high-powered light microscopes.   In fact, some biologists into the 1950's believed the brewer's yeast might not even have chromosomes.

diseases like cancer. He wanted to use a single-celled microbe as a model. On the other hand, maybe he was just doing whatever he had to, I thought. Researchers tend to be a resourceful lot.

Yet the scientific highway is littered with the careers of those who have gotten too overconfident, took a wrong turn early on, drifting too close to the edge of the scientific method in favor of their own version, creating a method so perverse as to not even be scientific anymore. The luckier ones figure it out early and get back on track while the unlucky never do, at least not until they've wasted decades, or even entire careers, not to mention all that money. "The easiest person to fool in science is yourself," as the physicist Richard Feynman was fond of saying, and King Tut's tomb may have contained a pair of blinders for his horses, but the truth is, blinders aren't just for animals. Scientists can wear them too sometimes.[cxix]

After a miscommunication with the person running the projector, Hartwell was soon under way, beginning by explaining how he came to be working with brewer's yeast in the first place.

There were many respected people who at the time believed the only way to understand what happened when a cell divided in two was to have all the pieces of the puzzle in place first...what amounted to a "master parts list" of all the genes and proteins involved. This is sometimes called the "bottom-up approach". The strategy is similar to compiling a program for a baseball game, where you have all the names and numbers identifying each player, his stats, and what each one's job is on the field. But Hartwell believed he could shed light on cancer by coming at the problem of mitosis "from the middle" so to speak.[147]

Thanks to Winge and the others over at the Carlsberg Brewery that had worked on the brewer's yeast for decades, it was possible by the 1960's to do some interesting experiments, mutagenesis experiments designed to damage only some of the yeast's several thousand genes, and then see what happens next. It's like removing a single part from an automobile at random – say the windshield wiper blade – and then seeing what happens to the car, or doesn't happen as the case may be. When the car doesn't drive as well in the rain, you might guess that the removed part had something to do with water and you might even speculate that the part was involved with removing water from the windshield. This is basically what gene knockout experiments are all about. They're designed to give you clues about a particular gene's function; leads you can later follow up on.

The brewer's yeast has about 6,000 genes (blueprints for how proteins are made), and these protein parts do just about everything a yeast cell needs to do, whether it's fermenting grape juice into wine, making bread dough rise,

---

[147] Howard Carter came to Egyptology "from the middle". Lacking a university education in the subject, he was not as respected and therefore had to learn Egyptology as he went along.

mating, or even copying its DNA before dividing into two new cells.[148] In the 1960's, Hartwell and his colleagues set about randomly knocking out each of the brewer's yeast genes with a chemical called a mutagen. If his hunch was correct, then it might be possible to identify key players in eukaryotic cell division by homing in on and identifying mutant yeast cells, hobbled ones that could no longer divide the right way, like Boveri's sea urchin cells so long ago that ended up with an unusual number of chromosomes. That was the idea at least and it must have seemed a long shot at the time.[149]

But an important thing Hartwell had going for him was that microbes are so much easier to work with than animals. Still, most researchers in the field, leaders in cancer for example, thought of microbes as far too simple to yield useful information about a complex human disease. But the advantages were too tempting for Hartwell and this new generation of researchers coming up to ignore.[150] A yeast cell has perhaps five times less genes than a mouse cell, and yeast aren't picky eaters and so can be grown on Petri dishes as colonies. Each yeast colony is distinct and made up of clones (identical cells) and so each colony can have its own mutation in one particular gene.

On the surface of the agar in one dish it's possible to have as many as 300 different yeast mutants, each with a different gene knocked out of commission, while that many mutant mice would take up so much more space and require care and feeding, which of course adds up. Alternatively, a dozen or so Petri dishes can fit all stacked up neatly on top of one another on a shelf in some out of the way incubator. [151]

Hartwell wasn't too concerned about how similar or different yeast and humans were at the time. As he explained it that day, he was far more interested in finding specific genes inside a eukaryote cell involved in cell division. He was also gambling on the notion that even after a billion years of evolution that separated the brewer's yeast from humanity,[152] many of these same genes used for undergoing cell division were the same, still recognizable in other words, and that the yeast genes used for doing its cell division would be similar enough to shed light on our own reproductive genes. Rather than the painstaking bottom-up approach, as so many in the field were

---

[148] By comparison, the parasitic bacterium *Mycoplasma genitalium* has just 517 genes, only 300 of which appear to be essential for life.

[149] Yeast has about 200 times less DNA than a human cell and can be grown in shakers and spread onto Petri dishes like ordinary bacteria. It's much easier than raising mice. And the animal rights people don't seem to mind, either.

[150] "Necessity is the mother of taking chances." Mark Twain.

[151] Some other advantages of microbes like the yeast are that they don't cause disease so they're safer to work with; they reproduce every hour and a half simply by budding, thus producing an exact clone (i.e. copy, or "genetic replica") which can then be grown into a separate colony (millions of identical cells) on a dish; they have simple nutritional needs including sugar, some nitrogen salts, and water, which is why people were able to cultivate them for thousands of years just by making bread, beer, and wine.

[152] The length of time since we last shared a common, single-celled ancestor.

advocating, he was basically turning over rocks randomly to see what lay underneath; coming at the problem from the middle. It's probably always a good idea to drink upstream from the herd if you can.[153]

Within days of mutating the yeast, Hartwell's team already had some results to follow up on. Many of the cells appeared under his microscope to be stuck in various stages of the budding process. Some yeast cells had tried to reproduce but had formed, like a bulge in the inner tube of a bicycle's tire that's about to pop, only a portion of a round new bud on its surface. And then that bud had malfunctioned and simply stopped growing for some reason. These new yeast buds (daughter cells) never detached from its mother cell like they were supposed to.

And what was perhaps most intriguing was the fact that there appeared to be specific steps during mitosis, and that these steps relied on previous steps in the process, almost as if the cell had quality control checkpoints along the way and these checkpoints could communicate with one another to find out what was going on globally inside the cell. Cell division was looking more and more like some kind of an assembly line, one that was hierarchical, and controlled by just a handful of key players. It seems there were more important genes and the proteins made from these genes were the foremen, or managers, and they went around deciding whether cell division could start, continue, and even stop if things went wrong. This would help account for the clock-like timing people had noticed for the past 100 years.

One mutant in particular called *cdc28* seemed to work at a critical point called the "Start Site" in the cell division cycle of the brewer's yeast.[154] Wouldn't it be interesting, Hartwell speculated, if the Start Site turned out to be the same place human cells had to cross when they too divided? And what if something went wrong there? Could it lead to cancer? [cxx] The finish line in a race is often the same place as the beginning. Perhaps cancer cells simply became deaf to the "stop" commands shouted at them near the Start Site, the same way a race car driver just keeps going around the track if he doesn't see the checkered flag. Maybe the cancer cell just keeps going through the cell division cycle, dividing in two endlessly even when they shouldn't…ignoring important signals. Cancer couldn't be that basic, Hartwell thought, but then again, maybe it could.

The next step was to find out what kind of protein the cdc28 gene made in the brewer's yeast cell and what it was that made this protein so important. It was more difficult to do this kind of work back in the 60's and 70's – perhaps years would go by with little to show – but eventually they did characterize it

---

[153] Now that I'm older and more well read, I believe my grandfather borrowed that one from Will Rogers.
[154] Which stood simply for *cell division cycle number 28*. Note that cdc28 is the name of the *gene*, while the *protein* made from the gene is called Cdk1. For the sake of simplicity, I'll use the term cdc28 to describe both the gene and the protein.

and the protein coded for by cdc28 turned out to be a member of a class of enzymes that had already been discovered, a class of proteins called a *kinase*.

This was a ready made clue for they could then speculate that the cdc28 protein was some sort of a messenger molecule that went around adding little chemical tags (called "phosphates") onto other proteins to get them to change their shape and thereby alter their function (adding phosphate tags is like adding tiny on & off switches to proteins). Enzyme reactions in cells are often cascade-like events, meaning that a change in just one single "upstream" protein like cdc28 can have dramatic effects on the rest of the proteins within that same assembly line but further downstream because a single cdc28 protein could, in theory, go around tagging so many other kinds of proteins with phosphates, which could in turn tag so many others, and so on.

When Hartwell published his findings on cdc28 in the journal *Science* in 1974, others including Paul Nurse in England, took an immediate interest. Inspired by Hartwell, in 1982 Nurse would go on to use a cousin of the brewer's yeast to show that the same cdc28-like protein existed in it too. Then Nurse helped make the remarkable discovery in 1987 that something similar existed in human cells, and that the cdc28 protein from the yeast could take the place of the damaged cdc28-like protein inside the human cell... step in for it in other words.[cxxi]

He found that the parts list for doing mitosis was interchangeable between cells of completely different species (like the way windshield wiper blades can sometimes be swapped between different makes of cars) even after a billion years or more of having gone their separate ways during evolution. They may not work perfectly, but close enough to restore normal functioning in the other cell. On a fundamental molecular level, our cells were looking more and more like the brewer's yeast than few had ever allowed themselves to imagine. [cxxii] [cxxiii]

By the late 1980's thanks to Hartwell and the others, the details of mitosis were finally coming into focus, and it was possible to understand what it was that controlled the process that had so baffled the early cell biologists. Mitosis was a carefully controlled event. It has to be, especially in higher animals, so cancer won't result, and most remarkably, just a handful of genes controlled the whole process, and they did so in response to sometimes very different factors ("stimuli") in the cell's environment.

In the case of the brewer's yeast cell, cdc28 normally receives a signal from its surroundings to tell the yeast to divide when conditions are just right, for example when there is an abundance of sugars to be eaten. This stimulus, like the sound of a pistol beginning a race, triggers the cell to grow and then (with the help of enzymes) the chromosomes to form duplicates of one another.

But in the case of human cells, they normally pass their "start site" and begin growing and copying their DNA whenever they need to replace other cells in the body, for example when making new skin cells if a wound needs to be healed, the replacement of worn out red blood cells, or the expansion of white blood cells to fight off an infection.   Both humans and yeast were putting to work many of the same genes for controlling mitosis.   Yeast cells turned out to be, in a surprisingly fundamental way, a very good stand-in for human cells because many of their genes turned out to be so similar.   In 2001, Hartwell and Nurse were awarded the Nobel Prize in Medicine for using brewer's yeast to shed new light on human cancer.[155]

I sat in the back and waited through the entire question and answer period that always follows a stimulating talk like this one was.   The whole thing was out of my area since I was trained as a bacteriologist.   Bacteria have been separated from human cells three times longer than a yeast cell has and bacteria divide by a different process called "fission".   They don't have their DNA inside a nucleus like yeast and human cells do so the whole cell division process is a lot different for bacteria.

But being simple is a good strategy and has worked well for bacteria. There are some that can reproduce in as little as 10 minutes...which is a mere 600 seconds!   Human cells take hours.   At one point, Hartwell did notice me, I think.   Throughout the talk I had been sitting in the very back wearing a pair of sunglasses...the only prescription ones I owned at the time.   I had left for Seattle so quickly that there wasn't enough time to replace the ones I'd left back at the Hofbrauhaus so my only other option was to wear the prescription sunglasses.   What with my sporting a sunburn in a city famous for its lack of sunshine, I couldn't help but wonder if perhaps Hartwell had assumed I was someone who had wandered in off the street high on drugs.[156] [157] [158]

Today, Hartwell's work with the yeast has influenced work in labs all around the world and it's now recognized that most human cancers are caused by mutations in the genes that control mitosis...defects in genes similar to cdc28.   In fact, successful cancer drugs we now know often work by inhibiting this "runaway cell cycle" during mitosis and there are at least 23 different genes in normal human cells that can lead to cancer when they're

---

[155] The third recipient, Tim Hunt, used sea urchin eggs.   In the years since, it's been found that the brewer's yeast uses upwards of 800 different genes to accomplish mitosis.

[156] It turns out every human cancer cell has been found with a defect in at least one of its genes that – like cdc28 – control the cell cycle and mitosis. / There are now new cancer drugs that target specific proteins in the cell that control division, as Hartwell had hoped.

[157] Unlike many wines, the yeast doesn't age very well.   As a yeast cell gets older (they divide about every 90 minutes if conditions are favorable), the yeast even takes on some similarities to a cancer cell, for example its DNA chromosomes become more unstable, more easily mutated.

[158] Yeast uses the same 4-letter code in their DNA as our own cells use.

damaged and perhaps most remarkably of all, each of these important, cancer-causing genes has a relative found deep inside a brewer's yeast. [cxxiv]

ALMOST HOME

Montana winters can be so long and cold that the hens have to lay standing up, as my grandfather used to say. I sometimes think I spend more time looking forward to summer, than actually enjoying it. After catching the redeye back to Missoula, I was as exhausted as I'd ever been working on the ranch and yet I still found my mind drifting back to the new project. When the ancient Greeks finished a symposium, they would parade through the streets of Athens singing together. I, on the other hand, had arrived home alone towards the end of an unexpected snowstorm. They seem to come out of nowhere this time of year, and this one brought with it the sense of dread that it was going to be a particularly long winter. It snowed on Lewis & Clark's men in 1805 as they traversed the Continental Divide not far from my family's ranch...and it was only September 3rd.[159]

Snow is the great equalizer, and accordingly a thick blanket of it lay on top of everything when I got back – cars, roads, houses, trees, and even lampposts – blending the living with the non-living into graceful ivory curves, the wind smoothing out all the sharp corners into diminutive drifts. It was all the tangible evidence I needed of what can happen when a dry Canadian cold front coming down from the north merges with a warm, moisture-laden Pacific one with just enough energy to make it up and over The Divide from the west. Thistle and the Lewis Valley were in its crosshairs. Seeing the snow reminded me of my grandfather explaining how the early settlers made their apple brandy – or homemade applejack as they called it – during the wintertime. In the autumn, after allowing the yeast to "harden" their cider into wine, they would leave it outside in the cold, whereupon the water would freeze solid, leaving behind just the alcohol as a liquid; a convenient way to concentrate alcohol and avoid the hassles of distillation.[cxxv]

Another hour of driving and I felt the warm sensation one often gets after they've passed the place they consider themselves home after a long journey. For me it was just after finishing the climb up to the top of the final ridge and before beginning my descent down into the valley, the point I usually considered myself home these days anyway. I thought back to the poster

---

[159] It was so cold, there was no game and their salt pork was gone. Before getting to the western side, they ate some of their horses and tallow candles. Clark wrote "I have been wet and as cold in every part as I ever was in my life."

session while gazing across at the scene laid out before me in the soft, sparkling moonlight.

One of the researchers had mentioned that it was possible to buy through the mail a panel of yeast clones, each with a single one of its 6,000 or so genes knocked out ahead of time, including the same genes Hartwell had discovered. Another panel I could buy from the same company included each of the yeast genes tagged so that the proteins they produced would be easier to purify, or perhaps glow green, almost like each protein having its own bar code attached, the kind of barcode that makes keeping track of merchandise easier in a supermarket. We're so focused on knowing the structure of proteins these days, I lamented, stopping the truck alongside the road to take care of some personal business that couldn't wait. A yeast panel might be just the place to start when I got back.

Looking out over the guard railing at the rolling foothills beneath me, Thistle appeared as a miniature village, the kind you see under a Christmas tree behind a department store window. Barney was still at the ranch, asleep next to my mother's bed, no doubt protecting her in his mind, so I drove the rest of the way to the college, managing to find a space next to the microbiology building for a change. The entire lot was empty and the fact that the space belonged to the chair of the department made it that much better. Being a lowly, untenured professor often means parking the furthest away, like having the smallest office or drawing the least desirable committee assignments.

On the walk up to the building the lower cuffs of my trousers turned bright white as I blazed a trail through the new snow, but I hardly noticed as I was still too busy thinking about what would happen if I wasn't here in a year from now. Where would I be and who would be in my parking space? Losing out on tenure would be the first major defeat in my life and I had no backup plans yet.

I slid the snow away from the door with my boot, the sidewalk underneath still damp, as if the storm had taken even the ground by surprise. I pried open the heavy steel door and stopped just long enough to rid the powder from my boots by knocking my heels against the door, then walked carefully down the stairs so as not to disturb Charlie if he should be working on a night like this. There was some consolation in knowing that the ground was warm enough that the snow would all be gone by morning.

As I got closer to my office, I could just make out the outline of a package, a tall thin one in the dim light. It was leaning against my door and I as I got closer I could see it was a cylinder. I picked it up and placed it under my arm, fumbling around with my keys in the dark, then opened the door and turned on the lights, seeing now the colorful yellow and black stamps were in Arabic. I looked closer at the address. It was from Tahany.

I pried open the metal lid with a key and slid the contents out. The paper gently unfurled, curling up at the bottom just before reaching the floor. I held the smooth papyrus up to the light. It was a replica of one of the scenes I'd come across in the museum that day with Gamal, a colorful scene depicting an ancient Egyptian family baking bread together. The scene was meant to insure the deceased pharaoh would have enough bread for the afterlife. They hadn't forgotten even though I had. Looking back, it was hard to imagine that just two short weeks ago I hadn't even met Tahany, or Gamal, or Wolf or even knew who Lee Hartwell was. Research can make for some unexpected acquaintances.

With the papyrus back inside its canister, I walked the short distance across the hall to my lab and began rummaging through my freezers and refrigerators, looking for what it was that had drawn me back this late at night. They had to be here, I reasoned, but where? There would be no sleep if I failed to see them again. But they weren't in any of the obvious places, so I checked the glass dehydration chamber, gently sliding the heavy, jelly-smeared lid over to one side to peer down through the jar's frosted glass but they weren't in there either. It wouldn't be until I had given up and sat down next to the computer in the back room, my head resting wearily in my palms, that I noticed them sitting on the bench right in front of me. The foil package of dried brewer's yeast I had turned to that night two weeks ago were right where I had left them.

I thought about how my search for the yeast had taken me halfway around the world and was just now bringing me home again, and about how Hartwell and the others have since gone on to show that over a quarter of our own genes have a counterpart deep inside the brewer's yeast. We're 25% alike, genetically, human cells and yeast.

I reached for a pair of tweezers, grasping some of the delicate crumbs, the prongs held lightly between my thumb and forefinger, managing to lift a few of the golden nuggets out and then I dropped them into a Dixie cup. Next, I added a teaspoon of water, agitating the mixture until it gradually disappeared...as if the crumbs had been consumed by the water. Then I transferred a small drop onto a glass slide, fixed a cover slip, and focused the fine-tuning knob on the microscope until I saw their faint outlines in the circle of light, as if a multitude of tiny actors caught in a bright spotlight. They were still there all right, round and intact, bobbing up and down in the water as I squinted. Each of the yeast cells a self-contained laboratory, floating right in front of me, each one a far better biochemist than I could ever hope of being, tenure or not.

I thought about how amazing it was that they had sat there in the same place night after night without any refrigeration, or nutrients, or even any water for that matter, as if waiting for me to come back and give them something better to do.   The freeze-dried yeast cells hadn't changed in the slightest in the last two weeks, but I certainly had.[160]   And it was also true that where I found myself at this particular point in my life wasn't the Mecca for scientific research every graduate student hopes of going to when they get their degree and go out into the world.   And it was also true that I wouldn't be working on any glamorous diseases like anthrax, or TB, or the Bubonic Plague anytime soon.

But then I though about Tahany, and what she was accomplishing on her meager budget – of finding new and better ways of diagnosing schistosomiasis – and what a difference that would make to so many who didn't even know her, and probably never would.   And then I thought about the yeast cells settling to the bottom of the water in front of me, and the opportunity I had been given to add to an already growing body of knowledge, of understand this amazing, versatile little fungus a little bit better.   And as I leaned back in my chair, my feet propped up even with my body on the bench, I took a moment to savor the notion that this was the very first time since coming here as an assistant professor three years ago, the very first time that it all seemed like it might be enough for me now.

"God made yeast, and loves fermentation just as dearly as he loves vegetation."

- Emerson

~ ~ ~

---

[160] Since drinking beer or wine was safer than river water, modern humans have evolved to metabolize alcohol better than our pre-civilized ancestors could. In this way, the yeast is said to have altered the distribution of genes in the human population, allowing us to live in cities necessary for the Industrial Revolution.   (i.e. a modern human could probably beat Neanderthal in a drinking contest)

PART 2: WINTER SEMESTER

TWO MEDICINE & JEFFERSON

Like a handful of leaves floating in a mountain stream that come across a rock in the middle, after losing out on the anthrax grant, my group could have gone either of two ways – scattered in so many directions as often happens – or remain together and try and make the best of things. Fortunately for me, things flowed the second way and within three months after getting back from Seattle the yeast project was already underway and even yielding some results, so at least we had that to look forward to, even though making human proteins in yeast is far from glamorous work in microbiology these days. Still, we had produced our first ones – insulin and interferon, both of them hormones – using the genetically modified fungus as our "biofactory" and so I was gaining confidence we could do the same again but this time with more interesting and challenging proteins.

I'd like to give my pep talk some of the credit for keeping my small group together but the truth is that if my students had quit, they would have had to join someone else's lab and for them it would have meant starting new projects all over again... several months of hard work for the most part wasted. So the incentive was on them to stay put with me as well.

As for me this early March morning as I poked my fingers between the blinds to have a look at what was left of the snowline up on the Bitterroot mountains in the distance, I had to remind myself that it was still technically winter outside. It's a given this time of year in the Northern Rockies that life is a challenge and yet it's also true that even a blind hog finds an acorn once in a while because for the last two days we had been experiencing a true gift from above, or more accurately, from below.

A current of warm moist air called the Pineapple Express had been finding its way up from Hawaii and across the Pacific and then over the Continental Divide and as a result, Barney was tugging at the tip of my sock desperately trying to get my attention to let him outside. Still, I had no idea what this particular day held in store for me, not even the slightest hint whatsoever that I would be offered a chance to work on another disease so soon, but that's the way it works in research sometimes.

Microbiologists may find ourselves working on projects for any number of reasons. It could be something that interests us to the exclusion of all else, maybe sparked by a chance talk we took in from some visiting professor and we simply "caught the bug" from them. Or maybe the drive comes from

146

someplace deeper and farther away, a place going all the way back to our childhood that even we don't fully understand.[cxxvi]   Or maybe it's just that our project is a continuation of an earlier one, something our mentors worked on before us and since we learned to *test the waters* from them as graduate students, well maybe we just never felt the need to give it up.   Several in the department fell into this third category including me when it came to anthrax.

I often remind myself it's nothing to be ashamed of, as this type of project is the most common one.   Staying put is usually the best thing to do in research just as it is in other walks of life.   To up and change horses in the middle of a stream can be a challenge on so many unexpected levels. Technology changes more rapidly than ever before; and then there's the unforeseen hurdles in the form of new procedures to standardize and test the limits of, fresh contacts to be made in the field and that's not always easy either.   After a while it can seem as if all the experts you used to admire and think so highly of are actually pretty territorial in a subtle sort of way – even the ones you might have once looked up to – and that can be a problem too.

One semester back in graduate school I found myself working in this guy's lab on a virus, the microbe that causes hoof-and-mouth disease if I remember right.   And as it turned out there was this other researcher, one of the leaders in the field no less, but at a different university, and he had his name on just about every major publication concerning hoof-and-mouth disease, and in good journals, too.   But do you think he would ever return any of my calls?   It was a question I had concerning some minor detail in a procedure he had pioneered (minor details can be really important in research), and yet he refused to even acknowledge my existence.   It finally dawned on me that he was dodging me because he didn't want any more competition, especially from some young upstart at another university who had the audacity to get his feet wet in *his* pond.   These are big fish, after all, and they can be found inhabiting all bodies of water both large and small.   It was of little consolation when I found out later on that others had had the same problem with him.

The unvarnished truth is that there are a certain percentage of researchers in science who operate under the guiding principle that there is a finite amount of slop in the trough and they want to get all of theirs no matter what, even if it means squeezing all the others out.   Sad, but true I'm afraid, but in their defense it does seem sometimes like there are too many pigs for the tits in research.   I'm not trying to be a glass is half empty kind of person; I'm just letting you in so you'll know what new professors like me are up against.

No, changing fields in science is definitely not for the faint of heart.   In a way it reminds me of how, whenever I go fly-fishing someplace new, like up on the Madison or the Jefferson rivers, and how I'm mostly familiar with the Two Medicine or the Musselshell from my boyhood, common sense says it's always best to take a guide along the first time out…someone who knows the waters better than you do because there's no substitute for *local knowledge*. It's not possible to get everything you need from journal articles, even peer-reviewed ones, so in research as in fly-fishing, it's always best to line someone up ahead of time. Just a few words of advice here and there from someone who's been there can save a whole lot of time with your procedures.[cxxvii]

Take my specialty – protein chemistry for instance – and how it can seem as if each protein in a bacteria cell, and there are literally thousands of different kinds, how each one comes in its own shape and size and has its own "chemical personality" if you will. There are idiosyncrasies in the way each one needs to be handled, coddled sometimes even, during purification and subsequent tests for both its purity and ability to function the way it's supposed to.[cxxviii] You might get lucky and the protein you're interested in turns out to be as abundant as brown sparrows in summertime and maybe they're indestructible too, able to withstand almost anything you can throw at it during purification in the form of liquid chromatography and centrifugation. On the other hand, your protein may be so sensitive that even the slightest bit of agitation with a tabletop vortexer – the kind used to gently stir up liquids in test tubes – will cause your protein to unravel and lose its activity as if it were a delicate ball of yarn being tugged at by a playful kitten. You have to somehow know all these things in advance to make it in research. It's what we call having "good scientific intuition" and it's a real intangible and you're really lucky if you've got it.

Still another reason researchers might find themselves fishing in unfamiliar waters is when new funding becomes available from someplace that didn't exist before, which is what happened with the arrival of HIV, the virus that causes AIDS, in the 1980's. There were two or three of us in the department that fell into this category as well. As for me, the project that would be dangled in front of me over breakfast this particular morning was by a senior investigator in our department by the name of Stanley Beufort. And the bacterium Stanley works with causes an infectious disease with an entire rap sheet of crimes and misdemeanors throughout human history. The bug is called *Vibrio cholerae*, and nearly everything I describe here happened on this same day, one of the more interesting twelve hours or so I've had in my career, but as my grandmother still likes to remind us boys from time to time, no man ever drowned in his own sweat.

THE GREAT SALT LICK

The ancient Greek philosopher Aristotle believed that the natural world was put together in such a way as there were just four elements worth considering and that, in various combinations, these four were the building blocks that made everything else up. The "periodic table of elements" for the ancient Greeks, therefore, was extremely simple, consisting of just fire, earth, air and water, and in fact Aristotle wasn't even the first to believe that water was an element.[cxxix] The Egyptians and the Hindus all held water in this high of regard, and some Greeks even before Aristotle – Thales for instance – believed that water was the only element worth considering, so important that it actually formed the basis of all other things. But it was Aristotle's views on the subject that held sway for the next two thousand years or so, throughout the entire Middle Ages and right on up until the late 1700's and while their beliefs may seem strange to us now, it's worth remembering that these folks were taking the very first steps away from relying on supernatural forces to explain the everyday world around. Some, like Democritus, even speculated on the existence of atoms 450 years before the birth of Christ.

On the other hand, I had come to view water from a slightly different perspective than Aristotle had. As a product of a 20th century educational system, I had learned that water was not really an element but a molecule, a combination of two elements, hydrogen & oxygen, and that these two could be joined together as one. And even though water still plays a central role in so many religions as it has since ancient times – whether taking the form of prophets walking on it and turning it into wine – or the use of water in purification of the spirit while bathing in sacred rivers like the Ganges, or being baptized as in Christianity – science has nonetheless relegated water to a relatively mundane category of matter today called a *compound*.

This view of water as a mixture of two different elements rather than a single one has been accepted as fact in all of science for it has stood the test of time the best and since nearly all of modern science is based on experiment, and all experiments concerning water ever since the Enlightenment have reinforced this view, whether it was splitting water molecules apart into hydrogen & oxygen gas using a strong current of electricity,[161] or joining these two gasses back together again by igniting them with a flame to form water anew,[162] or whatever experiment one could conceive of, the notion of water as a 3-atom compound – two parts hydrogen and one part oxygen – holds up extremely well and I had therefore given it my full endorsement.

---

[161] 1800 by Nicholson and Ritter.
[162] 1783 by Henry Cavendish. He also discovered hydrogen but mistook it for philogiston.

A central oxygen atom extending out its "arms" to bond with two smaller hydrogen atoms, in the course of which tracing out a nearly right angle all within the backbone of the molecule, is the overall view we have of water today and indeed one of the first things you notice when seeing a water molecule sketched out onto a piece of paper for the first time is that it has this bend right smack dab in the middle of it. And this boomerang-like shape has rather profound consequences for the chemistry of a water molecule, helping to explain its many peculiar behaviors.

And it's hard not to think about water on any given day when working with microbes for a living, which probably makes me a bit of a water chemist at heart. Water has been called the *matrix of life* in biology for good reason. It is, plainly and simply, our solvent, the very thing our molecules are dissolved within. Practically everything inside your cells is either suspended in water or touching it, and if that's not enough, consider how water also cools us in the form of sweat on our skin, lubricates and cleanses our eyes in the form of tears, helps us swallow food as saliva, and even transports digested food around the body as a major component of the blood. Water is the solvent making up our urine, allowing us to flush away toxic wastes like urea and other poisons that would eventually build up and kill us. Astrobiologists usually won't even look for life on a moon or some planet if there isn't at least the slightest hint of liquid water having been there at one time or another.

And according to geologists here on Earth, there's another ocean's worth of water right beneath our feet, and its presence has had a profound influence on what happens up at the surface. Water can change the melting point of rock, the place where it goes from being a solid to a magma, therefore water plays a role in the creation of new continental crust thanks to volcanic eruptions and the lubrication of continental plates along their fault lines. In fact, ours is one of the only places we know of in the solar system with a constantly changing surface like this.

And it's also true that over half of the ninety-two naturally occurring elements listed in the periodic table will dissolve in water without too much coaxing. Water percolating underground picks up salts and other minerals including copper and gold, depositing them in concentrated veins within the crust of the Earth after they have precipitated back out of solution upon cooling. This fact has helped shape not only the terrain, but also the history of Montana, by attracting gold-seekers here in the late 1800's.

Water is so good at dissolving other compounds that you will always consume a bit of the glass each and every time you take a sip of water from one. It's because the glass itself is made of silica molecules and the water will simply eat away at them...in the process dissolving some of its own container. The Earth's surface is the only place we know of in the solar system where the compound can exist naturally in all three states, as a liquid, a solid, and a gas (vapor).   This is extremely fortunate, for if the Earth had been just a few percent farther away or a few percent closer to the sun in its orbit, our supply of water would have either gotten completely locked up as ice, or boiled away as a gas and lost to outer space long ago.[cxxx] [163]

Our planet's oceans, and indeed all its water, is a 50-50 mixture drawn from two different sources, half having been delivered on the backs of asteroids and meteorites, the other half emerging from deep below the surface and if you examine a drop of seawater closely enough, you will find more than a thousand cells and viruses all floating around within it.[cxxxi] In fact, it's possible that the salt inside our blood and in our cells testifies to the same relative proportion of salts that were once making up the seas back when life first began there.   Homer was the first to call salt the "divine substance" and in a way a poet could claim that we never really left the sea because our ancestors took a small droplet of saltwater along with them for the journey inside each and every cell of their bodies.   The Earth is a constantly changing place and since our ancestor's 4-finned journey onto land, the oceans have become progressively saltier thanks to erosion.[164]

But before moving onto land permanently, life needed a way to hold onto water more tenaciously.   For land plants, the solution lay in evolving not only roots but better insulation…thick outer walls around their cells and waxy cuticles to protect their leaves.[cxxxii]   With vertebrates like reptiles, a tough, oily skin was called for, as well as a hard shell for its eggs (later on for birds too), and an efficient kidney that could concentrate wastes like urea without losing excess water.

Humans and nearly all land animals reproduce by "internal fertilization" because this method guarantees our reproductive cells will remain moist. Human egg and sperm cells are not meant to spend time outside a watery body while creatures like fish and amphibians reproduce in water and so can get away with fertilizing their eggs externally, yet we lost this ability millions of years ago.   Sex, it seems, is part of the price we paid for leaving the seas behind.

Like all mammals, each of us began life floating in our own ocean of water inside our mothers called the amniotic fluid, which protected us from stress during pregnancy.   Mammals evolved from reptiles about 200 million years ago, reptiles that developed thick hair to lessen the evaporative loss of water

---

[163] The continent of Antarctica is about the same size of the United States; yet only 3% of its surface is free of ice.
[164] About 4-times saltier. This rising concentration of salt in the oceans has been suggested as a reason life eventually moved onto land.

from their skin, while birds were also reptiles at one time except that their ancestors developed feathers to accomplish this same task.[165]

Once freed from the ocean, our ancestors had to find a way to get salts back in their bodies, but this time from the land.   Unlike plants, which have roots to take up minerals, or by drinking from springs as herbivores do, we humans get all the salt we need in our food these days.   And there seems to be an endless number of ways water has influenced life here on Earth.   A cactus has narrow spines for leaves so it can keep water loss to a minimum while surviving in the desert.   These sharp appendages also keep animals at bay intent on stealing the water stored within their stems, while the elephant's trunk and the giraffe's neck look the way they do because their ancestors needed a way to remain upright while drinking from a stream or puddle.

So as a green-as-grass high school student back when I first began learning about microbes, one of the things I remember doing lots of was adding water to things, whether it was to the dry powder used to make a microbial broth, or to agars so they could be heated and spread out onto Petri dishes and allowed to cool and solidify.[166]

Microbial food in the laboratory – white like sifted flour and composed mostly of simple salts, sugars, and amino acids – is light in its consistency, so much so that some of it always seemed to make its way up into the air against the force of gravity whenever I had to weigh any out onto a piece of paper, whereupon it would then gently find its way back down into the passageways of my eyes and throat.   But as nutritious as it was for microbes, this store-bought powder was worthless without any water.   In fact, keeping food dry is why we can send microbial media through the mail by parcel post and store it in plastic jars with screw top lids on shelves for years just so long as it never gets wet.[cxxxiii]

Ancient people knew that drying and salting food kept it safe from spoiling, in fact drying was one of the few ways people had of preserving food before refrigeration was invented.[cxxxiv]   Genghis Kahn and his soldiers carried dried yak milk in the Gobi desert during the 12th century and re-suspended it with water from their canteens to make a yogurt they ate around their campfires.[167] And the reason adding salt to food works so well at preserving it is that salt makes whatever water is in the food unavailable for microbes.   In other words, if heavily salted, the water may as well not even be there.   Pickling, for

---

[165] Today some of the best evidence we have that all vertebrates had a common ancestor at one time is that the salts dissolved within the blood of a lizard, fish, or human, are so similar in chemical profile as to be identical.

[166] It's possible to "select" for specific microbes depending on the exact ingredients of the broth.   For example, anthrax bacteria grow well on SBA (sheep's blood agar) but not on MacConkey agar, which contains bile salts that inhibit gram-positive bacteria like *Bacillus anthracis* yet encourages gram-negatives like *E. coli* to thrive.

[167] They would also ferment mare's milk into an alcoholic drink.

instance, is done using brine water because the salt inhibits stray yeast cells from fermenting the plant's sugars into unwanted alcohol. It's like one of those old Westerns on TV where someone poisons the waterhole so no one coming along after them can take a drink. Microbes aren't able to use water if it's too high in salt because the salt raises metabolic havoc inside them. Neither can we consume too much salt, which is why even the most experienced sailor will never be able to drink untreated seawater to quench his thirst.[168]

Another interesting example of osmosis is when a gardener accidentally "burns" a plant by giving it too much fertilizer. The salts in the fertilizer can suck the water right out of the plant's roots, in the opposite direction the plant needs the water to go in, so the leaves brown up and die. Like plants and animals on land, microbes need to keep control of water too, which is why *Bacillus anthracis* will produce over two-dozen different proteins while making its spore coat: so it can remain impermeable to water.[169]

And it's not just salt that can increase osmosis, in fact any small molecule will do the trick. When I was a kid I remember my grandfather canning leftover fruits and vegetables from our garden every summer, and one of the main ingredients for jams was always sugar...lots and lots of it. Without knowing about osmosis, what he did wouldn't make much sense since sugar is usually a rich carbon source – a food in other words – for microbes. But anything dissolved in water in high enough concentration makes water unattainable, in fact poisonous, for life. Sugar can actually be toxic because in high enough amounts – like salt – sugar will steal a cell's water.[cxxxv]

Honey is so high in sugar that it's the reason the Egyptians could use it as an ingredient in their salves and ointments. They knew honey helped in healing when applied to flesh wounds, while the Babylonians used honey to preserve dead bodies for the same reason: it keeps bacteria from growing. When I was at the Mummification Museum in Luxor, I saw a demonstration on the way natron salt was used by ancient embalmers to preserve bodies. The Egyptians packed 600 pounds of it all around the deceased for 40 days and nights and – because of osmosis – the salt was able to mimic the dryness of the desert, drawing nearly all of the water out from the body, preserving it for thousands of years. When his bandages were removed by Howard Carter in the 1920's, King Tut still had white spots on his face from the natron salt used in his embalming over three thousand years before.[cxxxvi]

---

[168] When Lewis explained to the Indians on the Missouri of their intentions to cross the Rocky Mountains to get to the Pacific, the Indians couldn't understand why anyone would want to travel such great distances just to get to what they called a great lake that was "illy taisted".
[169] If your city uses rock salt on the roads to reduce ice, you may have noticed brown patches near the sides of the road. Too much salt kills grasses too. / Single-celled microbes like brewer's yeast and protozoa use salt to regulate the amount of water they take in or expel.

Yet paradoxically, all living organisms need salt; in fact chemists have a special name for these important salts. They call them *electrolytes*. Too much water when pure can actually cause *water poisoning* because, in an effort to rid our blood of this excess water, our kidneys have to use the body's salt to encourage water to be flushed away in the urine. The net result is that this extra water ends up leaching away our electrolytes. Drinking just three liters of pure water at a single setting can kill a fully-grown adult, proving once again that too much of anything – even ordinary water – can be hazardous to your health. [cxxxvii]

Salt has been so important throughout history that cities were named due to their proximity to salt mines. Salzburg in Austria and La Salle in France as well as in England all towns ending in "wich" were located near ancient salt deposits. Monasteries in the Middle Ages were sometimes founded on top of salt so they could use the commodity as income for their order. [cxxxviii] Taxes have been levied on salt and some societies even used it as a currency. The Great Wall of China was financed by Chinese emperors from the sale of salt and even our word *salary* is a reminder that Roman soldiers in ancient times were sometimes paid in salt. [170] The Union Army during the Civil War fought battles to keep the Confederacy away from salt supplies and later on in the 20th century, Mahatma Gandhi would protest the British tax on it in India by bringing his followers to the sea, whereupon they simply boiled seawater to obtain salt for free. More than 100 years before Gandhi, Lewis & Clark's men boiled seawater too in an effort to extract salt after reaching the Pacific – 7 bucket's worth – so they could cure a ton of elk meat and trade the leftover salt with the Indians on their way back to St. Louis (they were almost broke by then). [cxxxix]

Maintaining the proper balance of salts in the blood is also important for keeping your blood pressure right, which is why doctors tell their patients to go easy on the salt shaker and may even prescribe diuretics to increase salt's removal by the kidneys. Too much salt in the blood means water from the tissues will rush into the bloodstream (by osmosis again), which increases the pressure inside the vascular plumbing (our arteries), possibly leading to a leak in the brain (a common type of stroke) or a heart attack. [cxl] People with a condition in the inner ear called Meniere's disease are told to reduce their salt intake for this same reason: too much salt in the inner ear can lead to increased water pressure in that confined space, which can, in turn, cause dizziness. [171] Not just animals but plants also need to maintain the right

---

[170] Cleopatra imposed a salt tax in ancient Egypt / *Salad* gets its name from the first salad dressings, which were salt.
[171] Certain salts like potassium chloride & sodium chloride are also responsible for transmitting messages in nerve cells (salt is composed of both positive & negative ions and all ions carry a net electrical charge). When Sacagawea was treated by Clark after she fell ill in Montana, Clark inadvertently lowered her body's salts by bleeding her, which may have

balance of salts and water. Plants rely on pressure inside their cells to create a *hydrostatic skeleton*, which is why without enough water they will wilt.[cxli] The point is, where salt goes water follows, and this is one of the most fundamental tenants in all of biology.

MERCURIAL LIFE

So the phenomena of osmosis, whereby water flows from a place with a smaller amount of salt across a barrier to an adjacent space where the salt is higher, without having to put any energy into making it flow, is pretty influential in all biological systems, and is why a sailor cannot drink from the ocean without getting rid of the salt first, and why a tree's roots have the ability to split a solid granite boulder, and why slices of raw vegetables immersed in water will lose their crispiness after awhile.[172]   Now having explained a little about osmosis, I can go into what my meeting with Stanley Beufort was all about that morning.

Some scientists seem to be born under a lucky star, others prove their salt by doing careful, detailed work their entire careers, while the rest of us fall somewhere in between.   It took the methodical Howard Carter ten years to number, photograph, and pack up all of King Tut's belongings.[173]   And perhaps like Carter must have been, Stanley Beufort was the hardest working scientist around, on that everyone would agree.   In fact, Stanley had made a long career out of studying the cholera bacterium from the point of view of its mechanisms.   He was among the first to show how the microbe uses salt to cause cholera's symptoms.   I was surprised he wanted to talk with me about a new project on account of how he was already the most well known microbiologist in the department.   I first met Stanley when I had to give a talk at one of our departmental meetings my first year in a situation that rolls around once a week called journal club.

Journal club is a chance for all the graduate students, a few ambitious undergrads, postdocs, and faculty within the department to read an article that's been published in a journal recently – perhaps because the author used some new technique or procedure everyone might want to know about, or maybe the article is interesting simply because it has generated a lot of buzz – and to get together for an hour or so and put their two cents in on it.   When I was in graduate school, Dolly the sheep became the first mammal cloned so

---

caused one of her symptoms: "twitching fingers".   Then Lewis gave her mineral water (which has natural salts) from a nearby sulfur spring and she quickly recovered.

[172] Water being able to expand wood is also why wine bottles are stored on their sides: so the wine will touch the cork, otherwise corks can shrink and allow oxygen inside.

[173] Because Carter could only work in Tut's tomb during the winter due to heat in summer.

of course Dolly was among the topics we discussed that semester. Journal clubs are about as common as house dust in academic departments and are a good chance once a week to feel like you're a part of something because in research it's easy to spend too much time cooped up in your own little world.

So during breakfast in a corner booth in the basement of the student union cafeteria that morning, I mostly sat and listened while slowly coming to the conclusion that Stanley was trying to make me feel better about my recent setback. He'd no doubt heard about my losing the anthrax grant. I watched as he painstakingly picked apart his grapes with a knife and fork to dislodge the seeds, thinking about how much time he must have spent in the lab during his long career. Yet there were never any acid burns on his shirts, no distinct whiffs of microbial broth I could detect in the air when we passed in the hallway. He always had on a lab coat while doing bench work, and he may have thought I had some time to spare on a new project. Perhaps that was it, I remember thinking. Maybe Stanley just wants to help me get my name on a paper.[174] But I still wasn't sure and I didn't bother asking that morning. Mostly I just sat there, looked down at my food, and listened.

Stanley did all the talking and somewhere between my first sip of orange juice and my last strip of bacon it dawned on me he was giving me a crash course on cholera. By the time I finished mopping up the last of my yolk with a crust of toast, I had learned that although the disease dates back to ancient Greece,[175] far from a thing of the past, there have been seven worldwide pandemics of cholera just in the last two centuries. Cholera can spread so rapidly in contaminated water that the only disease known to outrun it is influenza, which travels through the air.[176] At times the only way to defeat cholera was to give in and leave; in fact entire cities have been abandoned due to it. In 1849 cholera killed a tenth the population of St Louis while a few years later in Paris 150,000 would also die of it.   James K. Polk fell victim to cholera in New Orleans in 1850 while another president, Zachary Taylor, may also have succumbed to the disease.

During the Civil War, by far the bloodiest conflict in American history, 60% of deaths on both sides were caused not by bullets but bacteria in drinking water (cholera, dysentery, typhoid), something the soldiers referred to as "Virginia Quickstep". Towards the end of the war, it was noticed that military camps having their latrines dug further downstream from where they drew

---

[174] Scientists sometimes get their names added onto the list of other authors of a scientific paper by contributing rather minor details.   It can be a kind of a "you scratch my back I'll scratch yours" arrangement to help each other build up resumes in science (science is competitive and people will do all sorts of things like this).

[175] Hippocrates, the father of medicine, and Galen, the Roman physician to the gladiators, both described it.

[176] Roman soldiers sometimes used dead animals to contaminate an enemy's water supply. / Microbes killed 1 in 13 soldiers during the American Civil War.   In contrast, only 1 in 65 died in battle.

their drinking water rather than upstream had lower death rates but no one ever figured out why.[cxlii]

Cholera reminds me of something the biochemist Isaac Asimov once said about microbes: "Everything about the microscopic world is terribly upsetting. How can things so small be so important?" A cholera victim can lose four times their own body water in under a day and their skin can take on the consistency of clay, even appearing blue and wrinkling up at the extremities. Victorian-era physicians referred to this phenomenon as "washer woman's hands." The victim's blood thickens until it rivals molasses and if not treated quickly enough, the cause of death will be complications due to extremely low blood pressure (called hypotension).[177] This disfigurement of a cholera victim was so unsettling to 19th century Victorians that cholera was their most feared disease.

By comparison, TB (then called "consumption") tended to kill more slowly and leave behind an attractive corpse, while with cholera at the other extreme, physicians described having to step over pools of a patient's fluids just to reach the body. In short, Stanley explained how the cholera bacterium had acquired a way to manipulate a person's salts (and with it their water) inside their intestines, just by using its toxins, and these toxic proteins have the ability to direct water right out of a patient's blood and divert it into their lower digestive tract, where the bacterium is able to reproduce and then escape back into the water supply, reminiscent of a child sliding down a waterslide. The bacterium uses our own body's ability to accomplish osmosis against us in an effort to perpetuate its mercurial lifestyle.

Stanley explained that morning how cholera moved so quickly between towns in the 1800's that doctors assumed it had to have been spread through the air. During outbreaks, neighborhoods would burn so much tar trying to ward off cholera that the smoke could be seen for miles. And when it wasn't air or bad sanitation that was suspected, it was leading a life of sin to blame. A cholera victim's death certificate before 1850 might have read, "Died of alcohol" or perhaps "Gluttony".

Stanley is originally from England and so instead of telling people he works in a lab, he calls it his la-BORE-ahh-tory. Well one of the things about listening to someone with a British accent all morning is that I often feel educated afterwards even if I don't remember everything, which is probably why so many nature documentaries are narrated by Brits these days.[178] Yet there was one thing that kept coming back to me. Time is likely the most important resource an assistant professor has and I simply couldn't afford to

---

[177] Which can lead to multiple organ failure, especially the kidneys. Also blood flow can be so reduced that not enough oxygen reaches the brain, causing coma and death.

[178] In fact, about all I remembered of cholera from my undergraduate days was adding a pinch of salt to the Petri dishes to grow the bacteria in. It's because cholera-causing bacteria apparently evolved to live inside a marine invertebrate.

spread myself too thin just then with a new project. It's easy to spend too much time chasing up blind alleys and turning over rocks with nothing under them to show for your efforts in research. I guess what I'm trying to say is that you need to pick your battles carefully in science, just like everywhere else in life. And so I listened while thinking about what difficulties competing in an entirely new area would entail. With well-funded already established labs – not to mention Big Pharma – as competition, to enter mainstream cholera research as an assistant professor at a small college would be difficult, if not impossible for me. So what a microbiologist like myself needed was a new angle, but I just sat there and listened that morning, mostly because I couldn't think of any.

Like I said, Stanley had spent most of his career working on the mechanism *Vibrio* uses to cause cholera. He was truly fascinated, almost like a child with a new toy, on what was likely a far deeper level than I could ever be, by how the toxins released by the bacterium caused this massive water loss that results in dehydration and death so quickly.

You hear a lot about mechanisms in science these days, and for good reason.[cxliii] Knowledge about mechanisms can prove invaluable. If, for instance microbiologists could learn the exact mechanisms bacteria in hospitals use to share antibiotic-resistance genes with one another, it could make diseases like MRSA a thing of the past. Discovering a mechanism is also an important way scientists get their work accepted by others. The reason you've heard of Charles Darwin and not Robert Chambers, for example, even though Chambers came up with the idea of evolution first, is that only Darwin provided a reasonable explanation (or mechanism) for evolution; something he called "natural selection".

Then there's the phenomenon of the drifting continents. The theory of continental drift was first proposed before 1911 but had so little support among geologists that it was forgotten about; that is until sea-floor spreading was observed using submersibles in the 1960's. When there was visible evidence of a mechanism whereby continents could spread – magma welling up from beneath the ocean floor and forcing the adjacent plates apart – it all made sense. Because people could actually see the mechanism, continental drift was easily accepted after that.[cxliv]

When water is transferred throughout the human body (often right through and in between cells) there are no pumps for accomplishing it. What happens is that the skin cells lining the interior of a particular body cavity – whether it's along the intestines, a sweat pore, the lung, or a tear duct – releases salt through numerous tiny holes in the cell membrane created by ion channels (proteins) opening and closing. This allows the power of osmosis to move water along in the body, following a trail set up by salt ahead of it. Almost like a traffic cop at a busy intersection, this salt basically directs the flow of water inside the body…by traveling there first and then encouraging the water to follow.

One of the striking details Stanley had uncovered during his long career, first in industry, then in academia, is that the cholera bacterium opens up salt pores (ion channels) in the intestines using its toxins, allowing the water from the patient's own plasma to leave their bloodstream and flow out into the intestinal cavity, in the process causing massive diarrhea and, of course, the accompanying fatal water loss.

Stanley needed some kind of an answer by the end of the day and so as I was on my way out the door I rolled my jacket up into a tight ball and tucked it under my arm, for the first time in months indulging in the simple but unexpected luxury of being able to walk outside without a coat on.   With a warm stiff breeze against my face, I didn't resent the walk back to the microbiology department like I usually did this time of year, in fact the air carried with it a welcome sweet smell, perhaps due to decomposing vegetation, yet I still couldn't help wondering why Stanley had asked me of all the professors in the department he could have turned to.

True, I knew something about toxins having worked with anthrax back in graduate school, and it's also true that in some ways, proteins are proteins whether they function as toxins from a microbe that causes anthrax or another that causes cholera.[cxlv]   Toxins are all made out of amino acids because they're…well…proteins.   But the differences had to far outweigh any similarities.   Different bacterial toxins will employ completely different mechanisms for causing a disease.   And there was another thing that bothered me too about the whole thing that morning that just wouldn't go away.   Senior faculty members don't usually go around seeking out collaborations with junior faculty.   Usually it's the other way around.

Back in my office with the door closed, I tossed the handful of articles onto the top of my desk while accepting that I needed get down to some serious reading.   I had to get up to speed by afternoon if I was to make any kind of a decision.   Before going further, I should explain what a review article is because these were the tools I was planning to use to learn about cholera, and all in a matter of a few hours.

Review articles are usually only a few pages in length and published right alongside regular scientific articles in journals, except that their purpose is to review work that's already been done by others.   It's a way of summarizing the achievements of 10's or even 100's of different researchers all in one place.   Review articles are "one-stop shopping" for researchers.   They are also the tools we use to knock down barriers other scientists inadvertently put up in the form of unfamiliar terms (i.e. jargon).   Review articles are valuable for anyone wishing to enter a new field without having to go back and gather up every important article that's ever been published on the subject. They can save a lot of time and I had carefully chosen a few from the stack Stanley had presented me with that morning. So with this in mind I slowly sat down and began reading.

Some forty minutes later I finished the first article, then looked over at the clock on my computer. Only 9:31 with no eyestrain on the horizon and just four more articles to go with the rest of the day to do it in. Feeling slightly groggy from all the heat building up in the basement and a large breakfast on my stomach, I took my sweater off and allowed my mind the freedom to drift a little. To stall for time, I reached over and changed the clock on my computer from digital to analog, then leaned back to contemplate the notion that all clocks move in the direction they do because clockwise is how the shadows on ancient sundials moved. This thought must have been sparked by a documentary I'd seen the night before about some archeologists that found a large sundial in Pompeii at the Temple of Apollo.

Feeling more confident having tackled the first article, I thumbed through the small calendar on my desk, trying to remember why it was I had gone out of my way to leave the bulk of this particular afternoon free.[179] Looking more closely at my handwriting it suddenly occurred to me. We were to be paid a visit by an eminent physiologist by the name of Oskar Mosbacher. Mosbacher was a lung specialist who, in spite of what I had assumed by his name, would turn out to speak perfect English with no noticeable accent, and he was coming to address us on diseases of the lung, which are an all too common problem in western Montana due to a history of mining, one disease in particular called *asbestosis*.[180]

SANITARY REFORMER

I picked up the second review article, recalling how Stanley had mentioned something about cholera having a silver lining. In fact, it's why cholera is sometimes called "the great sanitary reformer". It seems that, back in the mid-1800's, once it was realized cholera germs were carried in water, cleaning up drinking water became a primary concern of governments. As hard as it may be to believe today, before the Enlightenment most folks never even considered the notion that they had the same right to drink clean, fresh water as rich people had.

In India they knew about boiling water 2000 years before Christ and one of the benefits of being an ancient Roman was the expectation that you could

---

[179] We have the ancient Egyptians to thank for our 365-day calendar.
[180] While not a microbial disease, our department hosts speakers on occasion that benefit the community.

bath in fresh, clean spring water carried down from mountains by aqueducts stretching all the way into the heart of the city. At its height, engineers were able to divert some 200 million gallons of fresh water each day to Rome's baths and fountains, and they even used it to flush the city of its waste, which was then washed into the Tiber River.[cxlvi] It took invasion by the barbarians to break the flow of Rome's aqueducts just before her collapse, whereupon Rome's population was reduced from over a million to just 20,000 – its citizens having been forced to get their drinking water from the polluted Tiber instead.

The need for water in biology is absolute, which has also provided for some interesting history. Our ancestors took to walking upright because of drought in Africa reducing the number of trees and creating savannahs while the Egyptian civilization came about much later only after wetlands there began drying out, forcing people to move closer together near the Nile River. Lack of water equates with death when you're surrounded by deserts so the ancient Egyptians actually feared leaving the Nile Valley, avoiding traveling out whenever possible, and since there were no camels in Egypt until the 5th century BC, crossing a desert was always taking one's life into one's hands. The need for water has set the stage for battles and even caused wars. It's no coincidence Custer found Sitting Bull's village near the Little Bighorn River in Montana because that's where the buffalo were,[181] and consider today how the Golan Heights is contested by both Syria and Israel in the Middle East in part because, whenever snow melts from its peaks, forming rivers further below, this water helps create badly needed farmland for irrigating grapevines.

In the 1600's, London was supplied with clean drinking water carried in hollowed out logs leading from springs several miles to the north. Like ancient Rome, London's water supply was also fed by gravity. As London grew, its water wasn't as clean, which is also how the very first epidemiological study in history came about during a cholera outbreak in a neighborhood called Broadstreet in 1854. The source of cholera would be traced back to a single pump by a tenacious doctor (the MD kind) named John Snow, who used a map upon which he carefully plotted the addresses of each recent victim. Another clue Snow came across was that workers at a nearby brewery managed to avoid cholera because they were allowed to drink beer by the owner and so didn't need the contaminated well. The only person to get the disease was the mother of the brewer and it turned out she had her servant fetch water from the contaminated well because she'd grown accustomed to it. The unprecedented amount of information Snow compiled was enough to convince local authorities to remove the pump handle. But

---

[181] The approximately 8,000 Indians had 20,000 ponies that needed water too.

old ideas die hard and even in light of Snow's new evidence, cholera still wasn't accepted as a waterborne disease for several more years.[182]

Still, by the 1860's cholera had managed to go from being a disease widely regarded as a result of one's own personal failings to a full-fledged sanitation problem.   A report on cholera led to the formation of England's Board of Heath and fear of the disease encouraged Parisians to rebuild their decaying underground in the late 1800's.   This timing was fortunate because cities were expanding in the late 19th century due to the Industrial Revolution.[cxlvii]

One of the things researchers worry about these days is doing a lot of work and then not getting credit for it.   It's one of the reasons we publish as often as we can and in fact the history of biology is littered with examples of work never making it into the mainstream for various reasons.   In 1854 the Italian anatomist Filipo Pacini described the curve-shaped bacterium he'd seen in the intestines of a cholera victim, but was largely ignored.[cxlviii]   Later, Robert Koch got the credit by isolating the microbe and studying it in more detail using his culturing methods for growing bacteria.   By 1883, teams of both French and German microbe hunters were searching Alexandria, Egypt for the cause of cholera, trying to isolate and describe the germ responsible. Because the Suez Canal had opened recently, Egypt was now an important link in a much shorter chain of cholera transmission between India and Europe.   When word spread in London or Paris that Egypt had gotten cholera from India, it was assumed that Europe would be next.[183]

The tenacious Koch finally caught sight of the bacterium – describing it as comma-shaped – in a tissue slice from the intestines of a victim, but he wasn't able to culture it (i.e. expand its numbers) so he couldn't be absolutely certain it was the cause.   When the epidemic in Egypt finally receded, the French team returned to France while the Germans stayed on, following the disease all the way back to its source in India.   There, in a water cistern in Calcutta, Koch finally came upon the microbe he'd caught sight of, but only briefly in Egypt, and he even managed to grow it in some agar and beef broth on a plate.[cxlix]   One of the surprising things about the cholera bacterium for him was how easily it died without sufficient water.   This stood in stark contrast to Koch's earlier work on anthrax spores, which were able to withstand drought for years yet could still cause disease.   Koch eventually came to the conclusion that the cholera bacterium was producing some sort of toxin to bring about its symptoms, but he was never able to prove it.   The field of protein chemistry was still a good half-century away.

---

[182] Snow died of a stroke at the age of 45 before he saw his theory widely accepted. In fact, government officials put the Broad Street pump handle back on soon after the epidemic subsided.

[183] Estimates are that 40 million died of cholera in India during the 1800's.

London's sewage system now empties downstream into the Thames, and by the 1890's several large cities were avoiding cholera simply by using sand to filter drinking water.   Koch would someday pay tribute to cholera as "our old ally" because cholera became the first preventable disease in the history of medicine, the first time that widespread governmental measures were taken that actually made a difference in extending the lives of everyday folks.[184]

I put the second review article down on my desk and rubbed my eyes, nursing full-on eyestrain, then looked over at the clock again.   Only 10:30 with two more articles to go, plus a talk in the afternoon to attend.   I thought about how, for years, microbiology was scoffed at as "stamp collecting" by other scientists.   Microbiology had been about finding and classifying microbes in those early days, in fact it's the way it was done for over a hundred years.   Doing experiments on bacteria in detailed and systematic ways was slower in coming.   It wasn't really until the advent of genetic engineering in the late 1970's that microbiology has taken off and gained the respectability it deserves.   In Koch's day genes were unknown and so folks were still scratching their heads as to what proteins were exactly, let alone what the mechanism of heredity was, so of course no one knew that different kinds of proteins have different shapes and sizes, which means they can do jobs based in large part on what those shapes are.

The 3-dimensional structures of the cholera toxin proteins wouldn't be seen by human eyes until 1995, and it turned out then that the toxin was not just one but a series of six proteins in all, bundled together as five identical ones and one completely different protein per package.   Yet these six proteins worked as a team and it was even learned that the five identical ones bind to the intestinal cell and combine to allow, like a shoehorn slipping on a shoe, the 6th protein to travel inside and invade the human cell, and it turns out it's this 6th protein that does all the damage.   It's an enzyme, or catalyst, that commandeers the human cell's metabolism to cause salt to flow out.

These days, it's possible to not only change (or knock out completely) any of the genes in a disease-causing bacterium like Vibrio cholerae, but we can even alter the genes of the animal that the bacterium infects… its host in other words.   In this way it's possible nowadays to look at the complex interplay between a host and its pathogen on a fine-grained genetic level never before imagined, not even in the 1970's.   As Stanley Falkow has called it, "a genetic choreography of a pathogen in contact with its host".   And it does seem to be a kind of dance, between two different organisms, one large, the other tiny, a dance that's turned out to be surprisingly elaborate at times too, with host

---

[184] In 1895, H. G. Wells published a short story called The Stolen Bacillus, in which an anarchist steals a vial of cholera and puts it in London's water supply.

tissue and bacterium "talking" to one another, almost like they are having a conversation, but with a chemical vocabulary of course.

I started in on the next article but only got about halfway through before the eyestrain returned. But by then I'd picked up on a few details I could at least mull over. The genes used to make cholera toxins actually come from a virus that infects the bacteria cell. The cholera toxin genes therefore are viral in origin, which the virus transfers to the chromosome of the bacterium. These virus genes literally stitch themselves into the bacteria's own DNA and at this point the bacterium is capable of causing cholera. Prior to this, it's possible to drink an entire glassful of *Vibrio cholerae* without getting so much a stomachache afterwards. The virus confers the ability to cause the disease to the bacterium via its genes and it's now realized that the bacterium hides out in plankton, specifically a small marine copepod, somewhere out at sea, until it gets ingested by a larger animal. It's a bit like the proverbial minnow getting swallowed up by the fish, which then gets swallowed by the bigger fish, and so on all the way to the whale. A virus that delivers its gene into a bacterium that lives inside a plankton, which is then taken up by a shellfish and finally consumed by a person. In fact, cholera outbreaks in Bangladesh coincide with plankton blooms, as well as changes in ocean salinity and increasingly warmer temperatures, which is why scientists using satellites can now track these variables with the goal of predicting future outbreaks more quickly and reliably.[cl]

Vibrio cholerae is classified as a gram-negative bacterium related to an interesting and diverse group of others within the same genus (*Vibrio*) and this genus happens to be widespread across all of Earth's oceans. Some are able to live inside the digestive system of fish and a few *Vibrios* even glow in the dark because of special genes they have, which is why beachcombers sometimes report seeing dead fish washed up at night that give off a green light. Scientists thought for years that free-floating *Vibrio cholerae* in Chesapeake Bay were dead because they couldn't be cultured (grown) in the lab. No one could encourage them to reproduce no matter how fresh the nutrients were (similar to how Koch had cultured *Vibrio* in the past with beef broth). Then someone gave rabbit gut a try (as a growth medium) and the bacterium from Chesapeake Bay resurrected itself and began reproducing again, proving it was not so dead after all.[185]

I also learned that morning that no one knows why the bacterium lets down its guard, allowing the virus to put its genes inside the bacteria's chromosome. One theory is that these toxins help protect the tiny plankton the bacterium lives inside of, protect it from "grazers" in the ocean, for

---

[185] This ability to go dormant may account for its ability to survive feast-or-famine cycles.

example a fish that might happen along and then consume the plankton. And of course what's good for the plankton is going to be good for the bacterium since *Vibrio cholerae* uses the plankton as its home. Protection from grazers like bison and cows are also the reason so many medicinal (and recreational) drugs come from plants. What we view as drugs are actually the plant's way of protecting itself, on account of how plants can't simply run away and hide from danger like an animal can. Plants invest in chemicals for weapons to ensure their continued survival.

The last item I came across was how quickly *Vibrio cholerae* moves, inspiring me to flip the article over and do a calculation on the blank side. The article mentioned that the cholera bacterium has a flagellum on one end that rotates about 100,000 times every minute, whipping itself at speeds of up to a millimeter in ten seconds. Now a millimeter may not sound very fast, but when you consider that to a microbe swimming in water would be like for us swimming in a vat of honey, and that a millimeter is the same distance as 300 cholera bacteria lined up end-to-end, it's pretty darn respectable. And once the bacterium in a victim's intestine makes it to where it needs to be, it can shed its flagellum and change its shape – go from comma-shaped to looking like a round dot – which probably isn't all that surprising. Always striving for economy means bacteria have gotten good at shedding things they don't need, even genes on occasion.[186] One can't tell just by looking at the cholera bacterium whether it has already been infected by the virus and can cause the disease, so investigators usually have to resort to using techniques like PCR, the same one Tahany was using to spot schistosome worm genes in human urine.

AMANDA

I soon gave up on my eyestrain getting any better and began wandering the halls of the dungeon until I came across one of my more recent recruits, Amanda, carrying a rack of test tubes under her arm out in the hallway. I could tell she had just gotten the tubes from the cold room as the broth inside was still clear and yellow, with no foam on the tops. Not having any white froth meant they hadn't been on the shaker so she had just gotten them. The cold room is what our community cooler is called except that it reminds me more of a walk-in meat locker. Instead of sides of beef hanging up, we have benches all along the walls with a wide assortment of equipment, everything

---

[186] *Vibrio* is also unusual among bacteria in having its genes distributed over not one, but two circular chromosomes. This may give it an advantage in reproducing – by allowing it to copy its DNA in half the time.

from small vice grips and screws lying next to heavy tabletop centrifuges, shakers, and water baths useful for doing enzyme reactions in. Purifying proteins often has to be done at cooler temperatures to inhibit these proteins from denaturing (unfolding). And just like food in your refrigerator, microbial broth keeps longer when it's colder too.

On this particular morning, Amanda had been performing routine maintenance on her yeast cultures, taking the opportunity to change their food while in between classes. Seeing her in the hallway reminded me of the first time I'd ever noticed her. She liked to hide her light under a bushel basket back when she was an undergraduate, which explains how she could have been a student in one of my classes for an entire semester without me even realizing it. Thin and on the tall side, perhaps ungainly some might say, especially when wearing her platform shoes, her long brown hair still streaked with blonde in enough places that it was possible to imagine what it must have looked like as a child, but now that she was older it was as if her hair had gone behind the clouds and grown darker along with her spirit. I realized she was only in between classes and wouldn't be in the dungeon longer than it took to feed her yeast on account of how her hair was still down. She hadn't tied it up into a bun like she usually did when getting down to serious bench work.

To say Amanda was on the quiet side is like saying Montana winters can be a bit chilly. When she first joined my lab most of the time she opted for expressions of stoicism rather than the exuberance one usually expects in a college student. It was obvious she had a lot on her mind lately, like back when I first met her, for she had returned to reminding me of a spiritual apparition floating in and out of the lab rather than the carefree graduate student she might have been. We all go through ups and downs, and it didn't interfere with her work as far as I could tell and I also noticed she always made a point of washing her own glassware in the sink. In fact she even washed other people's glassware when she had the time, and stayed out of their business, too, which can be a real trick in itself because lab politics and office politics are pretty much the same thing, with only slight variations, the way British and American spelling of certain words are different, yet it seldom changes their meaning.[cli] And it turns out that Amanda hadn't been very fortunate in finding an advisor her first year of graduate school and I imagine having a child with medical problems didn't make things easier for her. Joshua, her son, had been born with cystic fibrosis.

CF is an inherited disease caused by a single defective gene that makes a protein normally put to use by cells deep inside the passageways of the lungs and other organs. Even though the location of the CF gene within the human genome has been pinpointed since 1989, there is still no cure for it. Finding a defective gene and knowing how to fix it are two completely different problems in biology. The protein responsible for sickle cell anemia – a defective hemoglobin – has been known since about 1948 and yet just knowing its identity hasn't changed treatment for sickle cell very much at all as far as I know.

A mere half century ago, a child with CF like Joshua would have died before reaching his fifth birthday. But today, with the help of modern antibiotics and other treatments that help keep the waves of invading bacteria and accompanying inflammation in the lung at bay for periods of time, a child with CF may live into his or her thirties or perhaps even longer if they're fortunate. In America, CF is the most common lethal inherited disease in children of European descent and Amanda once explained how she would hold Joshua over her knee and rap him on his back in the middle of the night with the heel of her hand to loosen up the mucus that was always threatening to build up deep within his lungs.

Joshua's lungs were becoming a deathtrap for pathogens, not so different than the way flypaper catches flies. Normally, it is this same sticky mucus that helps rid our lungs of foreign dust and pollen grains that float around in the air. But in CF patients this flow of mucus can't get out of the lung quite so fast...not on its own, anyway. It doesn't flow well because it doesn't have enough water in it. Invariably, the mucus gets backed up and anything caught within its grasp – like bacteria – rousts the macrophages and neutrophils and these watchdog cells in turn sound the body's alarm that is our immune system.

Neutrophils and macrophages are potent scavengers that travel around the body killing and digesting invaders and can sometimes turn on us too if we're unlucky enough to get caught in their crossfire. Healthy tissue, like soldiers on the battlefield wounded by friendly fire, can wither and die when they've been attacked, forming a scar tissue, which eventually will interfere with breathing.[clii] Today many CF children slowly succumb to this damage caused by their own immune systems, unable to breath on their own, so they end up attached to a ventilator waiting for a lung transplant.

So after about a month in my lab, Amanda quietly confided in me that she wanted to work on CF and perhaps help find a cure for it someday. It was, she said, one of the reasons she had changed her major and joined my lab. But there was nothing much I could do except listen as I knew there was no one working on CF at our small university.

At this point I need to shift gears and explain one of the problems I have with research. I don't like the way it's all set up. Laboratories can seem more like dictatorships than places of discovery, the lead investigator more in a position of a despot than scientist or teacher. All this stands in stark contrast to my own preconceived notions I brought with me into science, deep-seated ideas rooted in summers lying underneath the shade of an apple tree reading books like *The Microbe Hunters* and biographies of famous scientists like Louis Pasteur and Robert Koch. I had somehow gotten the idea that everyone in research was dedicated to doing just one thing... curing diseases. Later on this notion would be severely tested, in the end leaving me feeling somewhat naïve. It took graduate school to show me the error of my ways, after which I soon came to my next erroneous conclusion... that research is set up the way it is because it's meant to inspire competition, sort of like a capitalistic country's economy.

But after having run my own lab, I've since decided that research is done the way it is more often than not because it is just the way research has always been done. And the people the system most benefits are often the ones charged with changing things, so of course hardly anything changes. For one thing, there are a surprising number of barriers between labs, some obvious, others subtler, put up presumably to keep our work from being stolen by others. Research is competitive and so, rather than being done like some sort of joint "Manhattan Style Project" as you might expect with all the best minds in a particular field working together with the same goal, research is a lot more willy-nilly than that, I'm sorry to say. Most taxpayers would be surprised at where their money goes sometimes, to be honest. In fact, it's often a wonder to me that the whole thing works at all.

But then again, secrecy is hardly new. The ancient Egyptians never wrote down how to build a pyramid or preserve a mummy, and the Chinese successfully kept porcelain a secret as well as how to unravel a silkworm's cocoon all the way into the middle ages. Even Leeuwenhoek liked keeping under wraps his method for fashioning all those tiny lenses in his microscopes, in part because he enjoyed being the only one who could see so many different kinds of life.[187] And Jefferson and Lewis corresponded in code sometimes to avoid Spanish agents finding out about the planned expedition across the continent.

But all this secrecy and barrier-building in science these days can give someone who doesn't deserve it an awful lot of influence over someone younger and more vulnerable and while many use this opportunity benevolently, with an eye towards doing better science, and in the process molding the minds of future scientists, others see it as an opportunity to pad their own egos, plain and simple. I've experienced both kinds working my way through the system. A few professors are so cleaver at cranking up the pressure on their students that, even though they have won numerous awards from their peers, at the same time what is most talked about behind their backs at seminars is how they have had more than their share of students slide into serious depression (or worse) before anyone stepped in to stop it.

---

[187] To this day no one knows for sure how he achieved his 275-fold magnification.

In graduate school I remember this one professor who would set his pager to go off ahead of time right in the middle of someone's presentation during journal club, I assume so he could walk out and in the process look more important than everybody else.   And this was only journal club!   It came as little surprise later on when I found he had the unhappiest students in the department.   Anyone who's been to college has at least one story to tell of a professor where one wonders how they ever got their degree in the first place, let alone thought they could teach.   Having a Ph.D. doesn't necessarily make someone a good professor any more than being born in a stable makes someone a horse.   Or as my grandmother still likes to remind us from time to time: "Just because a cat has kittens in the oven, don't make 'em biscuits." Absolute power can corrupt whether it's a dishonest cop on the take, or a priest entrusted with the welfare of someone younger and more vulnerable. Some professors use their position to help students, while others see it as an opportunity to feather their own nests.   And word in the department was that Amanda had, by all accounts, been abused.

I have just let you in on one of research's dirty little secrets.   But even the greenest horse has something to teach the most experienced rider and accordingly, the longer Amanda stayed in my lab, the more she opened up, and as often happens the teacher became the student and so I found myself spending more of my free time in the lab after everyone except for Charlie the janitor had gone home for the evening.

Before changing research groups Amanda had been working for a scientist using lasers to understand the behavior of water molecules better and she explained how, of all the substances in the world, water is one of the simplest and yet least understood compounds.   We know more about the surface of the moon than we do the very substance that makes up over 60% of our own bodies.   And there isn't a single, all-encompassing theory to explain water's behavior, either, not even to this day.   Water can expand when it is cooled, precisely the opposite of most substances.   As a solid there are at least 15 different kinds of ice, each with its own unique structure and characteristics.

And it turns out Aristotle hadn't been so far off in considering water an element.   It's common enough.   In fact, over two thirds of the planet's surface is covered in water, even a good deal of the land as well.   If not for ice resting on top of Antarctica, the entire continent would float another 300 feet higher in the air.   Yet paradoxically, the water molecule itself is relatively small.

A molecule of water weighs less than a single atom of most other elements on the periodic table.   Something this light should, by all rights, be floating around up in the atmosphere as a gas rather than tied down here on the surface.   But fortunately for us, water is confined mostly as a liquid.   It turns out the water molecule – in part because of that bend in the middle, and also because of the difference between the hydrogen and the oxygen atoms making it up – has a strong attraction for other water molecules like itself. And this mutual attraction makes water rather "sticky", meaning it wants to attach to other water molecules in a way most substances don't and this affects its properties significantly.

It's why water beads up into a round drop as it's coming off your faucet or resting on the surface of a newly waxed car.   It's also why water seems to defy gravity when rising up above the rim of a glass or climbing paper towels and plant stems.   The narrow tubes inside plants that the water climbs are hollow because they're made of dead cells lined up in long columns.   Water will rise by capillary action even when the plant is no longer alive.   In fact it's the reason the ancient Egyptians were able to split massive boulders.   They simply drove a wooden peg inside a small hole drilled into the stone and then they let the water expand the wood enough to fracture their stone.[188] [cliii]

Water molecules are incredibly tiny, too.   If you could magnify a small drop of water you'd have to enlarge that drop all the way until it had a diameter of 20 miles before you could just begin to make out the individual molecules with your eyes.   Bacteria are the smallest life forms on Earth and yet a single bacterium like E. coli has perhaps 40 billion water molecules all crowded within it, jostling and competing for space with much larger proteins and other bio-polymers like DNA thrown into the mix.   And these molecules in cells are not just sitting around, either.   A single protein bumps into another protein or water molecule about a billion times every second.

And if that's not enough, consider what would happen if you took a glass of water and went to the edge of the nearest ocean and dumped it all in.   If you then waited long enough for that water to mix with all the water in all the oceans of the world, you could go back to the edge of that same, or any ocean for that matter, dip the same container back into the sea, and you would be holding in your hands about 100 of the original water molecules you had before.   In other words, there are more water molecules in a glass of water than there are glasses of water in all the world's oceans.

I didn't know as much about the physics of water as Amanda did, but I still managed to contribute a few interesting details to our conversations.   I recalled how people used water to keep track of time thousands of years before pendulums were invented.   The Egyptians had water clocks, as did the Greeks and Romans later on.[cliv]   They were simple enough, usually just two jars placed next to one another on two adjacent steps of a staircase.   The

---

[188] In building the Great Pyramid, the Egyptians used water to lubricate the sand on the desert floor so the blocks could slide easier, allowing as few as 10 men to pull a 2-ton rock. / The Egyptians didn't use nails to build boats, but instead sewed the planks together, relying on water swelling the wood to keep the boats from leaking.   Khufu's burial boat was based on this principle.

upper jar would be full of water and have a small hole in the bottom, while the lower jar would start off empty and collect water from the higher jar. When all the water had leaked from one jar into the other, that person's time was up. Water in this way helped keep things fairer in a democracy, useful for timing the length of speeches in Athens as well as the Roman Senate later.

Water has always been important for the stability of nations. It was lack of reliable flooding each season by the Nile that brought an end to the Old Kingdom in Egypt. 20 years of drought ushered in a period of starvation and eventual invasion by the Hyksos. [clv] City-states in Mesopotamia more often succumbed to decreasing levels of the Tigris and Euphrates Rivers (and the corresponding increase in salinity of the soils) than they did to invading armies and one of the advantages to being a Roman citizen besides peace during the Pax Romana was clean drinking water. When Roman soldiers weren't protecting the empire, they would have been put to work building more aqueducts. Eventually, there were eleven bringing water into the city from surrounding mountain springs, which meant water flowed continuously into Rome's baths and public fountains as well as houses of the wealthy. Nero, for example, had hot and cold running water in his palace, except that there were no taps to stem the flow into ancient Rome because the water's pressure was so great, its momentum having built up by its long journey into the city, it would break the pipes. One aqueduct carried 200,000 gallons of fresh water 60 miles into Rome every hour starting in 114 BC. [clvi]

Thousands of years before it would be used to cool nuclear power plants, moving water provided work by turning waterwheels attached to hammers, saws, bellows, and grinding stones alongside rivers and streams. The Domesday Book lists some 5,000 water mills, about one for every 400 inhabitants of Britain. And in time, many of the same water molecules that once turned waterwheels would do work during the Industrial Revolution inside the boilers of steam engines used in factories, railroads and ships. [189]

Water was necessary not just for drinking but also for transportation in a time before railways. The main purpose of the Lewis & Clark Expedition was to find an all-water route from the Missouri River to the Pacific, in hopes that trade from the Orient could one day flow east while European and American-made goods could flow west. In fact, before the Erie Canal was completed – opening up land west of the Appalachian Mountains to trade – it was often cheaper for someone living in New York City to buy goods made in London than to acquire that same item from Kentucky just on the other side of the

---

[189] The Earth's water is recycled; in fact you have some of the same water molecules that were once inside George Washington and Napoleon. / Machines evolve too, which is how waterwheels morphed into today's water turbines that "mill" electricity. Water-generated power at Niagara Falls is twice that produced by the nearest nuclear power plant. In fact, about half the water used in the US is by power plants.

mountains, so it's not so surprising that the digging of the Erie Canal eventually helped make New York City the trading capital it is today.

Like a lot of things, our knowledge seemed to complement one another and Amanda understood well the physics of water, having studied the molecule at the graduate level before joining us. She explained how moving water inside a typical thunderstorm can contain more energy than a medium-sized atomic bomb,[190] and the astronauts used the energy released when hydrogen and oxygen gases combine to make water molecules inside their fuel cells as electricity to run their spacecraft. They even drank the water byproduct, which is pure, but nonetheless leftover waste from the electricity generated.[clvii] She pointed out that the commonness of water shouldn't come as a surprise, either. Three quarters of the mass of the known universe is hydrogen, while the universe's third most abundant element is oxygen, water's other ingredient.

Water is never at rest, even if the temperature were to reach absolute zero, for its bonds would still vibrate somewhat. Moving water sculpted the mountains next to my family's ranch, not just as rain but also in the form of glaciers. Organized as a tsunami, water can transport ships to the tops of tall trees, its energy traveling at a surprising 300 miles an hour. In a glass of lemonade on a warm summer day, water molecules can zip around at three times the speed of a commercial jet liner, but not very far before colliding with another water molecule. On Earth, water regulates the planet's temperature, reflecting the sun's heat when frozen as ice, or absorbing it when in the form of vapor. If not for water's presence in the atmosphere, Earth would have frozen over like Mars long ago...in fact water vapor is the gas most responsible for the Greenhouse Effect. Venus, closer to the sun on the other hand, is nearly devoid of water, most of it having boiled away into outer space long ago, leaving behind sulfuric acid for clouds now.[clviii]

Each water molecule is like a tiny bar magnet possessing both a positive and a negative end. Therefore they can line up in long chains that form honeycombs extending out in three dimensions. This is the reason water can be filled just above the rim of a glass while seeming to defy gravity. Water likes to hold onto itself and therefore forms a tenacious "skin" on its outside due to this so-called *surface tension*.

When water freezes, its molecules form a more expanded honeycomb structure, which explains why ice floats (unlike most solids, which contract and therefore become more dense as they cool) and also why lakes and

---

[190] Looks can be deceiving. An average cumulus cloud can hold as much in water as 100 elephants would weigh. A typical storm cloud weighs in at about 200,000 elephants of water.

rivers freeze from the top down and ice cubes will always rise in your lemonade.[191]

One of the things that make me feel middle aged these days is that I can remember when biologists considered water inside a living cell to behave essentially the same way water does inside a glass of lemonade. It's now accepted that things aren't nearly so simple as I was led to believe. Water molecules can change their behavior depending on what's around them, and in the tiny confines of a cell, things can get surprisingly crowded. Almost half the interior of one of your cells is taken up by proteins (perhaps 200 million of them) and DNA (46 chromosomes), while the other portion will consist mainly of water, and these molecules all compete for space and can have an affect on one another in surprising ways. DNA has the distinct shape it does in part because of the way it reacts towards water molecules surrounding it, and the same holds true for our thousands of proteins. A saying among structural biologists is that "function follows form", which is a fancy way of saying that a molecule's shape has everything to do with what kind of job it performs inside the cell.[192]

Because of their different charges at the two ends, negative near the oxygen atom and more positive towards the two hydrogens, a water molecule can act like Velcro, attaching itself onto the surfaces of other biological molecules. And as you can imagine, this affects the shape of these biomolecules, as well as what other things they interact with. The most well studied protein in all biology – hemoglobin – has 60 water molecules that attach, and then detach themselves from the protein's surface every time it binds to, and then releases, a molecule of oxygen in your blood. And it's one thing to have a certain gene, quite another to activate it, which is why a disc-shaped red blood cell can look so different than a flat skin cell or a long, narrow neuron in your brain even though they all carry the same exact DNA instructions.

It turns out water is a major contributor to whether or not a gene is switched on inside the cell because water spontaneously binds to DNA all along its length, shielding the chromosome in some places while leaving it more exposed in others. In the exposed patches of DNA, proteins can then bind and act as switches, helping turn genes on or off. Water molecules inside a cell are beginning to look to biologists more and more like they're

---

[191] When Lewis and Clark were over-wintering with the Mandan Indians along the Missouri River in South Dakota, the temperature dropped to - 45° F and in the evenings they could hear what sounded like cannons in the distance. The sound was actually water trapped inside the cottonwood trees expanding as it froze and then shattering the tree's trunks.
[192] Our enzymes require water to digest (break down) fats & proteins in our intestines. Water also degrades DNA, which is why some paleobiologists believe the upper limit for recovery of ancient DNA will probably be around a million years. DNA is too hydrophilic, meaning it attracts water strongly, almost like a piece of paper attracting its own scissors.

directing traffic rather than randomly zipping around in a glass of lemonade like they used to back when I was a student.[clix]

This ability of water to bind DNA and alter its shape has even played a part in the history of biology. When the structure of DNA was being worked out for the first time in the 1950's, surprisingly there were just two or three groups competing for the honor. In one laboratory, researchers were using an x-ray photograph of DNA that was more *hydrated*, meaning it had more water molecules surrounding it. This structure they called beta-DNA. Another lab used a photograph of DNA that was less hydrated – more dried out in other words – which gave that DNA a more compact shape. This second type of DNA they called alpha-DNA. Both were the same DNA, chemically, it's just that their shapes were altered because of the amount of water surrounding them. By chance, it turned out the beta-DNA was easier to work with because it yielded less data in the photograph, and so was more easily interpreted using x-rays. In fact, it's more like the way DNA is normally found inside the cell.

Dr Rosalind Franklin (the Ph.D. kind) and her student were working with the alpha-DNA, while doctors Maurice Wilkins, James Watson, and Francis Crick used photograph 51, the famous snapshot of beta-DNA that Franklin took.[193] Working with the beta-DNA allowed Watson & Crick to arrive at the correct structure of DNA first and publish their results in the journal *Nature* in 1953. Therefore, it could be said that, in this case, water molecules and their affect on giant polymers like DNA helped lead the way to a Nobel Prize.

TO KILLIFISH

Amanda handed me some graphs and so I took my glasses off to have a closer look. They were the activity profiles of the endostatin (an angiogenesis inhibitor) she recently made using our yeast. This test protein looked good, which meant our system for making human proteins in the brewer's yeast checked out. After she had purified some endostatin, we had it sent off to a government lab, which injected our yeast-made protein into mice with human tumors and the tumors shrank as would be expected of the human form of the protein. We were now officially a "protein expression lab" and so things were looking up for a change.[194]

I suddenly remembered my promise to Stanley and so hurried off upstairs to get my snail mail from the wooden box in the department's office, then

---

[193] All were Ph.D.'s except Crick, who didn't earn his until 1954 at age 37.
[194] *Expression* is a molecular biology term for the production of a particular protein from one cell in another kind of cell.

came back downstairs to settle in for some more reading. But an uneasy feeling crept over me as I was going down the stairs because I was still bone dry in the idea department, still clueless as to what I could possibly contribute to a cholera project.

While in the upstairs office I couldn't help noticing the icicles on the roof becoming clearer and smaller due to the heat, and there were only a few patches of snow left where the stuff had been piled up all winter around Clark Hall by the snowplows. It was really warming up outside. I'd even overheard someone say that the mercury might reach seventy. Cabin fever was settling in and it was on days like this I felt a twinge of envy thinking about my brothers working outside on the ranch. Being a scientist and working indoors has always seemed at odds to me. Maybe that's why Galileo told his students to "always seek truth directly from nature" and not, I've always assumed, from some dusty old lab.

Back in my office, I couldn't resist checking my e-mail again, anything to put off having to read another review article and risk worsening my eyestrain, so I clicked on the latest one from Tahany. It seemed she was updating me on what had gone on in her lab since I was there several months ago, and about Gamal's book, and even some things I'd forgotten all about, like her friend across the hall whose project involved stocking estuaries in the Nile Delta with killifish.

Killifish are an interesting little creature because it lives in saltwater marshes along the delta and yet this fish also has a choice as to what kind of water it swims in. They can remain in the marine estuaries along the Mediterranean where it's salty, or they can swim upstream into fresher water, dining on mosquito larvae as they go. And from a practical standpoint, this is extremely useful because their appetite for mosquito larvae has been shown to reduce the incidence of malaria.[clx] So killifish are *euryhaline*, meaning they can live in either salt or fresh water. Most fish die when the concentration of salt in their surroundings is too different from that of their blood.

I thought back to a demonstration I saw once at a municipal aquarium in Billings. One of the workers there put a freshwater guppy in a saltwater tank to feed an extravagant-looking purple sea anemone, the kind that looks like shag carpeting. It's a surefire way to get an elementary school kid's attention by the way. But before the anemone could curl up and wrap its tentacles around the fish, the poor thing began fluttering and struggling, swimming around in circles as if lost. We watched with that strange mixture of fascination and horror as the tiny fish slowly lost consciousness and began floating helplessly with the current of the aquarium.

Euryhaline animals like the killifish and the salmon have a gift provided by their genes. They have the ability to regulate the flow of salt through their gills, which means they can adjust the amount of it in their blood pretty quickly. The poor guppy, on the other hand, couldn't. This information was just the excuse I needed to procrastinate a little further, so I did an Internet search using the keyword "killifish"…soon coming up with a list of hits.

Besides providing the killifish with the ability to find new sources of food, it seems being euryhaline is also a clever way of escaping from danger, including getting rid of those pesky parasites if you happen to be a fish because most predators and parasites can't make the adjustment to vastly different salt concentrations like the killifish can. I found other information I hadn't come across before either. The Internet is great for that, its strongest point in my opinion. The killifish it turns out is just three inches long yet it can eat 2000 mosquito larva a day, plus it has the distinction of being the first bony fish to go into outer space, aboard Skylab back in the 70's.

It seems killifish also have a long history in high school biology labs here on earth because their eggs are transparent, meaning it's possible to see beating hearts and red eyes as they begin to form in embryos. There are several other euryhaline animals (and even some plants), including the European shore crab, lamprey, and bull shark. But there was nothing on how the killifish or any of the others accomplished this surprising feat, which for some reason was what I was most interested in. I started my search over again, this time typing in the words "killifish" and "mechanism" together, and then waited. But nothing of any value came back. I knew someone had to be working on it, though. It was too interesting of a mechanism *not* to have someone on it. Maybe they just hadn't published, I reasoned.[195]

And so it was in this mood that I reluctantly picked up the last of Stanley's articles, tucked them under my arm and headed off to lunch, leaving my coat on the back of my chair, unnecessary for the first time since September, as I headed across campus with a warm wind at my back. It seemed almost sacrilegious to be in a hurry walking between buildings on a day like this. Something else I didn't know that day was that, thanks to this rare confluence of factors we were experiencing, bringing the Chinook and its dry winds up and over the edge of the Rockies, nudging out the usually damp cold air that tends to hang over the Lewis Valley this time of year, the whole situation would soon be cleared up for me by one final visit to my office later in the afternoon.

It's an interesting thing about pathogenic microbes like *Vibrio cholera*, I considered as I scanned the final article in a booth all to myself. The more quickly they can get back into the water supply, the deadlier they can become. If the cholera bacterium is able to cause massive fluid loss in people, it can

---

[195] Even the methodical Howard Carter never got around to formally publishing his discovery of King Tut's tomb in a scientific journal (though he apparently planned to).

then go on to infect its next victim that much quicker, which sets up a cycle. The microbe doesn't have to concern itself with how badly it harms its host because it doesn't need them to be walking around anyway, or even alive for that matter.

Consider cold viruses, which get spread around more easily if they *don't* make us too sick. They have a vested interest in *not* killing us...of keeping us on our feet. Rhinoviruses that cause the common cold make you just miserable enough to cough and have a runny nose, and as a result the bug gets to come into contact with a lot more hosts because we're still up and spreading it around. With cholera, being a waterborne bacterium means not having to travel through the air, so it's not as much of a problem moving, meaning that you can make your victim sick enough to land him or her on their backside and still get spread. Once the microbe gets back into the water supply, the water does all the work of transportation...no more need for a host...in fact the host becomes expendable. A cholera victim can release up to two trillion *Vibrios* back into the water supply in a single day thanks to this ability to cause diarrhea and vomiting.[clxi]

And of all the bacteria that cause disease, none kill quicker than cholera. It's a record holder. The microbe's presence in the body may be felt as just some dizziness in the morning, then by afternoon the hands will be cold and clammy, and by evening the microbes have already done their damage and moved on. There are cases where someone infected has left home in the morning feeling perfectly fine and by suppertime were stretched out in the morgue...cold, cyanotic and very dead. "Like a superannuated hag with black eyelids," is how one Victorian physician described his patient, a young girl who died of cholera in 1831. [clxii]

The only slave on the Lewis & Clark expedition, York, survived the entire 8,000-mile journey, yet died from cholera after gaining his freedom. Mark Twain was still a child when cholera reached its high water mark in Missouri. During Twain's childhood cholera traveled up the Mississippi every summer like clockwork, carried by paddlewheel steamboats beginning down in New Orleans. In fact, it was due to a rumor in 1835 of cholera in St. Louis that his family decided to settle in nearby Florida, Missouri, where Twain would be born two years later. In 1849 cholera killed over 4,000 inhabitants of St. Louis, which accounted for more deaths that year than all other diseases combined. Cholera also traveled with the gold seekers (forty-niners) heading west from the Mississippi River, where they picked up and carried the bug all the way to the Sierra Nevada Mountains to the west. It's been estimated that 5,000 died of cholera in contaminated watering holes along the trail, while another 7,000 along the more northerly Oregon Trail perished because of cholera, more than accidents and other diseases put together.[196]   Twain wrote in his autobiography:

> "Those were the cholera days of '49...the people along the Mississippi were paralyzed with fright.   Those who could run away, did it.   And many died of fright in the flight.   Fright killed three persons where cholera killed one.   Those who couldn't flee kept themselves drenched with cholera preventatives, and my mother chose Perry Davis's Pain-Killer for me."[197]

In the late 1800's and even well into the 1900's the assumption was that the bacteria were doing massive and irreversible damage to the intestines.[clxiii] Few believed a cure could turn out to be something as simple as fluid and salt replacement.   Given a boost early enough, the human body actually recovers on its own.   Even antibiotics often don't need to be given because the immune system takes over and does the rest, ridding the body of the microbe itself.   Called ORT for oral rehydration therapy, it was developed in the 1960's when it was realized both water *and* salts needed to be replaced in a cholera patient, not just their water.   Today relief agencies use little packets of dried salts and sugars that remind me of Kool-Aid.   ORT has saved the lives of about three million people every year, mostly children that otherwise would have died of bacterial diseases like cholera.   This simple intervention has cut the cholera death rate from 50% down to about 1% where it is today. But getting information, clean drinking water, and ORT to people in time can still be a major obstacle.

GUEST SPEAKER

I finished the last article with a sense of relief and only a tad of eyestrain, which isn't too bad.   Once I get it, eyestrain usually sticks around for the day. Getting out of the dungeon can be a remedy for just about anything it seems. But I needed to get back as I still had a meeting to attend.   Whenever a visiting speaker comes to Thistle, they usually want to stop and talk with the faculty one-on-one in their office.   Weeks ago I had placed my name on a roster circulated among the faculty as one who wished to speak with Dr.

---

[196] The forty-niners also inadvertently spread cholera to the Northern Cheyenne, killing about half.
[197] A patent medicine consisting of alcohol and cayenne pepper, Twain's mother used it as a preventative for cholera.

Mosbacher but apparently I wasn't on *his* list.   I wasn't too bothered by it. After all, I wasn't even tenured and besides, there were others doing more interesting work and so I probably wasn't even on his radar.   The reason I wanted to speak with him had more to do with Amanda's son than it did me anyway.   I vaguely recalled that Mosbacher at one time or another had worked on cystic fibrosis.   But perhaps it had been a while and my memory was never that good anyway when it comes to names.   I looked at the handbill again on my desk.   The title of his talk was "Miner's Lung and the American West".   We have a long history of mining in these parts and unfortunately asbestos is a not uncommon mineral here.

During the walk back to the department, my shadow stretched out ahead of me on the sidewalk like one of those comic book characters, the kind that can tie themselves up into knots.   The sun in January, even at high noon, barely nudges above the peaks of the mountains to the south, almost as if the sun has decided to go south for the winter too along with everyone else that fled for Southern California.   Due to the Chinook, though, the snow had been steadily melting up at the ski resorts, forcing Dr. Mosbacher and his party to cancel their plans for the slopes.

A Chinook is a weather phenomenon set up by unusually warm, dry winds created when moist Pacific air is forced to rise quickly over the tops of the mountains, producing warm dry air (as the water condenses out; condensation being an exothermic process) and this now-drier warmer air then feeds down the eastern slopes, filling up the valleys below.   The name itself comes from the Chinook Indians who used to live to the southwest, down near where the warm air was thought by early fur trappers to have originated.   Chinooks are also sometimes called "snow eaters" due to the sudden increase in temperatures they bring.   Chinooks have been known to melt a foot of snow without even giving it a chance to turn into a liquid first. The greatest temperature difference ever recorded on Earth was right here in Montana back in January of '72 when Chinook winds increased the temperature of the air from minus 54° F all the way up to plus 49° F, an almost 100-degree temperature swing in less than 24 hours.

While I'm sure Dr. Mosbacher and his party weren't happy about the Chinook – as we have world class skiing less than an hour from Thistle – just the same it didn't take away from my pleasure of seeing grass on campus for the first time in months.   And more importantly, I now had the chance to meet with him individually in my office before his seminar began this particular evening.   Apparently my name had been on his backup list.

Which explains how I came to find myself precisely at 2 p.m. in my office, watching and waiting for the great Dr. Mosbacher, wondering if he'd live up to my expectations.   He wasn't due to arrive until 2:30 but just in case he was early I wanted to be waiting.   I needn't have worried, though.   2:30 rolled around, then 2:45, after which I began to feel silly. It's always easy to spot a guest speaker out here, though. No one wears a suit and tie in Thistle unless it's to church on Sunday, a wedding on Saturday, or a funeral the rest of the week. Finally, almost 30 minutes late, the fully attired Dr Moesbacher arrived at my threshold, briefcase in hand and a smile on his face. Apparently, the Chinook winds hadn't tarnished his outlook any.

My first impression of Moesbacher was of one of those people in science who make a name for themselves by relying on their forceful personalities rather than putting in the long hours in a wet lab. In fact, these types seem to avoid labs as if they harbor the plague or something. The giveaway was his suit coat, noticeably lopsided, having been buttoned up in haste, each of the buttons in the hole belonging to the one above it, and as a result one half of his suit coat was riding higher than the other half.

I knew a professor back in graduate school, the kind who always managed to have a firm grasp of the "big picture" in biology yet couldn't peel an orange with his fingers or tell you something as simple as the definition of a Brownsted base if his life depended on it. They have a natural brilliance that outshines everyone in the room, and they seem to know everyone, too, by name, from the department head on down to the janitor who sweeps the floors and empties the trashcans. We used to place small wagers on this one as to whether he would know Dr. So and So at the Moscow Institute of Physics and Technology or wherever. More often than not, he would. I still had a hard time with every researcher's name at my own university while it seemed he knew the whole research community personally. He used to say, "We scientists are one big family." I never saw it that way, but was soon wise enough not to bet against him.

We shook hands and he closed the door, unfastened his coat and sat across from me in the only other chair I could squeeze in. I also had one of those plastic dry erase-boards useful for explaining to students various concepts they couldn't get in lecture, while the rest of the room was taken up by my computer, a printer and a long row of books on a shelf that spanned from one wall to the other. Obviously, I wasn't a big name around here. He knew it the minute he saw my little niche, yet he still acted as if I were the most important person on campus that day.

"I seem to recall you used to work with Donald Bonadurer on cystic fibrosis," I began as he sat down, still smiling.

"Yes. That was me. We had some good times, Don and I. But lately it's been miner's lung solamente." Apparently he knew some Spanish, too.

"Well you see...I have this student whose son has CF..." I began, still hoping to get the conversation started in the right direction. But instead he headed me off at the pass, holding his hand up and turning his head to one side as if directing traffic.

"I'm just not up on it like I used to be. Sorry, Ben. Fields move so fast these days. You can ask me anything you want to about lung physiology and the theory of CF, though. That I can handle. Physiology doesn't change too much."

We both had to laugh because it was true enough. Projects come and go as do professors and their students, but the basic physiology of the human body always stays the same. I decided then and there to concentrate on figuring out his legendary teaching philosophy. We all have one whether we realize it or not. I pick up pointers in the most unexpected places sometimes. One of the best teachers I ever had on the subject of teaching was a young associate professor in organic chemistry. In fact, he didn't even have his own office yet, or a class to teach for that matter, but he'd hold study hours in the library...open to anyone. And he used to make it a habit of never giving us the answers, but lead us to them instead so that we felt as if we'd arrived at the solutions ourselves. A small tip, but it made a huge difference in my own teaching.

"What exactly causes CF?" I asked, "I know it has something to do with sticky mucus but..."

Without waiting for me to finish, Moesbacher leapt out of his chair as if there was a hot rock underneath him, in the process reaching for the marker on my desk. He then began scribbling in wide circles and short lines on the whiteboard.

"It all comes down to a faulty gene, Ben, a gene that makes a protein called CFTR.[198] You see, to move water around, our bodies have to set up a salt gradient in the lung. Take this desert toad for instance."

I could see now it was supposed to be an animal of some sort. Yes, I decided, if I squinted hard enough, it could just be a frog. It's a good thing he'd never become an artist.

"The spadefoot toad likes to bury itself in the sand."

I found myself trying to make sense out of what looked like a lump of coal sprouting four appendages.

"During the heat of the day, this little guy stays buried, yet he won't miss out on getting a drink." Moesbacher stopped scribbling, leaned back, and then tilted his head, apparently admiring his artwork. "How does he do it?"

I had no idea and merely shook my head no, while noticing that for an older man, Oskar still had a full head of hair. It was completely grey and his upper lip was curved in the very middle. The gap this produced combined with two oversized front teeth gave him the appearance of a large friendly

---

[198] Proteins and the genes that code for them are often given letters (or combinations of letters and numbers) as names. CFTR in this case stands for the *cystic fibrosis transmembrane conductance regulator*, a real mouthful.

dog, maybe a Saint Bernard or a basset hound. It made him look permanently affable and I wondered if he worked it to his advantage.

"By using osmosis," he then drew a series of arrows from the toad's surroundings, all pointing inward, towards the lump of coal with legs. "To get its water the spadefoot merely increases the concentration of salts in his tissues." He next drew a cluster of dots close together inside the toad that apparently represented salt. "This little guy will even increase the urea inside his body. The buildup of ions and molecules causes the water in the surrounding soil to flow inward, right into the toad...right through his skin. This very same principle of osmosis is what allows our kidneys to reclaim water before we expel it as urine. Recycling water with an efficient kidney is how our ancestors were able to adapt to the land so well."

"My father had a ranch hand who was part Blackfoot Indian, and he used to say that every animal knows more than you do. Increasing your salt is a nice trick to get a drink of water in the middle of the desert."

"And without leaving the safety of your burrow, either. You see..." and here is where I mentioned earlier how things this particular afternoon would all begin to converge, "...the cholera bacterium gives off toxins that turn our own proteins against us...in our intestines. They make our own proteins overactive there. Well, it turns out the same protein we have in our intestines is also in our lungs; and its job is the same. It's CFTR. In both cases, CFTR makes water flow. It's all about, location, Ben...location, location, location," he began tapping the board with the closed end of the marker for emphasis, "Because CFTR in the lungs makes our mucus wetter there. CFTR allows salt to pass through it, like a doorway, from the bloodstream straight through the cell and into the mucus in the lung. And the water molecules will follow this trail of salt. When the mucus is wetted by water traveling from our bloodstream, the mucus can be removed...coughed up a lot easier. But if CFTR doesn't work properly, when the gene is mutated in other words, then Holy Moses the whole thing gets thrown off. The mucus gets drier. So it doesn't get removed as quickly. The lung slowly becomes a trap...a sticky trap made of mucus."

"How does CFTR come into play with cholera?" I wanted to know. Actually I needed to know at that point. There was a chance I could finally make sense of everything if I could just tie it all together somehow. I can usually remember details a lot easier if I can find a way to tie them into some kind of a coherent story in my mind.

Oskar began drawing again, the edges of his latest masterpiece now spilling over the first. He had uncapped the opposite end of the marker, which was green, and his mind was going so fast, he didn't have time to erase apparently.

"The intestines – like the lungs – need to use water to keep things flowing too. Water lubricates all living things. But with cholera, too much lubrication is the problem. *Too much* activity from CFTR causes water to be released into the intestines, which of course comes from the blood, and this water loss leads to diarrhea. It's how the bacterium makes good its escape. Too much water down there in the bowels...it's just the opposite problem in the lungs where the CFTR *isn't* working in a CF patient like your student's son. What's his name by the way?"

"Joshua," I replied, "So the cholera bacteria uses its poisons to make CFTR work even harder than it otherwise would...but in the intestines?"

"Yes, Ben. That's it exactly. It's the same protein doing the same job, but in two different organs of the body. On the one hand, in the lung of a CF patient, CFTR *doesn't* work so well so the water doesn't flow...follow the trail of salt in other words...while in the intestines when a person who doesn't have CF gets exposed to cholera...well the protein works too damn well."

I nodded in amazement.

"It's quite a thing, really. We used to think organisms couldn't survive at all without water. But cryptobiotic organisms do exist apparently...and they don't seem to age either."

"Cryptobiotic?" I asked, only vaguely having heard the term before.

"Cryptobiotic is when an organism becomes completely anhydrous, loses all its water, even inside its cells. Some can apparently survive this complete loss of moisture. Some plant seeds, worms and brine shrimp if I remember. Lotus seeds over 1000 years old have been grown. I seem to recall the world's record being something like 2,000 years...for a seed in the Middle East somewhere. It's possible they can even survive in outer space. Pretty exciting stuff. Some cryptobiotic organisms without any water have survived experiments down to about a degree above absolute zero. Even brewer's yeast seems to be able to survive this kind of extreme desiccation."[199]

It was interesting this new tangent we had somehow drifted off onto, but I kept thinking about Amanda and wanted to steer the conversation back to Josh's problem if I could. He was extremely knowledgeable and seemed to have a firm grasp of the larger picture, and I wanted to take full advantage of the meager 30 minutes I was allotted. If only his mind didn't wander so much. Scientific papers and review articles are great for picking up details, but not so good when it comes to putting things together like a real person can. There's something about a person's body language or the emphasis they place on certain words, the intonation perhaps, that a journal article will

---

[199] In 2012 Russian scientists germinated seeds buried by a squirrel in Siberia 32,000 years ago.

never quite capture.   It's because we're all human, I suppose, so it just comes with the territory.

"I didn't realize CF was so closely related to cholera.   It's kind of surprising, really."

"Yes.   In fact, one theory as to why CF is so common is that having a defective CFTR gene may actually be a *good* thing.   A single defective copy, that is.   It's possible having only *some* working CFTR in the intestines gives a person a leg up against cholera.   Most of us have two working CFTR genes…and so both genes produce good CFTR, and that lets salt flow properly."

"A carrier would be protected from cholera?"   Someone like Amanda and her husband, I was thinking.   Both had to be CF carriers in order to give Joshua two mutated versions of the gene.   It's often possible to get by with just half the amount of normal protein if one gene that makes it is still good while the other backup gene is mutated.[200]

"Less CFTR in the intestines could be a good thing, theoretically, if you happen to get exposed to cholera or some other microbe like it.   Less functional CFTR in the intestines would mean less diarrhea and therefore less water loss."   It was beginning to make more sense

"Oh.   And there's one more thing."   Without bothering to pick up the eraser he used his sleeve to wipe away the board, making room for something else, apparently.   "I almost forgot.   The killifish."

"The killifish?"   I exclaimed, and for a brief moment thought I might actually fall right out of my chair.

"Yes.   It turns out some fish have CFTR too…but in their gills.   And the killifish is one of them."   He flipped the pen over and drew in black ink again what could be taken as the outline of a fish but with gills exaggerated so wildly that they looked like a pair of gloves the fish was wearing on its head.   "It uses CFTR when it swims back and forth, between saltwater and freshwater."   He then drew two parallel lines on either side.   Apparently this fish was supposed to be swimming in a stream.

"When it's in freshwater, it needs more salt so it takes in salt from the water…in through its CFTR situated in the lining of its gills.   CFTR is the hole…the passageway that the salt flows through, the same way CFTR allows salt to flow in our lungs.   But when the fish enters the sea, it has the opposite problem because then the fish has to get rid of all that extra salt it drinks in. Too much salt in the blood is bad for a fish just as it would be for you and me. In the ocean, CFTR lets chloride flow through the fish's gills the opposite way, back out into the ocean and thereby escape the blood.   It's a pretty amazing talent because it means CFTR allows for two-way traffic.   People and fish last

---

[200] Approximately 1 in 29 white Americans has a mutation in one of their two CFTR genes and are therefore carriers of CF.

shared a common ancestor some 375 million years ago. You probably knew that. Back when air-breathing fish left the water. Some of their offspring eventually became amphibians, and then reptiles and eventually us. The thing is, the CFTR gene, it stayed around the whole time... in both lineages. But in the killifish, CFTR was kept because it could move salt back and forth between the fish's blood and the surrounding water, by way of its gills. In humans, CFTR was kept around because, when made in the lung, it could move salt from the blood into the mucus and thereby help keep the mucus moist, and the same holds true for CFTR in the intestines. Pretty amazing how this all fits together, isn't it?"

I was still speechless as Oskar went on to explain how the fish's CFTR is now known to be 60% identical to human CFTR, which means that, of the 1,480 amino acids strung together to make up a single molecule of human CFTR, 6 out of ten amino acids along the protein's chain are going to be the same amino acid in both humans and fish.[clxiv] This is a remarkable indication of relatedness... of having a common ancestor at one time.

"Nature is the true experimenter, not us. And she's good at keeping things around when they work rather than inventing something new... because it's easier. She reallocates genes, assigns them different duties slightly. Well, a lot differently depending on how you look at it. People certainly aren't fish. But we're a lot more like fish than most of us would care to admit sometimes."

I sensed the meeting was coming to an end yet I wasn't too disappointed as I already had a lot to think about by then. There was just one more thing I wanted to know before he left and since time was running out I went ahead and asked it.

"Out of curiosity, why did you choose to speak with me today? Was it because I published a paper on inhalation anthrax back in graduate school?" For once he hesitated before answering, as if weighing the pros and cons of whether he should let me in on something or not.

"No... err... well," he began fidgeting with the pen, playing with it between his fingers...then looked in the direction of the door, apparently lamenting his failed plans for an escape. I almost started to wish I could take the question back when he must have decided to trust me.

"Well you see yesterday I was talking to Stanley Beufort. We go back a long ways... all the way to Cambridge, in fact. We used to call him Stan the Man back then. He was so damn good at everything." Moesbacher paused and then smiled, apparently thinking of better days. "I never could go in just one direction like Stanley could, he was like a bloodhound... always on the trail when it came to doing science. Nothing could distract him when he was like that. While he was experimenting with rabbits and cholera, I was experimenting with the girls. Needless to say who had the better career.

But we're both happily married so I guess," his voice trailed off and an awkward silence took its place. Mosbacher would finally be the one to break it. "Well, anyway, I was talking to Stanley, and he seems to think you're a pretty good protein chemist, by the way."

For a brief moment I thought about letting him in on our recent work with the yeast, but not that many MD-PhDs are interested in brewer's yeast, so I just kept it to myself.

"It seems Stanley is trying to get into studying the cholera toxins and their structures this time around.[clxv] Well, it's new territory for him and that stuff you've been doing with anthrax, the toxin genes I mean, well all that's pretty amazing from what I hear. I just thought that maybe you could help him out a little."

The man standing at my vinyl board was in his late 60's or maybe even early 70's, and yet he reminded me more of a proud schoolboy who comes home with a frog in his pocket expecting the full approval of his parents, yet can't understand why they aren't as enthusiastic as he is. I didn't want to disappoint him, so I didn't say anything. I still couldn't think of a way I could help Stanley out.

Finally I broke the silence by adding reluctantly, "I'd like to, but I can't seem to think of anything, really. I've been at it all day, Oskar. I'm really sorry."

"Well maybe you'll come up with something, Ben. Give it a chance," he patted me on the shoulder. "Marinate on it for a while. Just between you and me and the wall, Stanley's been going through a rough time of it lately. We all do sooner or later. He lost two big grants this year, back to back ones he was counting on, and so he can't afford to hire a postdoc, or anyone who's worked with cholera toxin vectors before. He could use an extra set of hands right now."

And with that I knew for sure the meeting was at an end because the older man began looking around for his briefcase. I joined in the search until we both located it underneath his chair, the same place he'd put it when he first sat down. We said our goodbyes and I reassured him I was looking forward to his seminar later in the evening. As he was leaving, a sense of uneasiness descended over me, not so different than the way a Canadian cold front can bump a Chinook out of its sloping valley. Stanley Beaufort was, at least financially, in the same boat I was in. Times have been tough all over lately, what with the government cutting back and all. But at least Stan the Man had his tenure. At least he still had that.

"ALL THE WISDOM OF EGYPT"

I spent the rest of that afternoon in an odd sort of mood, not good or bad, more like restlessness, until I finally went for another walk with just my sweater on and my hands buried in my jean pockets. It's not often we peak out at 70 degrees in March with the sun so low on the horizon. The whole atmosphere had a calm, unnatural feel, as if we were about to be visited by aliens. And it cast an eerie spectacle over everything else the rest of the day, too. Chinooks make for some spectacular sunsets and this one was pulling out all the stops, currently turning the high clouds a bright pink, as if strings of cotton candy had been feathered against a flawless blue canvas. It reminded me of the rind on a genetically engineered watermelon.

The devil is in the details, I thought, as I kept on walking but to nowhere in particular. Napoleon's scientists assumed it would take just a few weeks to decipher the Rosetta stone once they got back to France. Instead it took them 24 long years. And finding where the Nile River ended was sure a lot easier for the ancient Egyptians than finding its source, which is why it took over three thousand years and was done by someone else. And it's true that details matter when it comes to meteorology, too.

Galileo was the first to realize back in the 17th century that if science was to progress any further, it needed more accurate instruments with which to measure the natural world. But I doubted Galileo had in mind cross barrier wind flows and changing dew point temperatures. Only one of the local TV stations had successfully predicted the Chinook's arrival. Even the ancient Egyptians knew the importance of details.

The casing stones of the Great Pyramid were cut to an accuracy of within a 100th of an inch...using just copper chisels. And then they placed these 15-ton stones so close together that today it's impossible to stick a knife blade between them. And it's also true that one of the reasons Anton van Leeuwenhoek was able to see so many different microbes while other scientists of his day – including the great Robert Hooke – couldn't, was that the cloth merchant had allowed the light from his candle to pass directly through his specimens and the glass lens to reach his eye, using what microbiologists today would call *transmitted light*. The others had been using *reflected light* and so they saw a good deal less. Scientific research can be so full of details it's often hard to see the forest from the trees sometimes. Oskar was at one end of the spectrum, always looking at the broader picture, while Stanley was at the opposite end, the closer one, always looking at the details and where they meshed together to become mechanisms. Yet both he and Moesbacher had managed to carve out nice respectable careers. I, on the other hand, was somewhere in the middle. What happens, I wondered, to all the ones like me, the ones who cast a small shadow in research? And then the little voice inside answered without any hesitation. We're probably the vast majority of scientists no one ever hears about.

Oskar's seminar was informative, as everyone expected, unfortunately I was still in a funk and sat in the back of the room so I could make a quiet exit if the mood should strike me, which it did. Throughout much of his talk my eyes were drawn to the bald spot on the back of Stanley's head for some reason. He was sitting in the very front row, and I couldn't shake the feeling that I'd let him down somehow. Because Oskar's talk had been advertised in the local paper the room was not only filled to capacity, but some folks had to sit out on folding chairs in the hallway and watch on closed circuit TV. There were retired professors I hadn't even seen since I was a kid, and whole families involved in mining copper and asbestos.

During the question and answer session I slipped out and picked my way back down the stairs to the dungeon, the dim light of the stairwell making me feel as if I were climbing down a groundhog's hole, a reminder that the sun had already set and we were still in winter no matter what the thermometer outside claimed. Sometimes it's hard to listen to a successful researcher when your own work isn't going so well. And it seemed unsettling that everyone thought I could help Stanley out but me. Stanley had had a great career. I, on the other hand, could barely get a protein expression grant from NIH. I silently feared what my grandfather used to warn me was coming true…that the surest way to stop a horse is to bet on him.

Mostly from habit I found myself in the back room of the lab next to the computer, staring up at the yeast stocks swirling above my head on a shelf. They were there inside, suspended within the amber liquid, bounded by the glass beakers on the mechanical shakers, the jars going back and forth, back and forth, over and over again, as if the machines were stuck, and yet they were working just fine. Getting results from experiments decent enough for a publication can happen at rates that make sweet grass growing on the prairie seem lightning fast by comparison.

Yes, the devil is often in the details, I lamented. It took Gaston Maspero a mere 15 minutes to unwrap Ramesses the Great's mummy in 1886 yet Max Perutz[clxvi] and John Kendrew had to spend six long years determining the 3-dimensional structure of the first proteins ever seen, to view their unique arrangement of atoms, the shape of myoglobin from a whale and hemoglobin from a horse.[201] They used the same technique – x-ray diffraction – that had been used to gaze at the atoms of ordinary table salt way back in 1914, over 50 years earlier. It took so long in between table salt and proteins because all the details had to be worked out. Proteins sure are a lot more complicated than table salt, I lamented.

---

[201] They shared the Nobel Prize for it in 1962.

I leaned further back and ran through the mental list one more time, trying to figure out any way I could possibly help Stanley. But I still needed to banish the elephant in the room first before I could concentrate on anything.

Things seemed so unfair in research these days. Wasn't it bad enough that it was always underfunded? Then there were all the various ways money found its way in and out of labs that involved politics. I'd even heard recently from a colleague who happened to know someone who was on the committee at NIH that reviewed my anthrax grant. He had come to the conclusion that my grant proposal had been reviewed too early in the morning. As incredible as it may sound, some of the same forces that influence Olympic judges to withhold all their higher scores for ice skaters who perform later in the day were at work even within the halls of the National Institutes of Health. My grant, my whole chance at tenure and a career in science, may have come down to the simple luck of the draw. My score may have been based in part upon what time of day my grant had been reviewed.

I hadn't expected the "you scratch my back I'll scratch yours" aspect of research, either. I'd been blindsided by that as well. Why hadn't I seen it coming back in graduate school, I wondered? Everyone else seemed to see it. I leaned further into my chair and stared up at the ceiling, noticing where the paint was beginning to fleck off in places.

The same professor who had been so nasty to Amanda, the one who had worked her like a rented mule, then left her to slowly twist in the wind without any help on her project, probably so she'd fail, maybe because she was a woman, he was still up there, walking around as if nothing at all had happened. But I couldn't shake the notion that perhaps I was letting myself off the hook too easily. Maybe the fact that he was still there, secure in his tenure, maybe the rest of us had more than a little to do with it. Perhaps what really kept him and others like him around in science was our collective silence on the matter, which seemed deafening at the moment. Was it possible that the rest of us were like the leaves floating in the stream when they come to a boulder and simply found it easier to go around?

It was true that one of us would probably have to collaborate with this guy someday, have to make use of one of his expensive lasers to obtain some preliminary data for a grant application. Then there's the fact that he could very well end up sitting on my tenure committee when it met again in the fall. He could even be the deciding vote on whether I stayed or whether I had to leave the university.

I took another slow look around the room, my eyes eventually coming to rest upon the glass jars again, topped with white foam, swirling around and around…endlessly and effortlessly…as if the yeast inside had all the time in the world. On top of the amber liquid the delicate white foam was reaching ever higher with each swing. The tiny brewer's yeast were all in there, oblivious to my concerns about cholera, or anthrax, or tenure…budding and dividing…making more and more protein…silently going about their business. They reminded me of the miller's daughter in the fairy tale Rumplestilskein…spinning gold out of ordinary straw. Except that our yeast were spinning valuable human proteins out of sugars and salts we fed it on a daily basis, and plenty of water, too.

I looked up at the jars again. It seemed as if the yeast were waiting – like Stanley and everyone else was – waiting for me to give them something better to do. It was surprising that the eyestrain hadn't come back though, I thought. Once I get it, it usually lasts the rest of the day. And it's also true what my grandmother says about how, if you wait long enough, the clouds will lift no matter how heavy the passing storm, for it was as if that simple act of resting my eyes on the glass jars of yeast had allowed the sun to shine in, albeit only briefly. The fog in my mind lifted for the first time all day, but only for a moment. Still, that may have been enough. When the ancient Greeks came up with a good idea they claimed it had "All the wisdom of Egypt."

There was at least one thing I could do and it wouldn't cost that much, either, not in the way of money or time or even effort for that matter. The physiologist who had paid me a visit, Oskar Moesbacher, had mentioned during our talk about a friend of his who had been working on cystic fibrosis in London. I couldn't recall his name but it didn't matter. I could still get it. Funny how the thought hadn't occurred to me at the time, though. His friend didn't have enough CFTR to work with. The protein was too expensive and he was holding off on doing some interesting work on it. He'd been getting his CFTR from human cells painstakingly growing them on tissue culture dishes, that is until his grant money ran out and he couldn't afford to work with human epithelial cells anymore.

We could try making CFTR inside the yeast. I was sure of it. If we got lucky and it worked, there would be enough protein for Stanley too, to work out all the mechanisms of the cholera toxins, all the mechanisms he wanted to, to his heart's delight. Yeast are incredibly cheap to maintain and there was no one working on that end of things as far as I knew. If all went well, Stanley Beufort could even have a brand new area in cholera research to call his own and my lab would have more collaborators than we would know what to do with.

A lot of researchers wanted to get their hands on CFTR and would be knocking down our doors if it we managed to get it. There was even a structural biologist I had gone to school with who recently published on anthrax toxins.[clxvii] I knew he'd like a crack at CFTR given half a chance. Cystic fibrosis is a hot area and has been for a while. Good journals like to publish in hot areas.

True, it might not be easy to produce human CFTR inside the yeast. The devil is in the details in protein chemistry. This was, after all, the very reason I had gotten the grant to work with the brewer's yeast in the first place There were still a lot of details that needed to be worked out with the yeast protein expression system in order to use it to manufacture large, complicated proteins like CFTR reliably.

But all I had to do to get the ball rolling was to ask someone for the gene, and there's no doubt they'd send it to us, in a clear plastic vial, sealed on ice and delivered by FedEx. All I'd have to do would be to sign for it and pay the shipping. We could have the human CFTR gene inside the brewer's yeast in about a week if everything went right. Starting would be the easy part. All I needed to get the ball rolling was to assign someone to get busy on the details.

And then, like so many other coincidences that day, I heard the door slam shut in the main part of the lab. It was probably Amanda returning from Oskar's talk to gather up her belongings before heading off home alone in the dark. I stood as I continued anticipating all the unforeseen hurdles and other details that would come our way in the following days and weeks ahead, the kind of things that always crop up when beginning a new project, and then I entered the main part of the lab, where Amanda's desk was. There was at least one thing I could do around here that would allow me to hold my head up for a change.

~ ~ ~

PART 3: SPRING SEMESTER

FRIED EGGS

Every once in a while Mother Nature likes to throw everyone a curveball no one saw coming. In science, there are the ones that come at you like fastballs and ultimately change the way you look at the world – the discovery of an expanding universe by Hubble, van Leeuwenhoek's descriptions of bacteria, the working out of the principles of genetics by Gregor Mendel come easily to mind[202] – and then there are the annoying little knuckleballs that seem to appear out of left field like when, for instance, a certain procedure that used to work just suddenly up and doesn't anymore.

I first became aware of the problem when I picked up my office phone early one morning in April. Our collaborators in England were on the other end. E-mails are for chitchatting about purification protocols and other details, phone calls for either really good or really bad news. I still hadn't had the benefit of my morning coffee, which made it all the more difficult to hear. The hardest thing about riding a horse is always the ground and so when you hear bad news in the morning and you're still half asleep there's that added sense of desperation because no back up plan comes easily to mind...at least none that you can think of.

The last batch of protein we'd purified and sent off to London had quit working apparently. In fact, according to them it wasn't working at all. The CFTR *was* functioning before we sent it off, I assured them. I knew it was because I'd seen the chloride transport graphs Amanda had printed off while using the *patch clamp* instrument, the one for doing the chloride flux assays (tests) over in the physics department.   Try and imagine how you'd feel if you were working in a lab that purified and sent away a protein like, say hemoglobin...and then all of a sudden other labs started clamoring because your hemoglobin didn't bind oxygen anymore like it is supposed to.   That's how I was looking at the moment...rather incompetent.   There would be no recall.   I only promised we'd send out another batch as soon as possible, which is what Amanda set out to do that very morning.

Maybe the real problem was that I had let my guard down; had allowed myself to become too optimistic again.   Just because a horse can jump a fence going north doesn't mean he'll jump the same fence again going south.

---

[202] When scientists at the British Museum in 1798 examined the first duckbilled platypus sent from Australia, they thought the specimen was a practical joke and someone had stitched together different animals.

Research is full of ups and downs and things had gone so well up until then with the new project.

We put the human insulin gene into the yeast and the yeast produced human insulin no problem. Then we did the same again and made endostatin and some other simple proteins. Then the human CFTR gene was sent to us from some researchers at the University of Michigan and using enzymes we clipped out the gene from its vector (a plasmid[clxviii]), purified the long stretch of DNA coding for CFTR in a single afternoon and transferred the gene to its new home, inside another plasmid, this one specially designed for making human proteins in brewer's yeast, and then Amanda expanded this plasmid's numbers using *E. coli*, purified all that extra DNA the bacteria made and simply added this DNA to her yeast cultures along with a quick jolt of electricity.

Just a few of the yeast cells took up the foreign DNA containing the CFTR gene and integrated it into their chromosomes. But that's all we needed. So far, so good because then all she had to do was select one or two of the yeast clones that had incorporated the human gene (from the surface of a Petri dish with a type of food missing only one particular nutrient) and expanded these survivors into new yeast stocks.[clxix] A little goes a long way in microbiology, that's one of the advantages to working with microbes; given the right conditions they reproduce like rabbits.

As George Elliot once said, a good horse makes for fast miles, and expanding clones that carry your gene can be almost pleasurable at times, in fact it only took a few more days to obtain enough protein to purify and start running tests on over in the physics department. As unbelievable as it seemed in so short a time, we had human CFTR being produced by our brewer's yeast in pure, functioning form. At that very moment, we may have had the ability to make more human CFTR than any other lab in the world thanks to the brewer's yeast.

Just about everyone else in the field was working with CFTR made by human cells, painstakingly growing them on the bottoms of tissue culture plates as far as I knew, which is tedious and expensive because this takes up a lot of space, not to mention providing maddeningly low yields.

The hardest part was to show that we could purify all this CFTR away from the yeast's other proteins, which my group did in just a few days. Then, relying on her physics background, Amanda checked the flow of chloride through the purified CFTR protein's open pore. The chloride flowed beautifully and at exactly the rate it was supposed to, in fact we had the graphs to prove it...showing the yeast-made CFTR working its magic in a test tube. We even added an inhibitor of chloride channels (a small drug known to interfere with CFTR's function) into the mix and the protein quit transporting chloride, just as it should have if it were indeed CFTR. We were in business and the whole thing had taken only a few weeks.[203]   But as Lincoln once said, "The wisest of all the creatures is the hen, for she only cackles after the egg is laid."

And in fact, the first two batches we'd sent off to London worked fine. The London people always checked with their own tests to make sure the protein wasn't damaged during transport.   But this time something had gone terribly wrong, that's all I knew for certain as I leaned against my office door probably looking rather shell-shocked.   Our CFTR didn't work anymore.   Too disappointed to stand around, I hung up the phone and walked across the hall to my lab, hoping for something to occur to me before I arrived.   God please don't tell me we sent them fried eggs, is all I remember thinking.

Fried eggs is a slang term biochemists have for denatured protein, the most obvious everyday example of proteins when they are no longer properly folded into their natural, compact little shapes.   We've all seen denatured protein, and even tasted it.   The protein I'm thinking of in this case is egg albumin and it's the heat of the skillet that vibrates these tiny egg proteins until they unravel and turn into an opaque-white solid.

Your kitchen is a surprisingly adequate laboratory for exploring the behavior of biological molecules, better than many of the tamed down chemistry sets kids get for Christmas these days, for your kitchen comes equipped with burners, weak acids and bases, even non-reactive glassware for doing all sorts of reactions in.[204]   Ordinary heat from a griddle can cause surprising physical and chemical changes to occur in biological substances, for instance change a transparent, raw, runny egg into a rubbery, white, easily digestible one in seconds.   Heat can kill the activity of the vast majority of proteins, in fact.[clxx]   But we'd added plenty of ice before shipping and besides, the London people would have mentioned something if the ice packs had melted, so the CFTR getting too warm clearly wasn't the problem.   This was truly odd because we hadn't changed anything in the procedure, either. If Amanda had I was sure she would have told me.[clxxi clxxii]   Or at least I was pretty sure she would have.

In the lab I found only Chris, a stocky premedical student who had recently joined our group.   Sitting at Amanda's desk he was hunched over a book, busily cramming for an upcoming test.   Biology majors are required to spend at least one semester in a lab of their choice and I wasn't flattering myself as I'm pretty sure he picked mine because of Amanda.   Premed students are always welcome because they tend to be bright, if somewhat...I

---

[203] Human CFTR was first produced (by genetic engineering) in bacteria shortly after the gene was discovered in 1989, however human CFTR wasn't functional when produced by bacteria. Bacteria often don't know what to do with large mammalian proteins like CFTR so they simply try to get rid of them by forming *inclusion bodies*, similar to 'fried eggs' where the protein is aggregated inside the cell.

[204] Acids include lemon juice and vinegar.   Bases are baking soda and many detergents.

don't know how else to say it other than to use the term "brown nosing". Getting into a good medical school can be one of the most competitive things they'll ever do in their lives, and a lot of premed students will do practically anything to get that ultimate recommendation letter. I remember this one in particular when I was a graduate student who was so good at sucking up to our mentor, he actually put his own photo right on the lab computer's monitor as wallpaper; a constant reminder to everyone including my mentor just who the real computer genius was. Lots of little things like this can cause friction, though. In fact I'd seen a few from Chris from time to time, but nothing worth making a stink about.

Premed students often look thin, almost malnourished, while Chris stood apart for his athletic build, yet he wasn't into sports as far as I knew. He looked as if he could have played football if he'd chosen to. Instead I found him gazing into an open textbook as I arrived. He could tell by the look on my face something had gone wrong. Premed students are nothing if not intuitive.

"Good morning Doctor Ketchum. Is everything all right?" he made a worried, half-hearted attempt to rise from his stool.

"Not really, Chris," I said as I rested my hand on his shoulder so he'd stay where he was. "By the way, I told you before. You can call me Ben. Like everybody else in here does." I had told Chris this several times, in fact. Maybe he'd forgotten, or more likely it was all part of the buttering up process. "I'm looking for Amanda. Have you seen her?"

"She's got a class now. Human anatomy was offered only once this semester. In the mornings."

That was definitely Amanda, a microbiology major auditing anatomy even though she didn't need it for her degree. I wondered how with Joshua to care for, her ex-husband out of the picture, what it was keeping her going these days. With me my only motivation had become fear...fear of having to explain to my family how I was not only unmarried but out of a job at age 40. I'd even heard someone saying recently how they'd seen Amanda waiting on tables at a truck stop just north of town. Her nearest lifeline was a couple hundred miles away, on a ranch up near the Canadian boarder in a town saddled with the unfortunate name of Two Dot, Montana.

"Do you know where she keeps her graphs," I asked with a tone that must have been urgent enough to send Chris rifling through the stacks of Amanda's papers. "That's okay. Don't worry Chris." I stopped him because I was afraid he would change everything so she wouldn't be able to find them when she got back. "Could you just tell her to stop by my office?"

Chris managed a puzzled nod, his arms spread over the desk in an attempt to keep the ceiling fan from blowing the now disturbed papers off onto the floor. He'd only been on the project with Amanda about a month, but he seemed like a nice enough kid. I could tell he liked her, though. Whenever Amanda was in the room it was the only time his attention wasn't fixated on gaining my approval over something. And I couldn't fault his tastes. She had a certain quality about her, nothing obvious, more in the facial expression and the way she carried herself than anything, a pout of some sort, the perpetual down turned corners of her mouth perhaps. It was definitely something intangible men sometimes look for in a woman... like need or something. Amanda seemed like she always needed things and most men in her life liked to imagine it was we she needed. We're always flattering ourselves that way.

After a few minutes alone in my office I got tired of squatting on my spurs and so returned to square one; silently ticking off a mental list of possibilities. CFTR was unusual to be sure. Even cystic fibrosis is unusual among diseases because it's caused by a single defective gene. Most diseases that involve genes are *polygenic*, meaning they have multiple genes involved, more than one signaling pathway damaged, like cancers seem to have. Defects in multiple genes are much more common than a defect in a single gene when it comes to diseases. CFTR was different in a lot of ways, though. But proteins are still proteins and the mental list could have belonged to any number I'd worked with all the way back to graduate school.

Maybe acid was causing the problem. But she and Chris had used plenty of buffers, which are dirt-cheap, not to mention effective. We always put them in the water to stabilize the pH, which is pretty routine during any protein purification. Buffers are also common over-the-counter medications for indigestion. Everyone's had one at one time or another. They reduce acid by neutralizing it, converting it into harmless water and salt inside your stomach. The effects of acid on proteins can be noticeable in everyday life, too. Cerviche is a Latin American dish where the fish is prepared not by cooking it with heat, like most food is, but by the natural acidity in the lemon juice the fish is soaked in.[clxxiii] The citric acid literally cooks (denatures) the fish's proteins apart.

Another common example of acid denaturing a protein we've all come across is when milk goes sour. As the bacteria in the milk increase,[clxxiv] they produce acetic acid – a waste product – while consuming the milk's sugars, and this acid in turn lowers the pH of the milk, causing the milk proteins (casein[205]) to denature, forming a curd as the proteins all stick together with one another. It's one of the ways we can get cheese from milk. These protein curds are solid and white while the leftover watery part is the whey.

---

[205] Researchers now believe that some of our proteins don't have well defined 3-D structures after all. Casein is a milk protein that has several floppy, highly mobile and exposed regions, probably so the infant can digest it easier. Perhaps a third of our proteins need to be partially or completely unstructured in order to do their jobs properly.

A favorite demonstration we do in the laboratory for non-science majors every semester is putting a whole raw egg still in its shell in a beaker filled with vinegar and then we let it sit for a couple days. When we return it's as if the egg became hardboiled on its own, taking on the texture of rubber. The acid in the vinegar not only eats away the calcium shell, but it also literally cooks the egg's proteins...but without any heat. Still another example of acid's ability to preserve once-living tissue can be seen in Europe's bog people. Because of the acidic nature of the earth in Denmark, one bog victim 2000 years old was found with a noose still around his neck from where he'd been strangled, apparently as a human sacrifice. Even his stomach contents were well preserved due to the low pH of the bog.

We also know that proteins unfolding at the wrong time and place can even cause diseases. Physicians more than 100 years ago looked at Alzheimer patient's brains, commenting on the stringy condition after death. It turns out to be a specific brain protein that forms an aggregate, kind of like scrambled eggs. Some proteins just prefer to join together whenever there's too many, or maybe it's because they're not folded properly in the first place. Something similar is happening in mad cow disease and Parkinson's too. The neurons in the brain are extremely sensitive to "fried egg" inside their cells. Neurons are the most highly developed, specialized cells in the human body, which is probably why they're so touchy, like a well-oiled machine you can easily bring down just by tossing a wrench into.[206]

CONTROLS & ROTATIONS

As I sat in my office waiting, I thought about how entire careers have been wasted in science due to overconfidence. Stretching a particular procedure – especially a new and untested one – beyond its limits can easily lead to errors that go undetected or perhaps worse...misinterpreted.

When electron microscopy was invented in the mid 20th century, people thought they had discovered all sorts of new structures inside cells, that is until they realized what they were actually looking at were artifacts caused by the metallic stain used in the procedure.[clxxv] Artifacts may be what archeologist's value most, but they have caused an awful lot of grief for the rest of us in science.

The better ones realize their mistakes and admit when they've been fooled. Others are either too naïve or too prideful so they never catch on.[clxxvi] There are countless temptations to cut corners in research and the spiral

---

[206] Disordered proteins can be dangerous to the cell because they disrupt the structure of other easily disordered proteins and form aggregates.

downward in someone's career can begin slowly enough. An early warning sign might be when other scientists begin to distance themselves, as if you have the flu or something. Invitations to meetings slow until one day you realize it probably wasn't just an oversight that you weren't invited to the annual seminar again this year. No one wants to reproduce your work anymore because, quite simply, they can't. And they don't usually come out and say so, either. Plus, there's always a chance they did the experiment wrong, too. But after a while, it becomes obvious enough who the real culprit is.[207]

About 30 minutes later Amanda showed up and after going over the graphs I had her begin a new purification. If we hurried, we could have a fresh batch of CFTR on its way across the Atlantic in a couple of days.

One of the most important ways scientists can run checks on themselves is by doing *controls*. Controls are something it's almost impossible to have too many of, and yet it never ceases to amaze me how often they aren't used in science. Controls are usually done right alongside the key experiments, so it's not as if they take up a lot of space.

An example of a control everyone's done is when batteries go bad. The radio (or some other appliance) stops working so you immediately suspect the batteries. Why? Because batteries are always going bad, aren't they? But you can't usually prove it until you get new ones. When you use the new batteries to show that the radio still works, and that the problem was, in fact, the old batteries, you've just run a positive control. If you didn't, then there's always a chance the problem could have been that the radio got wet or was dropped, etc. Controls are a tool, a valuable way of feeling your way around in the dark so to speak, something that helps keep good scientists honest and on the right path and they can save a lot of time and effort in the long run. When we first began the yeast project and made human insulin inside the yeast we were essentially using the insulin gene as a positive control to convince ourselves we could make other human proteins in yeast.

When Lewis & Clark first began paddling their way down the Ohio River from Indiana, they stopped at the place where the Ohio empties into the Mississippi because they wanted to determine the longitude and latitude using their new sextant.[clxxvii] They did this not to find the coordinates of the two rivers, as this juncture had already been determined by others. What they really wanted was to make sure they knew how to use the instrument correctly before they headed off into unknown territory. In this way, the coordinates at the mouth of the Ohio were used as a kind of positive control, to test their abilities as navigators. Lewis and Clark spent a week convincing themselves

---

[207] Reproducibility is a cornerstone of modern science. The historian Pliny the Elder wrote that he had lost faith with physicians in ancient Rome because he never got the same diagnosis from any two, calling their advice "the vacant words of intellectual Greeks".

they could use a sextant before moving on up to St. Louis and then into history.[clxxviii]

During the first two years of graduate school students are often required to do lab rotations. Rotations are a good way for each student to spend about a month in a different investigator's group...then switch to a new group run by another scientist. Rotations are a way to get a feel for whom to pick as a mentor during your second year, which all graduate students have to do by then. Rotations are also a good way to gain some experience in a lot of different fields in a short amount of time.

My very first rotation I found myself in this guy's lab who was so stubborn he'd have argued with a stop sign. This professor actually had convinced himself he could just sit around in his office all day long, sip coffee with his feet propped up on his desk, and come up with all sorts of theories to explain the way nature worked.[208] But whenever an experiment didn't back up his claims, he'd blame the person doing the experiment, or he'd change the controls. Trying to get him to see that he might wrong was about as much fun as trying to Baptize cats. He'd discovered a long time ago that "tweaking controls" is a really simple way to cheat in science. If you use an inappropriate control you can interpret your results almost any way you want to. Sometimes he didn't even use a control, to be honest. After about a month or so I quit reading his publications because they'd become as worthless as a Dixie dollar in my eyes. He'd lost all credibility with a graduate student, which is pretty hard to do since most of us are still pretty gullible at that point.[clxxix]

An interesting thing about him was that he was from abroad and normally his mastery of the English language was more than adequate. But if you cornered him on something – his aversion to using controls for example – all of a sudden his accent would become so thick you'd need a knife to cut it, he became that difficult to understand. I eventually realized he was using it as a sort of camouflage. Instead of "the protein is supposed to bind to the receptor" it suddenly became "ze proten es zuppoz ta bin a za recepter". It may have even been subconscious on his part for all I know. A way to gain an advantage in research is to sound like you're from someplace far away, which of course means you know more than everyone else in the room (the "Einstein effect"). It's human nature to think others from someplace else know more than you do. But probably more often than not, bad science isn't deliberate.

---

[208] As Mark Twain noted: "There is something fascinating about science. One gets such wholesome returns of conjecture out of such a trifling investment of fact." / Aristotle often took a hypothesis and made his observations fit his preconceived notions, which is one of the strategies 9/11 conspiracy theorists today will often use.

Negative controls work along the same lines as positive controls, except that they're designed *not* to work. Negative controls reassure the investigator that the results he or she is seeing in an experiment are real, and not due to some other cause. They're also a way of "subtracting out" the noise of an experiment, and there is always some noise in one form or another. Maybe it's something as simple as contamination in your buffer because you wanted to save a few dollars and bought it from some cheap supply house. The closer and more detailed you look at your results, the more noise there always seems to be, too.

Back in 1989 two well-respected scientists in Utah announced at a press conference they had found a way to harness the power of the sun in a jar at room temperature. Of course this was big news at the time. If it had worked as claimed, it could have solved the world's energy problems inside a decade. Referred to as "cold fusion", it should have seemed too good, and in fact it was, but not before the University of Utah asked the US government to invest 25 million dollars.[209]

After a while the scientific method worked, even if cold fusion couldn't, as red flags began appearing. It reminds me of those old black & white movies on TV where the star, I think it was Charlie Chan, says: "Truth like oil in water. Sooner or later must rise to surface". Or as the skeptical physicists said with a grain of salt: "Cold fusion could not be reproduced by any university without a good football team", in other words scientists at MIT or Caltech couldn't get it to work. At a conference someone finally asked Martin Fleishmann, one of the two scientists, what sort of controls he and his former student Stanley Pons had used. It turns out they hadn't used any.

Their "cold fusion" experiment required the use of *heavy water*, which is like ordinary water except that the two hydrogen atoms making up the water molecule are a bit heavier, having to carry around an extra neutron in their nucleus. If their theory had been correct, and they had accomplished cold fusion in a jar, by replacing the heavy water with ordinary water as a negative control, they shouldn't have been able to get cold fusion to occur. Very basic but apparently they didn't do it. When they did finally use ordinary water as a negative control, they still got positive results, which could have been a clue earlier on that something was wrong. In this case, a simple negative control might have saved a whole lot of trouble for these two scientists who, I believe, were basically honest.

---

[209] The state of Utah did invest several million dollars.

MIDDLE OF THE ROAD

Over the next couple of days, Amanda and Chris repeated the purification of CFTR with another batch of brewer's yeast, and we sent it off to London. Two days later we got the same phone call. Still no activity. Something was clearly happening during shipment, it had to be because there was no other answer. The London people knew what they were doing and besides, Amanda always checked the protein just before sending it off, it was her careful nature, always doubting herself that way. But just the same, we were back to square one. Our protein was no good if it didn't work. No one would believe any of the structure-determining experiments we had planned in London if we provided our collaborators with "dead CFTR" to begin with.[210] I retreated to my office and closed the door because in science, just like on the ranch, it's as important to know how to fall off a horse as it is to ride one.

And troubleshooting in research can be a lot like troubleshooting anywhere in life, whether it's trying to figure out why a calf keeps coming up lame in one of the back pastures, or why a protein inside a cell that normally transports chloride ions just up and quits transporting chloride ions all of a sudden. I thought about what my grandfather used to say, how if you ever find yourself in a hole, the first thing you need to do is stop digging. So I began with the easiest things to change first.

At least we had the problem narrowed down to shipping because the protein was working before we boxed the tubes up and sent them off. Which is good because if the protein had been damaged earlier on, during Amanda and Chris's purification for example, it would have opened up a whole Pandora's box of problems. Each step in the protein purification procedure would have been called into question and rechecked, and there were at least a dozen steps.[clxxx]

I leaned back in my chair staring up at the ceiling, grasping for anything that might help. Vaguely, I seemed to recall a group on the East Coast somewhere who had a similar problem with their protein, a cancer drug. To see if it was going bad during shipment they'd wrapped some of their protein up just as if they were going to send it off, but they drove around with it in the trunk of their car for a few days instead. I couldn't recall how their little

---

[210] In the back of my mind there was a growing concern that perhaps our yeast cultures had become *overrun* by a mutant yeast making faulty CFTR. I'd seen it before. Sometimes one bacterial cell carrying a human gene will undergo a spontaneous mutation, in which case it quits making the full-length, correct version of the human protein and instead begins making a shorter, defective version. It's like cheating. These mutant descendants will all be hardier than the other cells and can therefore overrun them because now they don't have to expend as much energy and resources making the full length version of the protein, meaning they have a *selective advantage*, similar to what Darwin realized drove evolution in nature, where one species is more fit and replaces another. I was worried about it, but didn't say anything. We had enough to think about these days.

experiment turned out, but it was better than anything I had at the moment. Chris was on an interview for medical school so I had Amanda purify another batch of CFTR with the help of my two technicians. I hoped I wasn't cutting into her schedule too much, with Josh and all, but desperate times call for desperate measures. My tenure was on the line and if I didn't get that, we'd all be out of a job (everyone except Chris, who would probably be in medical school by then).

So I spent the better part of the next two days driving around and thinking a lot, with the Styrofoam cooler squeaking away next to me on the front seat, Barney blissfully unaware that he was even taking part in an experiment, instead insisting on riding with me up in the cab and poking his head out the window as I drove up and down the steep mountain passes, the gravel roads at times so narrow that the side-mirrors brushed up against tree branches, picking up twigs and brown leaves and even a few mummified berries left over from the previous summer.

On the second day we came across an ankle-deep creek with enough water to have small fish swimming in it, cutting cleanly across the road, fed by the rapidly melting snow pack higher up. The fact that it was nearly summer should give you some idea of the elevation here. When the Lewis & Clark Expedition first caught sight of these peaks 100 miles off in the distance, it was the first time any of them had ever seen snow in the middle of summer. Mountains back east are a lot tamer by comparison. In fact, sometimes when I'm driving along a highway in Montana or Idaho all by myself, a mountain way off in the distance can seem to change its position so little that even after a half hour or so of driving I can convince myself I've really been sitting in my living room the whole time and the mountain is just another piece of furniture, maybe a dresser or a chair depending on how it's shaped. It's a strange sensation when you have to look down at the weeds alongside the road just to convince yourself you're really doing the speed limit.

Another thing I kept worrying about that I didn't let on was the possibility of oxidation. CFTR – like most proteins – can oxidize, or "rust", when exposed to air, not so different than an iron nail rusts.[211] One of the leading theories as to why we age has it that our aging is due to this kind of oxidation taking place on a molecular level within important biomolecules like DNA, lipid, and protein. Two of the twenty amino acids that make up all proteins contain a sulfur atom, and one of these sulfur atoms, if in contact with oxygen for too long, can form bonds with other sulfur atoms, producing what's called a "disulfide bond." Disulfide bonds occur naturally within all kinds of proteins destined to be exported from the cell.

---

[211] To create the first anthrax vaccine in the 1870's, Pasteur first weakened his *Bacillus anthracis* by exposing these microbes to oxidation. He eventually received a patent for the process.

The protein that makes up 90% of the weight of your hair, keratin, has many such bonds cross-linking its amino acids.[212]   They're so strong they can keep hair curly, which is why getting a permanent to straighten hair involves such strong-smelling chemicals.   The natural disulfide bonds in hair protein are so secure that it takes this harsh treatment to break the proteins apart so the curls can straighten.

When I was in Egypt, I saw the mummy Queen Tiye, mother of the "heretic king" Akhenaton, and her hair was 3,000 years old, yet it still held its curl because the disulfide bonds were still holding her proteins together.[clxxxi]   I was getting a little worried that maybe our CFTR was forming disulfide bonds on its way across the Atlantic and becoming inactive because CFTR isn't supposed to have them.   *Intracellular proteins* like CFTR don't get issued disulfide bonds by the cell.   They're mainly awarded to proteins that are going to be exposed to a hostile environment *outside* a cell, to function in the blood, or to be toughened up and secreted in the saliva or to support the skin and hair from the harsh environment that is our atmosphere.   These proteins need to be stronger because they're going to be exposed more.

But CFTR is a protein that stays in the cell and is protected by the confines of its fatty membrane…mostly a "reducing environment".   It doesn't get issued disulfide bonds by the cell because CFTR doesn't need them.   One of the problems with using the brewer's yeast to make human proteins is that the yeast is sometimes known to assign inappropriate disulfide bonds to a human protein.   This was one of the issues we planned to address with the yeast project and was a primary reason for getting the grant from NIH in the first place.

On the last day of the experiment, with the box containing the plastic vials of CFTR on the seat between Barney and me, something happened that's hard to describe even now, perhaps because it all happened so quickly that it still seems like a dream.   In fact, the main player in the drama came tumbling down off the side of the mountain from my upper right doing what I took to be cartwheels before coming to an abrupt halt in the middle of the road right in front of us.   Instinctively, I hit the brakes without realizing it.   Whatever it was must have become disoriented, but only for a second or two.

My mind began playing tricks on me, first registering it as a couch with four legs and a tail.   It all happened so quickly yet I distinctly remember hearing the box with the protein as it went tumbling off onto the floorboard of the truck.   Barney somehow managed to stay on the seat (I still don't know how).   After a couple more seconds my mind cleared enough that I was able

---

[212] Keratin is also what makes up a horse's hoof and a rhinoceros's horn. / Cholera toxins are held together by disulfide bonds and once inside the host cell these disulfide bonds are cleaved, freeing the enzyme, which can then cause diarrhea.

to tell it was a mountain lion, and while I sat there wondering what to do next, the golden fleece of the lion's coat was already disappearing down the side of the mountain to my left. The whole thing happened so quickly that Barney never even saw the lion. Except for the cloud of dust stirred up by my hitting the brakes so quickly and my heart pumping so fast, it was as if the whole thing had never even happened.

Up until then, I'd almost stopped believing there were any big cats in Montana…that they were actually legends the old timers made up to scare us kids into not wandering off on camping trips. Even having grown up out here, I'd never seen one before. My hands still trembling, I slowly released the brake pedal and the truck began to inch its way down the mountain back in the direction of campus.

Regaining my composure, I leaned over and picked up the box, placing it on the seat while remembering back to when I was about 12 and my grandfather and I set off to see my first lion. It was the coldest part of the coldest month, late January, and we picked up the trail in the freshly fallen snow just as the sun was disappearing behind the mountains, turning the sky a crimson color and the landscape a strange shade of purple. We followed the cat prints from the barn where the chickens were all the way out to the edge of the pasture and then up into the woods on the hillside for a mile through increasingly dense forest, the trail only visible towards the top by the gap between the trees. We never did catch sight of the lion that day, but on the way back down we looked for its tracks again, except that this time they were directly over top of our own. We didn't know it at the time, but at some point during our hike we had stopped being the hunters and quietly became the hunted.

"SOLID INTELLIGENCE"

On the drive back to campus I kept thinking about what it takes to be a good scientist these days. Sometimes bad work sees the light of day because someone finally realizes it's too good to be real.

Jan Hendrik Schon published a research paper on average every eight days, from 1998 until 2001, often in prestigious journals like *Science* and *Nature*, before raising enough red flags that someone finally checked his latest data against his older graphs from two research papers he'd put his name to. It turns out he'd used the same exact data in both papers. They knew he had because it's impossible that the background noise in two different graphs would be exactly the same. Noise by definition is supposed to be random from experiment to experiment. And besides, Schon claimed to have done the experiments at two different temperatures, so the noise should have been noticeably different. Yet there it was, exactly the same tiny jagged little lines in both graphs in two separate articles.

Since joining my department, I had just one paper to my credit where I was a *first author.* I'd been involved with two others where I was a collaborator and one where I was listed as a *third author,* which doesn't carry as much weight as being a first author. Two papers in three years isn't considered prolific by any stretch of the imagination, yet I was sure enough about my results that someone could repeat my findings anywhere. If I slept uneasily at night, it wasn't because of that.[clxxxii] Reproducibility is important; in fact it's one of the cornerstones of modern science. And once again, dishonesty is nothing new. No one's ever found a curse written on a mummy's tomb anywhere in Egypt, yet that didn't stop an enterprising journalist back in Howard Carter's day from starting a rumor that there was a curse, a rumor that still persists to this day.[213]

Sometimes bad science doesn't see the light of day because it fits so neatly into what everyone expected to see in the first place. Which was the case with Piltdown man, a fraud begun when a collector named Charles Dawson in 1908 claimed to have unearthed a skull that looked just like the missing link, the long sought after relative common to both man and his ape-like ancestors.

It was what everyone in the field expected so the skull went largely unquestioned in England for decades. In the end, Piltdown man was revealed as a composite, and not a very good one at that, of a human skull attached to the lower jaw of an orangutan sporting chimpanzee's teeth. The teeth were at an odd angle and had markings indicating they'd been filed down and this had raised some eyebrows in the 1920's, but still the Piltdown man was accepted by the scientific community.

Not until the 1950's when more exact tests were possible using radioactivity to accurately date it was the hoax finally revealed. Yet it didn't happen until after the fraudulent skull had set back the field of archeology significantly. Important new finds in Africa were being ignored for decades. Many scientists simply wanted to believe that the missing link would turn out to be from Europe or Asia, and the skull gave them that reassurance; a false sense of security. Over 250 research papers on Piltdown man were entered into the scientific record and the skull was even used as evidence during the Scopes Monkey Trial by defense attorney Clarence Darrow.[214] [clxxxiii]

---

[213] Greek art was in such demand in ancient Rome that many were faked and sold to unsuspecting buyers.
[214] Another example of preconceived notions causing havoc is the discovery of the ancient city of Troy. Schliemann was so certain that the city described by Homer in the *Iliad* would be nearer the bottom of the mound that he tunneled right past upper layers which contained what he was looking for. Priceless artifacts were destroyed in the process.

Of course, cherry-picking information and getting drunk on your own whiskey isn't only confined to science. In 2003 in order to make its case for the invasion of Iraq, then US Secretary of State Colin Powell would later admit to telling 26 separate "falsehoods" while addressing the United Nations, all of it based on sources presented as "solid intelligence". But these falsehoods (mobile weapons laboratories for example) weren't brought to light until after the invasion and the majority of Americans had already been led to believe that the regime of Saddam Hussein represented an immediate threat to their security. Being less than two years after 9/11 and the anthrax letter attacks no doubt added credibility in the minds of many to Powell's accusations which were explained in a careful, clinical, and by all appearances scientific manner complete with computer generated graphics, recorded phone conversations, and satellite photos. Even to an impartial observer, it all looked very convincing. The head of the CIA, George Tenet, sat directly behind Powell and on camera to lend unspoken credibility to Powell's claims.

Only later would the public learn that much of Powell's testimony was based on conversations with an Iraqi defector code-named "Curveball".[clxxxiv] The CIA apparently didn't inform the State Department of what they knew, but by then the damage had been done. The invasion of Iraq began three weeks after Powell's presentation at the UN. The Bush Administration ignored weapons inspectors who questioned Curveball's claims and I thought about how rewarding loyalty over skepticism is something a third year graduate student knows enough to avoid, yet this self-deception happened at some of the highest levels of the strongest government on Earth.[215]

Sloppy science can be costly for any number of reasons. A combination of improper statistics early on, mixed with the resulting hysteria from the public afterwards, led to more than 25 billion dollars being wasted to investigate what turned out to be a nonexistent connection between power lines and cancer. Thinking about twenty-five billion dollars all for nothing and yet I couldn't even get a decent-sized grant was about all it took to leave a sour taste in my mouth as I rounded the last bend down near where the gravel road flattened out, then merged with the paved two-lane highway leading back to campus.

Sometimes new discoveries can be made in science when different fields are united for the first time. Before statistics was used in polling, there were no scientific surveys of any kind. In 1916, a popular magazine called *The Literary Digest* sent out postcards to determine which candidate its readers thought would win the upcoming presidential election between Woodrow Wilson and Charles Evans Hughes. Their readers predicted Wilson and sure enough, he won. The magazine then repeated the survey and its readers

---

[215] Powell also used the word *nuclear* 22 times in his speech.

predicted the winner of the next four presidential elections.   By then most folks thought the survey couldn't possibly be wrong so when the magazine picked Alf Landon over Franklin D. Roosevelt by a landslide in 1936, it was considered a foregone conclusion.   Much to their surprise, Roosevelt was the one who won in a landslide.

What few had considered in 1936 was that it was the height of the Great Depression and the ones who could most afford to buy luxury items (like magazines) tended to be better off...those who would probably vote for Hughes.   By having a predefined demographic, the magazine inadvertently skewed the results of its own poll.   In that same year there was a less publicized poll done by a recent graduate in political science – Dr. George Gallup (the Ph.D. kind) – who used statistical analysis and a much smaller group of voters, just 5,000 compared to the magazine's two million.

But Gallup polled what he thought were folks more representative of the country.   Unlike the magazine – which was read by those who could afford telephones and automobiles – Gallup went door-to-door to make his sampling as random as he could.   It was the first truly scientific survey in the history of politics.[216]   Gallup even predicted ahead of time that the magazine would incorrectly pick Landon over Roosevelt.   One of George Gallup's guiding principles was that he never took money from political parties, which he believed could skew his results.   Today there are medical researchers that receive grants from the government while at the same time taking money from special interest groups like drug companies to test their products, an obvious conflict of interest I suspect Gallup would have known enough to avoid way back in the 1930's.[217]

---

[216] After predicting the wrong result in 1936, the magazine went out of business. / One of the most common forms of graffiti on the walls of Pompeii are campaign posters...some 3,000 located so far.
[217] Today, it's known that the way a question is worded (and even the order the questions are asked) in a poll can affect the poll's outcome, which is why pollsters will do repeat testing with different versions of the same poll over longer periods to correct for this.

LOOKING ANEW

Alone in my office I began wondering if the confidence I'd placed in Amanda so easily had been justified.[clxxxv]   I only had her word that the tests on CFTR were being done properly.   I wasn't even sure what kind of controls she had been using over in the physics department when she ran the tests. Maybe I should be more hands-on, like Stanley Beufort is with his students, I wondered.   But my mentor in graduate school believed in lending just enough rope for his students to hang themselves with... and I didn't think I'd turned out so bad.   His office door was always open, like mine was now, when it came time to come crawling back for advice.   Maybe Amanda was too proud to ask, though.   Maybe deep down she wasn't able to admit she might be wrong, maybe not even to herself.   Like me, she'd been raised on a ranch, the kind of place that breeds self-reliance.   In spite of the advantages, it's possible our similarities made us a poor match in important ways.   I tended to be gullible. Maybe I'd been looking the other way too much.   Scientists like anyone else can have their paradigm shifts, moments when they start looking at things in a completely different way.

Waterwheels became more widespread in the Middle East and Europe around AD 100 after someone realized they'd be a lot more efficient and easier to operate standing up instead of lying down on their sides in the current as had been the case for centuries.   European astronomers during the Middle Ages missed discovering supernovae in the nighttime sky right before their eyes all because they'd been taught – according to Aristotle's doctrine – that the heavens were supposed to be unchanging.   We know these stellar explosions did occur because Chinese astronomers who had no such inhibitions (since they didn't know about Aristotle) did record them.[218]

The Frenchman credited with deciphering the Rosetta Stone, Jean-Francois Champollion, had a moment of insight when he realized hieroglyphs were phonetic, like modern languages, and not individual words (as was thought for over a thousand years) after he did something as simple as adding up all the Greek letters on the Rosetta stone and finding that the hieroglyphs outnumbered Greek letters 3 to 1.   People mistakenly thought these pictures had to represent objects... yet Champollion realized this wasn't the case. Most hieroglyphs turned out to represent sounds and syllables, similar to the way letters in our own alphabet do.   Then there was Belzoni, the Venetian explorer who discovered new tombs in Egypt's Valley of the Kings in 1817

---

[218] General Terry – who in 1876 was Custer's superior and planned the ill-fated campaign – still believed in fighting Indians the way armies in Europe fought each other for centuries: by favoring pincer movements.   Custer hastily attacked Sitting Bull's village at the Little Big Horn because he believed the Indians would scatter, as they had done when the 7th Calvary attacked in the past.   Instead, Two Moon and his warriors used a pincer moment on Custer.

after realizing he could look for piles of discarded rock chips and use these to predict ahead of time where the original tombs had been dug.

Another interesting paradigm shift took place in hospitals as they grew in size during the late 18th century. It wasn't until physicians began grouping patients with similar symptoms together that it dawned on them these patients might all share the same disease, meaning it was possible to distinguish diseases from one another. Prior to this shift, it was commonly believed that diseases could change, and that one disease could even turn into a completely different disease.[clxxxvi] Then there was Luther Burbank, the inventor of over 800 different varieties of plants in the late 1800's all because, rather than seeing Darwin's new theory of evolution as pertaining only to the past, Burbank realized natural selection could shape the future as well. He enhanced his plants with a version of natural selection today called "artificial selection", which he did simply by keeping the plants he wanted while discarding the others, creating for the first time crops we ranchers find so useful today.[219]

When Lewis & Clark set out into unknown territory in 1804, Thomas Jefferson believed like many of his day that the North American continent would turn out to be symmetrical. Common thought was that the expedition would encounter a mountain range taking only a day or two to cross; the kind of barrier the Appalachian Mountains back east had presented. But the great height and width of the Rockies would dispel this notion. The range Lewis and Clark encountered turned out to be 60 miles wide, hardly the single day's portage Jefferson had hoped for.[clxxxvii]

Yet it's also true that science doesn't always have to be about cold, dry logic. Some of our greatest thinkers had great imaginations to go along with their intellect. Before he revolutionized science by claiming that the Earth revolved around the Sun, Copernicus did *thought experiments*, picturing himself at the center of the solar system, in order to make sense out of what he was observing in the nighttime skies, while John Snow used imagination to visualize cholera as some sort of a wild beast he was trying to track down in mid-19th century London, but by using statistics to lay his trap.[clxxxviii]

Yet too much imagination can cause problems as well. When William Harvey showed how blood gets recycled in the human body by way of the heart and blood vessels in 1616, scientists assumed plants had to recycle water and nutrients inside their trunks the same way. This, of course, turned out to be wrong. Plants don't have a circulatory system like we do because water flows mainly in one direction, from roots to leaves and then out into the

---

[219] It's interesting that the theory of evolution came about during the Industrial Revolution; a time when there was already interest in improving machines like the steam engine. This probably made it easier to accept the notion that living organisms could be improved upon with time as well.

air, but because of this misconception it wasn't shown that water travels straight up plants and out into the atmosphere by a process called *transpiration* until the late 1800's.[clxxxix]   Also during this time there were chemists who believed they could apply Newton's laws of motion to help explain why certain chemical reactions took place readily while other reactions didn't, a blind alley that also turned out to be unfruitful.[220]   And then there were the farmers in Egypt in the 1800's who assumed all the great monuments and statues from antiquity in their own backyards must be filled with gold and riches for why else would European archeologists travel such great distances and spend so much time trying to lug them back to faraway places like Paris or London?

An interesting outcome from "looking at things anew" happened in the 1950's in the Seattle area, and it began with a simple observation. Someone noticed their car windshield was covered with numerous tiny pits. Word spread and it wasn't long before others started looking at their windshields and sure enough, they saw these same curious little pits. Some were sure the government was doing atomic testing and that these holes were the result of nuclear fallout. Others believed they'd discovered tiny eggs inside and one person even claimed to have seen a small sand-flea hatch out of his pit.

Still other theories included meteorites and cosmic rays from outer space to explain the damage. At one point the President of the United States, Dwight Eisenhower, was kept informed. In time, things began to swing back to normal when someone at a nearby college ran some tests and determined that the dark particles inside the pits were ordinary ash, like that found floating around in the air. It was also noticed that these holes were always on the outside of the glass, and only in the front windshield of the car and not on the back window. The near-hysteria finally died down, but not before millions of dollars were spent, simply because the people around Seattle, instead of looking at their windshields like drivers usually do, from the front seat, began looking at them from outside their cars at an angle. Almost all windshields have these small pits if you look closely enough. They're due to rocks and other debris bouncing up from the road and hitting them at high speeds.[221]

I thought about the project again. Maybe we were fooling ourselves into believing we had good CFTR when we never really had any in the first place.

---

[220] The dividing line between modern science and pseudoscience isn't always a neat one. Isaac Newton was a practicing alchemist and tried to make gold out of lead by discovering the recipe for the Philosopher's Stone.   The father of classical physics, Galileo, cast horoscopes for himself and his children in his spare time.
[221] In cancer research, scientists were misled for many years using animals transplanted with human tumors, erroneously believing that the spread of cancer (metastasis) was not as important a factor as researchers now know it to be.   Metastasis didn't happen in the animal models yet today it's recognized over 90% of human cancer deaths are due to this kind of tumor spreading.

Doing good science is hard and time-consuming so it's easy to fool yourself into seeing something you want to believe in research.

Back in the late 1800's there was a horse with a rather bad disposition that had many, including his trainer, convinced he could do math on command, and could even tell time and what day of the week it was. His name was Clever Hans the Arabian stallion and he did it all by tapping out his answers with his front hooves. Darwin's *Origin of Species* had been published not long before this so of course folks were curious to know the extent of animal intelligence.

Hans was examined by scientists and found to be the real deal. One mathematician even concluded the horse had the intellectual ability of a 14-year old schoolboy. Finally, a more skeptical investigator came forward and did some experiments using carefully chosen controls. He did things we'd consider routine today, like using different trainers, shielding the horse's eyes so he couldn't get visual clues from his trainers, and asking questions the trainers didn't know the answers to ahead of time. It soon became obvious to anyone with an open mind that Hans was just reading the body language of his handler to get his answers. And as any rancher worth his salt knows, animals are really good at reading body language. Obtaining visual clues from each other is what horses do all the time in the wild.

Whenever Han's tapping reached the right number, the trainer would subconsciously change his posture ever so slightly and the horse would pick up on this and quit tapping. He was just doing what his trainer wanted him to do. Today this is known as the *observer-expectancy effect*, and is the reason bomb-sniffing dogs work with police officers who also don't know where the practice bombs are hidden, and why psychologists doing experiments with rats perform double-blind experiments whenever possible so even *they* don't know the expected result just like the rat.[222] [223]

---

[222] It's usually a good idea to collect more data if you can. When Clark was making his map west of the Mississippi, he included points of interest he himself hadn't seen. To correct for this, Clark interviewed as many Native Americans as possible to make sure their recollections agreed.

[223] It's a common misconception that a connection between two separate events must exist because they happened simultaneously. When Benjamin Franklin invented the lightning rod and the city of Boston began installing them, a rare earthquake occurred. Residents blamed the lightning rods believing that channeling lightning into the ground had provided enough energy to cause the earthquake.

WORKHORSES OF THE CELL

Before week's end, we heard back from the London people again. This time they had managed to take some of our CFTR and "run it on a gel" using SDS-PAGE. SDS-PAGE is a simple enough procedure commonly done by anyone working with proteins, usually in the final steps of its purification. We use SDS-PAGE all the time to determine the size of a protein after it's been purified, and to make sure that it is pure. It's one of about a dozen procedures we teach the undergraduates in the biochemistry labs every semester.

One just needs to expose a small amount of the protein to heat and detergent in a plastic tube. Boiling water is about the right temperature to get the job done because hot water will make just about any protein unravel (denature), so we float the tubes in that for a few minutes. When the CFTR unravels, it, like other proteins, then binds strongly to a detergent. These detergent molecules, cleverly enough, all carry a negative charge. So when you apply a positive electrical charge to the far end of the rectangular gel containing your proteins, you can get all these negatively charged proteins to migrate through the gel towards the positive electricity (the gel itself is made of poly-acrylamide, hence the PA in PAGE. PA can be thought of as a clear plastic net).

Thanks to friction, the speed that the protein moves through this dense water-filled matrix is an indication of how long the protein is. The longer the protein, the slower it moves through the gel.[cxc] If you also put several proteins of known but different sizes (called a "ladder" or "standard") in an adjacent lane,[cxci] when the procedure is finished you can turn the electric off and then match up pretty closely the size of your protein by how far it moved in comparison to the other "known proteins" in the nearby ladder lane.

Interestingly, the London group had found three small protein bands in the lane where CFTR should have produced only a single band (if it were still intact). In other words, CFTR's amino acid chain was being degraded, or chopped up, into three smaller fragments at some point between the time Amanda had tested it, and when it arrived in London, which meant we now had the problem narrowed down to an errant protease; an enzyme whose job it is to go around chewing up other proteins. This particular one had snipped our full length CFTR twice, into three separate pieces.

They're not uncommon, proteases, in fact we rely on them all the time whenever we digest our food; otherwise the protein in our diet would go straight through us without getting absorbed into the blood. Plants and all microorganisms including yeast have proteases too. Papaya juice extract, a meat tenderizer, contains proteases that chew up protein in muscle, making steaks easier to chew.

My theory now was that, during purification of the CFTR, one of the yeast's own proteases had somehow gotten through and was co-purifying along with our CFTR (since they're both proteins it's not hard to imagine they could both co-purify at some point). It doesn't take a lot of protease contamination to accomplish the task of chewing up, either, since they're enzymes, or catalysts, which means proteases get recycled.[224] Just a few of these enzymes could easily have gone undetected on the SDS-PAGE gel Amanda and Chris had run to check the CFTR for purity before sending it off.

Still, one thing bothered me about it, though. Why didn't this problem show up with the CFTR I drove around with for two days on the front seat of my truck? If this was indeed the problem, then the yeast's enzyme should have chopped up the CFTR while it was rolling around between Barney and me. And yet it didn't. The CFTR worked perfectly fine afterwards when Amanda tested it both with SDS-PAGE and the patch clamp test in the physics department. But since it was all we had to go on, we ran with it.

It wasn't long ago, just 50 years or so, that scientists weren't even sure if proteins had a defined structure or not. It's now taken for granted that most do. In fact, proteins are chains of 20 different amino acids (often called "building blocks") strung together and then folded up into specialized shapes.[225] Some proteins fold up to become long and narrow, like stiff rods hundreds or even thousands of amino acids in length. Collagen under your skin is this kind of protein, one that's used for support, kind of like iron rods support concrete, or the way the ancient Egyptians used straw to strengthen their mud when they made bricks for their houses.[cxcii]

But most proteins in a cell are compact and round so they can dissolve in the watery environment of the cell and move around to do their jobs unimpeded. This category would include enzymes like the brewer's yeast protease[226] we were now on the trail of. Other proteins chop up glucose so

---

[224] Some of the most deadly bacteria toxins we know of function as enzymes. The diphtheria toxin is one such enzyme whose job is to shut down normal protein production inside a human cell. Just one molecule of this toxin is enough to kill an entire cell. One of the cholera toxins is also an enzyme that, once inside its host cell, indirectly causes CFTR to become overactive.

[225] A typical protein in mammals is a few hundred amino acids in length, making CFTR relatively large as human proteins go, a string of exactly 1480 amino acids, joined one after another. Amino acids not only are building blocks for proteins, but can also lend flavors and scents to foods. The amino acid methionine is responsible for the distinct smell of fresh-baked potatoes while glutamate tastes like meat. In fact, meat became so important in our ancestor's diets that we evolved a taste receptor on our tongue to detect it (called umami).

[226] The schistosome worm Tahany was trying to detect with PCR produces a protease in its saliva to digest protein in human skin so it can gain entry into the body.

each of our cells can extract energy (ATP) from it.[227]   Still other proteins normally zip around the body as messengers.   Small hormones like insulin in our blood are an example of this kind.   Proteins do just about everything our body needs to do all day and all night.   In fact, about half the weight of a freeze-dried brewer's yeast cell on your grocer's shelf is protein. [cxciii]

Interestingly, proteins fit the definition of a robot and so some structural biologists have even taken to calling them *nanobots* because they're on the scale of nanometers, or 0.000000001 meters in length.   Proteins are similar to robots because the specific role each plays inside the cell is built right into the protein's structure.   Structure for a biological molecule is analogous to a software program for a computer because structure allows biological molecules to do what they do.   And the way they fold up happens because of the exact sequence of amino acids making up the protein.   So like a robot, proteins don't need to be "told" how to perform all the time.   Proteins just fold themselves up and go about their business.[cxciv]   All of life on Earth depends on proteins being independent this way.   If they weren't autonomous, life as we know it would be impossible.   For example, each of your cells has perhaps a thousand *different* chemical reactions going on inside them right now, in each cell every second.   And almost every one of these reactions is helped along by a separate protein (specifically, an enzyme).

Even in the 1800's, chemists knew proteins were a distinct class of chemicals found only in living things.   They also knew that proteins were made up of just four elements – carbon, hydrogen, oxygen, and nitrogen with a little sulfur and phosphorus thrown in – and they even knew the relative proportions of these elements with respect to one another.   But since they got all this information by burning their proteins (called *combustion analysis*), they never got to find out just how detailed and unique each protein they had really was.[228]   All they knew were the few elements making it up, and their relative ratios, and that these ratios were all similar between different kinds of proteins from whatever the source happened to be, whether the protein came from an animal or vegetable.

They still had no idea how exquisitely varied proteins would turn out to be at their structural level.   They never suspected that having different shaped proteins was the reason they could see light with their eyes (using rhodopsin, a protein that acts like a mousetrap set to be sprung, not by a mouse but by light), fight off an endless number of infections in the blood (with antibodies, which stick like glue to microbes), send signals to distant parts of the body from one tissue to another (hormones like insulin, which bind only to specific

---

[227] The brewer's yeast enzymes that perform fermentation – the same ones that Buchner freed from the yeast cells by squeezing them in chapter one – are an example.
[228] In 1840 chemists had just 2 amino acids in pure form.   By the early 1900's all 20 would be identified and available as pure substances.

receptors on certain cells so these cells "get the message"), digest nutrients in their diet, allow nerve cells to carry signals to and from the brain (ion channels similar to CFTR), duplicate their chromosomes during mitosis (DNA polymerase), or form the hard coat on the outside of an anthrax spore, nor would they come to appreciate that the toxins bacteria make to cause disease would turn out to be proteins, in short, proteins do just about everything life needs to do.[229]   Proteins have earned the right to be called the *workhorses of the cell.*

Back in 1901, the German chemist Hermann Emil Fischer was able to chemically link up, in a test tube, two amino acids and he repeated this reaction over and over until he had a chain of about 18 all strung together into what he called a "polypeptide."   It was far short of his original goal of making an entire protein from scratch, but at least he showed it could be done and that it passed in the laboratory as a protein using simple tests.[230]   It was an important first step in understanding protein structure... that proteins might indeed be made up of long chains of amino acids (usually hundreds or even thousands in length as we know today).   It finally became accepted in the mid-1900's that Fischer was right and that proteins were indeed polymers of amino acids.   Fredrick Sanger was awarded a Nobel Prize for figuring out the sequence of all 51 amino acids making up the first protein whose actual amino acid order was ever determined, a small protein called insulin, in 1955. What took Sanger several years of hard work could easily be accomplished today in a matter of hours (if not minutes) in a protein chemistry lab by either of my two technicians.[cxcv]

One of the more interesting surprises about proteins recently is that they can be infectious, almost like a virus or a bacterium.   In Montana and other places in the West there is a neurodegenerative disease among elk called *chronic wasting disease*, or CWD, which is caused by a single renegade protein called a "prion", and this prion can get picked up and passed on horizontally from animal to animal after being shed in the soil.   It's a misfolded runaway protein, one that's lost its normal shape, and they're practically indestructible, too.   Inside the animal this prion then travels throughout the body in the blood and under its malevolent influence can encourage other proteins like it in the brain to misfold and become "fried egg." This particular protein is normally found in all mammals including us and it's

---

[229] Some proteins called *chaperones* help other proteins fold up properly into their correct shapes.   The most common mutation in CFTR causes an amino acid to go missing.   This *deletion mutation* changes how fast CFTR folds, so that defective CFTR gets "thrown away" because the cell senses the protein is damaged (it takes too long to fold.   Cells can be impatient too).

[230] To this day, trying to string together just 300 amino acids synthetically into a chain is difficult and one of the reasons human CFTR in still made in yeast and other cells (insect cells, for example) rather than in a test tube synthetically.

related to the same one that causes mad cow disease in cattle and CJD in humans.[231]

Another relatively recent surprise was the discovery of tiny "trashcans" inside our cells. Cells use these trashcans for recycling proteins when they've outlived their usefulness or were never properly folded in the first place. Some proteins are meant to last for months inside the cell; others outlive their usefulness in a mere half hour or so. In hindsight it makes sense that our cells would have these tiny disposal units inside since proteins, like the human body, age too. Some diseases seem to result when these protein trashcans don't function normally and the proteins that should be recycled get all backed up, a bit like constipation for a cell…which makes sense since most metabolic pathways are like assembly lines (or disassembly lines, depending on which way you're going).[cxcvi]

Proteins, along with their genes, can have new portions added on and even evolve new functions just as CFTR did over millions of years. In fact, we now know CFTR has a recognizable relative in brewer's yeast, a protein called Ste2. The yeast uses Ste2 to sense whenever another yeast of the opposite mating type is nearby. CFTR even has relatives in bacteria including the same ones that live happily inside your gut.

LEADING A DOUBLE LIFE

I had plenty of time to think while waiting for the latest purifications results. At least the ancient Egyptians had a way of speeding up fermentation of grape juice, I lamented, by adding honey and dates whenever they needed to.

Amanda and Chris prepared a new batch of human CFTR from our yeast stocks, but this time they did everything they could inside the cold room, even when they used the chromatography columns for the separation of the CFTR away from the yeast's own proteins. Proteases function less efficiently at lower temperatures, like everything else does in nature. Protein chains become less flexible when it's colder, less able to do their appointed tasks, in the case of the yeast nanobot its task was chewing up our valuable CFTR before it got to London.[232] I also had Amanda increase the amount of protease inhibitors, which are small chemicals that target and shut down proteases, making their active sites no longer open for business.[cxcvii] If it did turn out to be a yeast protease co-purifying with our CFTR, then there were some measures we could take to at least slow things down if not stop it all

---

[231] Stanley Prusiner received a Nobel Prize in 1997 for proving prion proteins can cause an infectious disease without the need for genes (i.e. DNA or RNA).
[232] You may have noticed egg albumin, the main protein in egg white, has a tendency to become a translucent solid if kept too long in the refrigerator.

together.[cxcviii]   I continued resisting the urge to micromanage my best student and spent the rest of my day either teaching, grading homework, or sitting in my office contemplating how I wasn't the only one slow to figure things out.

Everyone in the entire field of microbiology had been fundamentally wrong about bacteria for a good 150 years.   And it all began with the early microbiologists, the ones who grew bacteria in nice, rich, liquid broth.   It made sense, trying to imitate an animal's blood…but in a test tube or on a Petri dish.   Anthrax – the first disease ever proven to be caused by a microbe – was found growing in the blood of infected cows.   So don't get me wrong.   I don't fault Koch and Pasteur.   There's no doubt it was the right place to start, growing bacteria in nutritious liquids.   The problem is that the notion of bacteria as only swimming and floating around went unchallenged for too long.   Finally, in 1977 a microbiologist up in Canada had heard of work being done by others down here and so decided to come see if this were all really true.[cxcix]

It wasn't unexpected that Bill Costerton would find bacteria in the waters of Montana's mountain streams.   Even the pristine rivers up in Canada can be clear as glass and yet harbor bacteria.   And in fact he and his colleagues did find bacteria in the waters, about a dozen in every spoonful on average.   But it's what they found living on the rocks that caught everyone by surprise.[233]

It wasn't as easy an experiment as you might imagine, counting bacteria on rocks, because to get their answer they first had to invent a whole new way of counting microbes.   Everything, even as late as the 1970's, was still geared towards doing things the way Koch and Pasteur had pioneered them back in the 1800's, by counting bacteria that first grew in liquid.[cc]   What Costerton and the others found was that an area of submerged rock as small as a dime could harbor one hundred million bacteria.   In fact, rather than the exception their finding turned out to be the rule…as they found bacteria living on all sorts of other surfaces.   Costerton realized that the bacteria in the streams were spending the majority of their time not swimming in the water, as everyone had assumed for so many decades, but anchored onto rocks and pebbles.

This meant that bacteria had been leading a "double life" all this time. Mostly they were *sessile* (anchored) and they spent only a small minority of the time being *planktonic* (swimming in the water).   This stationary phase is today known as a *biofilm*.[234]   And biofilms weren't confined to streams and rivers, either.   Biofilms turned out to be on all kinds of surfaces folks looked at.   Our mouths have at least 500 different species of bacteria all living side-by-side as a complex biofilm community, some of which aren't found

---

[233] Rocks with bacteria on them make these surfaces slippery.
[234] Costerton coined the term "biofilm".

anywhere else in the body, or in nature for that matter.[cci]   Many of the colorful reds, oranges, and yellows of Yellowstone's hot springs turned out to be biofilms...basically different groups of thermophilic bacteria living in close quarters.

In medicine, biofilms are a menace when they take shape while by growing on pacemakers, artificial joints, and even contact lenses.   They don't seem to be too picky about where they set up shop, either.   By the 1990's it was clear that the majority of bacteria in nature preferred life as a biofilm. And why shouldn't they?   There's always strength in numbers, which is one of the reasons our ancestors in the sea became multi-cellular in the first place: to get large enough so other microbes couldn't eat them.

Early on during an infection, for instance, if only a few bacteria begin making their toxins, the human immune system is able to react fast enough to get rid of them.   But if the bacteria have patience and wait...say until their numbers are sufficiently high enough...then they can produce all their toxins at the same time and simply overwhelm their victim...basically a coordinated attack (septic shock is an example).[235]   And when they join together in these assemblies, bacteria can form more complicated structures, which ensures better trapping of food and more efficient release of their waste products.

And it turns out that only about a fifth of a biofilm is actually bacteria.   The other four fifths is made up of sugars and water, producing a kind of sticky protective jelly they make to keep out viruses, not to mention the occasional grazing amoeba making a living off gobbling up bacteria.[236]   Bacteria in biofilms are constantly organizing themselves by communicating with each other using small molecules they make, and forming tunnels and trenches reminiscent of our own system that transports the blood.

And when food becomes scarce, the bacteria form intricate mushroom-shaped structures, the tops of which are fragile enough to easily break away, allowing some members of the colony to travel far and wide in an effort to pioneer a new biofilm somewhere more favorable.   And even more amazing is that we now think most bacteria in nature have this ability to form a biofilm, and the structures they create are easily recognizable.   There are some scientists who believe biofilms, whether growing on our teeth, inside the gut of a termite, or the rumen of a cow, are every bit as complex as the structures our own cells create when they join together to become tissues like the heart, kidney, and brain during our time in the womb.[237]

---

[235] *Yersinia pestis* waits inside the flea and reproduces until there are enough of the bacteria to make the flea hungry.   *Y. pestis* is therefore able to alter the flea's behavior, causing it take a blood meal from its mammalian host, thus spreading plague.
[236] Some even use the discarded chromosomes of their dead neighbors as physical support.
[237] Bacteria residing in a stationary biofilm use a different set of genes than when they are free-swimming.

Like so many new discoveries in science, this one also came down to the invention of new tools.   Advances in microscopes and computers allowed sufficient strides to be made in visualizing biofilms in new ways.   Lasers mounted on confocal microscopes could light up and scan thin sections of the bacterial colony, highlighting just one layer at a time as thin slices, and then hundreds of these slices could be joined into a three-dimensional image using computers, revealing the streamers of slimy matrix, a true division of labor among microbes taking place in all three dimensions of space.

Recent experiments have also revealed that bacteria living close to one another on solid surfaces can exchange DNA more easily.[238]   It's how bacteria acquire resistance to antibiotics.   The bacteria in the lower portion of the biofilm (the lowest *microniche*) will even change their metabolic state, slowing down their life's processes until they become dormant; like seeds lying in wait.   This can make them less susceptible to antibiotics; harder to kill in other words.

It's been known for decades that cystic fibrosis researchers could take bacteria from a CF patient's lung, grow these same bacteria up in the lab the old fashioned way, like Koch did with anthrax, and that they could kill these bacteria with antibiotics.   But when the same drugs were given to the patient as a treatment, they often had little or no effect.   It turns out the microbes were probably protecting themselves inside the patient by forming a biofilm, deep within the lung.   CF patients can make natural antibodies that bind to and remove bacteria in a laboratory test tube.   But these same antibodies often have little or no effect inside their lungs.   It's probably because the bacteria there are forming a biofilm.   In fact, it may not take that much to tip the balance.   Biofilms inside astronauts who are otherwise healthy yet stressed by space travel can often be found growing in their lungs after returning to the Earth.

There's even a species of bacteria that normally lives in the soil called *Pseudomonas aeruginosa*, which forms biofilms just about anywhere it happens to land, even in jet fuel and hospital mop water[239] and in fact *Pseudomonas* is the most common cause of hospital-acquired infections.   It's no coincidence that the lungs of CF patients often become colonized with *Pseudomonas* given time.[ccii]   It lives on plants, too, which is why hospitals may prohibit live flowers in a patient's room.

Even *Vibrio cholerae* forms biofilms on different surfaces including plankton and various aquatic plants.   According to Stanley, it's another way the microbe survives for so long in the sea without causing outbreaks of disease.   It turns out the bacteria are simply creating biofilms so they can hide out better.[cciii]   Another reason *Vibrio cholerae* forms a biofilm may be to

---

[238] By horizontal gene transfer (HGT).
[239] It even uses the disinfectant as a food source.

shield itself against the strong acid in our stomachs, acid that normally kills whatever we eat, which may be how *Vibrio* sneaks past the stomach to get down into the intestines and cause cholera.

It's possible being able to form a biofilm is so ancient that this ability goes all the way back to when the Earth was still in the process of cooling off and conditions were a lot harsher than today.[240]   Forming biofilms may have been a necessity three and a half billion years ago just for microbes to survive.

So after 400 years, Costerton and the others had finally figured out the answer to the wine-gauger from Delft's question so long ago, of why it was Antony van Leeuwenhoek was able to kill the bacteria he scraped off his teeth using vinegar, but not these same bacteria still in his mouth.   It turns out the ones in his mouth had formed a biofilm.[241] [cciv]

It was already dark outside, which only accentuated my despair as I made my way back down to the dungeon.   Amanda and Chris and my two technicians had gone home for the evening and a new batch of CFTR had been purified and sent off to London.   There was still some leftover Styrofoam on the bench where they'd packed everything up.   Seeing the material there took me back to Egypt and Gamal who had mentioned the way some of the papyrus scrolls in the museum had been found inside crocodile mummies, used as stuffing.   One scroll even had a poem by Homer and another a lost play by Sophocles written on it.[242]

BACK ON THE RANCH

All that was left to do now was wait, so rather than waste another weekend, I picked up Barney and the two of us headed out to my family's ranch at the western end of the Lewis Valley.   The Lewis Valley isn't your typical valley I should probably have mentioned by now so you could have appreciated it better.   Most valleys are river valleys because they are narrow and V-shaped, having been carved by flowing water over long periods of time. The Grand Canyon is like this, as is the Nile Valley, which is even more typical because it's a couple miles wide on average.

But structural valleys like ours appear as if someone's deliberately cordoned off a generous portion of the high plains not with fences, but with

---

[240] There is evidence some bacteria in a biofilm will commit suicide for the good of the colony. Some use toxins to kill their neighbors.   This way the survivors obtain nutrients shed by the sacrificed bacteria to help the colony make it through periods of starvation.
[241] Today, it's estimated 80% of all diseases in the developed world are biofilms.
[242] Papyrus is where our word "paper" comes from.

mountain ranges on all four sides. Our valley is so broad and flat that you can grow wheat or corn just about anywhere without a problem. Structural valleys are a geological phenomenon formed by the happenstance of fault lines combined with tremendous pressures produced by plate tectonics, almost like scuffing up a rug on all four sides. That's the Lewis Valley. There is a river that runs through it, but the river never had anything to do with forming the valley. Instead, over millions of years it acted as a conveyor belt, depositing about a mile's worth of gravel and other sediment washed down from the nearby mountains. And fortunately for us, it's great for growing things in.

My father passed away a couple years after the brucellosis outbreak that led to the culling of our entire herd, so I needed to touch bases with my two brothers, who were busy running things. Being close to home rather than on one of the coasts – which is where most biotech research is done in America these days – is another reason getting my tenure would help out, in fact it's about all I thought about lately. Besides, it's good to be reminded of where you come from once in a while, especially when things aren't going so well in other areas of your life.

When I wasn't helping brand calves or scare off wolves, I reverted back to the studious little brother, spending most my free time lying next to Barney on the bed, writing in my journal. I thought about how even Antony van Leeuwenhoek, when he wasn't relating discoveries to the Royal Society (he wrote 372 letters over the years), sometimes just described his dog, bringing up the curious image of Isaac Newton reading about the janitor from Delft's dog.[243] After turning off the lights, I laid awake a while longer, tossing and turning, thinking about all sorts of things in no particular order... first the yeast project, then the lost anthrax grant, and finally some of the changes science has undergone over the years.

For tens of centuries, no one knew that Khufu's pyramid at Giza wasn't a perfect square at its base and was never intended to be...not until airplanes were invented and someone flew over the top at the right time of day in 1940 and snapped a photograph.[244] It was a secret the pyramid builders kept all to themselves for thousands of years. For some reason they had made each of the pyramid's sides slightly concave towards the center, but nobody knew because the incline was so slight it could never have been noticed from the ground. Even after all the previous explorations since Napoleon, there was still something more to be learned...that the pyramid was actually 8-sided, rather than 4-sided as had been assumed.[245] My mind then shifted

---

[243] Isaac Newton wrote a chronology of ancient Egypt.
[244] Robert Koch was also an amateur photographer and in 1877 became the first person to take a picture of bacteria.
[245] To this day, no one's ever found a pharaoh's mummy in a pyramid.

gears and I thought about how we still don't know where the cholera bacterium comes from exactly, or where it goes when it's not causing epidemics, or why people with type O blood like myself are more prone to getting the disease than someone with another blood type.[246]

I was just about to fall asleep when one last thought entered my head. Not once since losing out on the anthrax grant had I added a new microbe to my collection. But I was home now, I realized as I reached for the alarm clock, setting the small glowing hand for what I estimated would be sunrise. I knew just where there would be plenty of fresh steaming cow pies one could rummage through with a fork and a pair of sterile tweezers before breakfast.[ccv]

IN SOLUTION

During the night a cold front had moved through and with the radio on the fritz there was plenty of time to think during the drive back to campus the following morning. The air was crisp and Barney was standing next to me on the seat without a care as far as I could tell, keeping an eye out for big game through the closed window. One never knows when a herd of elk or deer – hungry after a winter of near-starvation searching for grass on the valley floor – might be foolish enough to wander out into an open field on their way back up into the high country. It was late spring and the creeks were getting closer to the road's edge…all the proof I needed of the expansive snow melt taking place higher up in the mountains, even if all the trees hadn't leafed out down in the valley yet.

Summer was just around the corner and we had four months left to get something done in the lab, and perhaps if I was lucky enough, to submit a publication to a good journal, all before my tenure committee met for the last time in September, so I was at least still in the game. A lot was riding on what the folks in London could do for us – that is if we were ever able to send them viable CFTR again.

After dropping Barney off I headed down to the lab with cow microflora in a plastic vial under my arm ready to be diluted with sterile water and then streaked onto the surface of a fresh Petri dish, purified by colony expansion, and then lyophilized and stored on a shelf.[247] When I got there, Amanda was waiting and I could tell by the look on her face it wasn't good news she was about to deliver. Amanda wouldn't have made a very good poker player.

---

[246] Surprisingly, the actual number of human chromosomes – 46 – wasn't completely agreed upon by biologists until 3 years after the 3-D structure of DNA was determined in 1953.
[247] Lyopholization occurs spontaneously in nature and is how, for example, South American mummies from Incan sacrifices were formed high up in the Andes. Otzi the Iceman from the Italian Alps and mammoths trapped in Siberian glaciers are other such examples.

When the London people couldn't reach me, they'd called the lab and gotten her instead. It seemed we were back to square one. Our CFTR was still as dead as a doornail.

But at least we knew it had to be the enzymes, the proteases from the yeast, which were chewing it up. This time they had run the CFTR on an SDS-PAGE gel and found that the protein had been degraded even more thoroughly than the last time. In fact, it was chewed up into so many smaller fragments during the trip over the Atlantic that there were no longer any distinct bands left on the gel. Instead the protein had produced only a fuzzy blur at the far end, a sure sign that our protein was being fully digested into hundreds of tiny random pieces by the yeast's proteases on its way to our collaborators.

I had Amanda begin another purification that same day. By now they were at least getting good at it. In fact, she and Chris had managed to shave off half a day from the procedure just by having done it so many times.[248] If they and my two technicians got started right away, we could have another sample ready to send off the following evening. I scheduled a group meeting...a kind of "sending off party" is probably what I had in mind. For some reason, it seemed like a good idea. I had only a vague idea of what I'd say there, but it was time we all relaxed. One should always refrain from kicking a cow chip on a hot day, as my grandfather used to say.[249]

This was also the first Monday after spring break and so the students were back from Cancun, or San Diego, or wherever it was students headed off to these days. Which also meant I had a biology class to teach in the morning and a repeat performance in the afternoon, so at least I wouldn't have time to dwell much. It's just as well, I figured. I was almost out of ideas anyway. I hadn't even given Amanda any new instructions. I think it was Einstein who defined insanity as doing the same thing over and over again while expecting different results. Milking a dry cow is how my grandfather would have put it. My only advice was to increase the amount of protease inhibitors in the buffer. That's all. Just increase the amount of protease inhibitors. Geesh, I thought, how hard could protein expression be? Maybe my only real mistake was in believing I deserved my tenure in the first place.

Everything went like clockwork the following day and in the afternoon one-by-one we each made our way up to the conference room on the forth floor above the lab, the same room everyone took turns holding their group meetings in during the week. Small departments share a lot of things

---

[248] By now we were growing yeast continuously in the lab and inducing them to make CFTR, fully expecting to have to do another purification.

[249] My grandfather sometimes based his words on Will Rogers quotes I found out as I've gotten older.

besides gossip and cold viruses and I'd heard recently that Amanda's old boss – the one who'd been so abusive to her – had managed to get a major grant from the government, something about lasers and defense contracts. I was in no mood to hear more, though, not on a day when so much else was already weighing on me.

"So where's Chris," I asked to no one in particular as the group was all assembled except for the premedical student, "At another interview already?"

"He's coming," Amanda answered. "Probably wrapping up the protein now."

So we sat and chewed the fat, literally, careful not to spill any of the crumbs from the cupcake's festive pink napkins while I studied Amanda's patch clamp graphs of the newly purified CFTR. The CFTR channels looked good, as usual, with just the right amount of chloride ions – several million of them every second – flowing through each channel. Marjorie, my senior technician, had gone to school with my mother and so she took it upon herself to bring homemade brownies or anything else she happened to be baking at home to group meetings. Fortunately for us, today was no exception.

Since Marjorie could also have a conversation with a tree, I had the option of turning the meeting over to her, but instead, because a horse is only as brave as his rider, I went up to the board and drew on my knowledge of history, trying to think of a few things inspirational.

I reminded them of how the Pilgrims made an average of just 2 miles an hour all the way across the Atlantic for 66 straight days and how Marie Curie ended up discovering two new elements only after she was denied a teaching position in Poland.

I could tell that that hadn't done much, so I next told them about how it took Napoleon's scientists more than a quarter century after returning from Egypt to decipher the Rosetta stone, and how if a little known archeologist named Theodore Davis had just kept digging for 6 more feet instead of giving up back in 1914, then he would have been the one to discover King Tut's tomb instead of Howard Carter.

That didn't draw forth any smiles either, but fortunately I still had Lewis & Clark to go to since they had suffered many a setback on their trip. I mentioned how Lewis was shot at by a Blackfoot Indian and that the bullet passed so close to his head he could feel the breeze as it went by, and how his dugout was nearly swamped, in fact the pair's journals and scientific instruments went right into the river, and how Sacagawea had, with her newborn baby still on her back in another canoe, managed to retrieve all their belongings including the journals.[250]

---

[250] She and Clark's slave York were the only unpaid members of the expedition.

Still no discernable reaction, so I next described how Lewis spent the entire second summer going up the Missouri collecting new plants and then drying them out by the campfires at night after pressing them inside a book. To lighten their load for the trek over the Continental Divide, the expedition cached his samples underground near present day Great Falls, Montana, lining the hole with dry grass and sticks, and wrapping his books in oiled cloth and sealing it all inside wooden barrels. But the captains hadn't counted on the river rising as high as it did and so by the time they returned the following spring on their way back to St. Louis, water had gotten into the barrels and destroyed nearly everything. They managed to salvage the valuable chart of the Missouri River Clark had compiled but the months of hard work Lewis spent collecting plants were all for nothing. Still undaunted, Lewis managed to recover and do some more collecting on his way back, ending up with some 200 plants, many of them entirely new to science.

I was just choosing between mentioning Jefferson and how he spent 40 years building Monticello and yet never finished it, and the Old Kingdom pharaoh Snefru who had not one but three pyramids constructed because the first was too massive, causing the outer casing to collapse, while the second was too bent at the top when Chris entered with beads of sweat on his neck. He was breathing heavily as he sat down and so I turned to him an effort to keep what little conversation we had going.

"So where were you?" Everyone else turned to him as well. He had obviously just run up the stairs from the dungeon and was still out of breath.

"Me?" he looked around, surprised to be the center of attention. "I was just finishing up, Dr. Ketchum. I always make sure our CFTR is in the good water before sending it off. I don't know why more research groups don't use it, though." Everyone then turned to me in an effort to gauge my reaction to this latest revelation.

"The good water, Chris?" was all I could squeeze out, too stunned to question him further.

"Yes, the good water. If we're going to send a protein off thousand of miles, it ought to be in the best stuff we have. Don't you think sir? Don't worry, though. I added plenty of buffers…to compensate."

"Could you show us the good water, Chris?" I asked, and so without another word my entire group headed off single file behind Chris, the pied piper leading the way down to the basement and then into one lab after another, traveling through the side entranceways joining each lab until we reached the very last room, the one set aside for an elaborate water purification system all the research groups in the building could get water from when their water had to be especially pure. It was one of those contraptions exhausting just to look at for it resembled a plumbing experiment, having pipes running back and forth against the wall, various filters for the water to percolate through on its zigzagging course down towards the floor. The entire room was eerily silent with only the sounds of Charlie running a floor polisher somewhere overhead.

"So you see Dr. Ketchum? I always change the old buffer and replace it with this new stuff here. It's gravity fed." He flicked on the pipes with his fingers...smiling proudly, still unaware he may have done anything out of the ordinary. "I use it just before sending our samples off."

Marjorie stared openmouthed, first at the plumbing on the wall, then at Chris, who was finally growing uneasy. I think he half-expected a pat on the back and was taken by surprise when we weren't all forthcoming with approval.

"But you never marked any of this down in your notebooks, Chris. Why not?" Amanda demanded.

"I guess I didn't think it was that big a deal," he shrugged, hiding his neck probably in an effort to look less vulnerable, the way a turtle uses its shell for cover.

I moved closer to the contraption so I could have a better look, tracing each of its pipes from beginning to end. Distilled water entered at the very top from a spigot fed by gravity across the wall down through dust-covered plastic pipes to the first of the PVC filters, then a deionizer, then another water filter after that, and then a third, the plastic pipes zigzagging all the way like a switchback trail going down a mountainside, lower each pass until finally reaching the last filter by the floor. There was even an ultraviolet light that shone through part of a clear window in the last section of pipe, useful for killing any spores that, however unlikely, might have snuck past all the other filters.

It's then that the thought occurred to me to have a closer look at the nozzle. I vaguely seemed to recall something similar happening to another group back in graduate school. Sure enough, there inside the small diameter of the plastic tubing where the last drop of ultra-pure water fell out into a bucket on the floor, inside the spigot still exposed to the air, was a brown fuzzy film that looked like algae from someone's aquarium.

I cupped the end while bending down...noticing the unmistakable odor of ripe fruit...all the evidence I needed of a *Pseudomonas*. My pulse quickened. The implications were both good and bad at the same time. These bacteria made enzymes to consume all kinds of things and anything was better than what we had to go on the last few weeks.

"You see that film?" I asked, tapping the outside of the clear tube with my fingers as all moved in like football players forming a huddle to have a closer look. "This dark stuff is a biofilm."

I turned to Chris, who was inching away now, his shoulders slumped, making him look defeated. And he had good reason to feel that way. It was clear to everyone including him that his overreaching may have cost us a lot of time... and maybe even my tenure.

The expensive water purification system was usually worth the price of upkeep, but like every chain it too had its weakest link. And the bacterium had found it without any problem at all. Bacteria are good at finding opportunity... it's what they do all the time in nature. In fact it helps explain why they're billions of years older than we are and practically everywhere, too.

"*Pseudomonas* live on almost anything," I began.

"It looks like dirt," Marjorie added, kneeling down to sample the smell at the end of the tube for herself.

"But it's an opportunist that would put any scientist to shame. That colony started off as a single cell that blew in through the window riding on an invisible speck of dust. The dust landed up inside the nozzle and the bacterium simply took up residence there. Out of every ten infections someone gets in a hospital, one will be a Pseudomonas. They eat practically anything... even us."

"That would include our protein, wouldn't it?" Amanda asked.

"It could. A small amount of the biofilm could have been shearing off every time Chris got some water. It would explain a lot. It would have eaten our CFTR all the way to London, even consumed the protease inhibitors as if we'd packed them for a picnic lunch."

The silence in the room was noticeable and I reached out and patted Chris on the shoulder. He had by now lost some of his color. It was his first semester in a lab. "At least we know what the problem might be. That's a lot more than what we've had to go on."

I took another look at the system. Bacteria are adaptable and natural selection had shaped an organism that was capable of finding the Achilles heel in even the most elaborate device for purifying water that government money can provide. But as my grandmother still takes delight in saying from time to time, "For that kind of money, it that oughtta make cornbread too."

We threw away the latest batch of CFTR and purified a new one, this time leaving out Chris's improvised step and then sent it all off to London. As expected, everything worked fine. For the first time in four long weeks we had whole, functional, yeast-made CFTR safely in the hands of our collaborators who were now ready to do structural analysis on it. The ball was firmly in their court, not ours, so at least we could feel good about that.

That night back in my apartment I wrote in my journal again, the same one with the blank pages in the back because I'd stopped writing for pleasure in college. While getting a college degree late due to my father's untimely death (I had to help out on the ranch for a few years), the first in my family to get a degree, then a second one, I'd somehow forgotten how soothing writing can be.

I wrote down some of the other curveballs nature has thrown at us over the years while they were still fresh in my mind. The strongest sources of energy known in the universe since the Big Bang – Gamma Ray Bursts – weren't even noticed by astronomers until satellites were sent up during the Cold War to detect nuclear weapons testing by the Soviets. Dr. Peyton Rous (the MD kind) was largely ridiculed for claiming he'd found viruses that could cause cancer way back in 1910. He got a Nobel Prize for it, but not for half a century after he did the experiments with chickens to prove it…not until 1966. Fifty-six years is a long time to wait for recognition from your peers.[ccvi]

It's easy to get the wrong impression of how science is done by watching TV and movies and seeing only the success stories. It's like the way we tend to think of the ancient Egyptians as being obsessed with death when just the opposite may have been closer to the truth. Like many ancient people, they held festivals and banquets whenever possible.[251] One pharaoh even built a resort city. It's because most of what we know about the Egyptians comes from their tombs. That's the reason we think of them like that. Even now, Egyptologists still can't say for sure whether there are any more tombs like Tut's left to be discovered in the Valley of the Kings. There might be another, or an even better one just waiting for the next Howard Carter to come along.

No one's found Alexander the Great's tomb in Egypt either, even though most historians believe he's still in Alexandria somewhere. And I'd also read somewhere how amorphous silica was uncovered for the first time on the surface of Mars only because the rover Spirit's wheels had gotten stuck and then dragged along until it formed a shallow trench.[252] Or what if Lord Carnarvan hadn't driven his car into a ditch in 1901 and gone off to Egypt and its warmer weather to recuperate? He probably wouldn't have met and been impressed by a young guide named Howard Carter who was then working as an artist.

Which led me to my next thought, the visit Stanley Beaufort had paid earlier in the day. It was unusual for Stanley to come all the way down to the dungeon just to make small talk in the middle of the day and at the time I did find it unusual, even more so when he'd worked in a question about whether or not I had any backup plans…in case I didn't get my tenure. I'd had the rest of the day to ruminate on this, which had by evening grown into the nagging suspicion that perhaps Stanley knew something and it was just his way of giving me a heads-up about it. Perhaps he'd gotten wind of someone newly available, someone the department wanted to hire in my place.

---

[251] During the reign of Claudius, ancient Rome had 160 holidays a year on its calendar.
[252] Amorphous silica is a sign Mars once had hydrothermal systems, like Yellowstone's hot springs, places we now know life can exist in.

Oftentimes, faculty will join departments when someone already established puts in a good word. Maybe a certain someone on my tenure committee, a fellow scientist who smiled and made small talk with me each time we passed in the hallway, wanted to work with someone they usually collaborated with, someone far off at the moment, but out of a job...and if they could just get rid of me...

I tried to banish the thought from my mind. Still, there was no escaping the fact that times were tough, what with the government cutting back due to the Iraq War and Afghanistan and Katrina down in New Orleans. I'd even read recently, it was in either *Science* or *Nature* I couldn't recall exactly, about how universities were laying off professors with twenty years experience all because they weren't bringing in the grant money like they used to. Twenty years of hard-earned experience being deemed worthless almost overnight seemed too unbelievable to accept. Universities sometimes refer to grant money as "soft money" and have come to rely on it. NIH had been funding big labs for a long time now, big labs with ambitious projects like sequencing the human genome, and now the cancer genome, and for some reason they had forgotten all about those of us just starting out in science.

Then there was the gossip making the rounds lately about someone that had recently won a Nobel Prize using a protein from a jellyfish that glows green,[253] Martin Chalfie was his name, and how Chalfie had used this jellyfish protein as a tag to measure the output from the worm as the worm switched its genes on and off during its embryonic development,[254] how word had gotten around that Chalfie had initially gotten the jellyfish protein's gene from a fellow scientist named Douglas Prasher, someone who was currently working at a car dealership of all places. Scientists were tracking individual proteins around inside a cell, something that would have been undreamt of just a decade ago, while the Ph.D. who found the gene that made it possible was driving a van because he couldn't find a job in science again...or so I'd heard.

I looked over at the clock on the nightstand next to me by the bed, then out the window and made a quick calculation in my head. It was still dark in Thistle, but the sun would be coming up in Cairo soon. Tahany managed somehow to make ends meet...and on a much smaller budget, too. I had to be more resourceful that was all, I decided, and finally managed to put the whole matter out of my mind for the rest of the night. NIH isn't the only game in town.[255] There's more than one way to raise a hog or crack an egg.

---

[253] Called GFP, for green fluorescent protein.

[254] Almost a half million dollars from Nobel Prizes was awarded all together for work with GFP.

[255] Before governments began funding science after WWII, scientists relied mainly on businesses & benefactors. Galileo, for example, was financed by the Medici Family of Florence. But as my grandfather used to say, 'yesterday's river won't turn the mill.'

I put the pen down and turned off the light, resting my eyes as I thought about some of the other surprises nature has thrown at us over the years. It wasn't until a professor at the University of Illinois began sequencing and comparing the genes of what he assumed were all bacteria that a new kingdom opened up before his eyes, an entirely new domain of life separate from bacteria now called Archaea.[256]   Then there were the other no less intriguing discoveries made by DNA sequencing in recent years.   The hippopotamus turns out to be more closely related to, of all things, a whale than they are to a pig, and a trout is more closely related to you and me than it is to a shark, even though a shark and a trout are both fish that inhabit the water and have gills.[257]

There is even a species of amoeba that has 200 times more DNA inside its nucleus than our own cells do and Christian Graham discovered Gram staining all by accident it seems.   There had to be lots of things left worth discovering, I reassured myself.   In the 1970's an original journal from one of Lewis & Clark's privates turned up in someone's attic.   In fact, historians now believe there are other journals from the Corps of Discovery still waiting to be found.[ccvii]   Ancient Egypt had over 320 pharaohs, yet only 70 of their tombs have ever been located, and the Egyptians never did discover where the source of their Nile was, choosing to believe instead that its waters originated from somewhere deep inside the Earth.   The ancient city of Herculaneum was rediscovered when a farmer dug a well in the 1700's and came up with pieces of colored marble instead of water in his bucket.   And one third of Pompeii still lies buried beneath lava from Vesuvius.[ccviii]

Lastly before drifting off to sleep I thought about the past few weeks and Chris and the mistake that had cost us all so much time.   I was his first mentor.   Amanda had already been broken in by the time she'd joined me. How I behaved towards Chris could have a lasting impact on him the rest of his career.   Labs are always intimidating places when you first join in. Everyone walks around as if they know everything already, working with equipment that looks impossibly complicated.   It's hard to see the forest from the trees sometimes, especially when you're first starting out.

Ordinary tasks we take for granted, like pipetting liquids from one jar into another without contaminating or spilling any of it, and the tricks to keeping unwanted germs out of Petri dishes, even weighing powdered broth out onto a piece of paper, these were not so easy to master the first time either.   It's easy to forget that.   Anyone who's ever left a slice of bread out on a counter

---

[256] This discovery necessitated the establishment of "domains" to categorize life.   There are now 3 domains that all of life falls into: Bacteria, Archaea, and Eukarya.
[257] We all had gills at one time during our development in the womb.   Usually they close up before birth but sometimes things can go wrong and people are born with a gill slit or two.

top for too long knows that microbial spores are floating around in the air everywhere.   We inhale them with every breath we take.

Aristotle once said that "nature does nothing uselessly", and when you see the rainbow of colors growing on a stale piece of bread, you begin to gain a real appreciation for the diversity of the microbial world around you, and for the strength of your own immune system as well.   And besides, if not for Chris's mistake, I may never have gotten to see the mountain lion that day high up on the pass.   After all, both my grandfather and my father went their entire lives out without ever seeing one.[ccix]

~ ~ ~

PART 4: SUMMER SEMESTER

A MEASURED PACE

It was by now the middle of summer and the lecture I found myself preparing this particular morning for the *biology for non-science majors* class held special significance for me only I was aware of.[258]   Most professors turn their course notes over when they retire, and the professor who taught the course before me had done this too.   He taught mostly out of the book and his notes reflected it.   Without questioning things, I also went along with the book for the first couple of semesters.   But by the time my first summer rolled around it was clear I'd fallen into a trap.

I've always believed that a good professor supplements their coursework with personal experiences in life, but for some reason I was afraid to.   Maybe I was too nervous, I reasoned.   Overconfidence has never been a problem for me.   People in Montana don't like too much confidence anyway, maybe because all that optimism reminds them of "big city desperation".   At least part of it, I knew, was that I didn't want to make any mistakes.   No one wants to look foolish in a room full of other people, especially when they're your students, the ones who are supposed to look up to you.

It was on that day three summers ago when I was about to give this same lecture – an introduction to metabolism – that things changed.   I had decided to go ahead and "wing it" and so I took along just a single sheet of paper with some names and dates scrawled on it.   No heavy textbooks or elaborate slides, no precious minutes spent fiddling with the projector on this particular day, and I felt freer just walking out the door.[ccx]   That morning three years ago it would all be off the cuff.   I figured if anything went wrong – if I froze and the words didn't come for some reason – I could always say I had an important experiment going back in my lab and dismiss class early.   I doubted if anyone would have complained.   It was the middle of summer, after all.

In truth, I could have done the entire metabolism lecture without even mentioning the word energy more than a couple of times.   It's surprising how much you can learn about biology without knowing hardly anything about energy, maybe because the concept is so abstract to begin with.   Scientists throughout history including Newton, Galileo, and Copernicus knew next to nothing about energy and yet they accomplished far greater things than I ever would.   The word itself wasn't even coined until 1807, yet I had decided to

---

[258] The geology department has a similar class for non-science majors they call *rocks for jocks*.

make energy the focal point of the day's lecture, mostly because it presented such an interesting angle and besides, energy is common ground for so many experiences in life. Energy's pretty useful that way. [259]

And energy is a strange enough concept as it is. No one's ever seen it directly, we only know about it by what it can accomplish; and then only when it changes from one form to another, which is pretty strange when you think about it. If energy were a person, its favorite holiday would probably be Halloween, for energy likes to take on disguises, hiding out in different forms. A simple pendulum when it swings back and forth constantly trades some of its potential energy (obtained by virtue of its mass & height at the top of its swing) for kinetic energy, the energy of motion that it gains as it reaches the bottom and so this conversion goes back and forth like this, potential energy changing into kinetic energy and back again with each swing the pendulum takes.

As a student I had only a vague notion of what energy was even though I knew how to memorize the equations. Looking back, there wasn't anything to be ashamed of, though. Just about everyone doesn't understand energy. Even the Nobel Prize-winning physicist Richard Feynman claimed that energy is more of a mathematical concept, an abstract idea, nothing concrete like the mechanisms scientists usually prefer to study.[ccxi] And even Einstein wasn't able to come up with a mechanism to explain how energy works, or where his formula used to keep track of it came from exactly. One of the reasons the word wasn't even coined until 1807 was that accurately measuring changes in energy requires tools that weren't available to Newton and the others, simple instruments found in most freshman chemistry labs today like thermometers and bomb calorimeters. Before the 19th century, the idea people had of energy was a fuzzy one, not quantitative like a solid mathematical framework can provide.[ccxii]

Heat from fire is an everyday example of what happens when energy changes, in this case from chemical into thermal energy. Heat is often referred to as thermal energy…the energy of motion…yet heat is not really a noun so much as a verb, one that's kept our ancestors safe for thousands of years, allowing them to spread out during the last ice age some 40,000 years ago.[260] But by the 1800's, heat would become the most important window scientists could use to look into the strange world of energy and form some

---

[259] The term "energy" can't be found in any of Newton's Laws of Motion. The physicist Neils Bohr once said, "nothing exists until it is measured" and after Thomas Young defined energy others set about trying to measure it. Young is the same scientist credited with helping decipher the Rosetta stone.
[260] More correctly, heat is the *transfer* of thermal energy. An object cannot possess heat (since heat is a process, not a thing). On the other hand an object can possess *thermal energy*. / The first good evidence for humanity's controlled use of fire is from South Africa about 1.5 million years ago.

kind of a picture of it. And as it turns out, ordinary heat wasn't so easy to understand either, yet by the mid-1800's they could at least keep track of it, and pretty accurately even by today's standards.

Ancient people saw fire more as a purifying influence. Hindus cremate their dead because they believe fire is still necessary to liberate the soul from the body, while trial by fire was a common ordeal the gods in ancient Greece proved themselves worthy. Dionysus was cooked by the Titans...Hercules struck by lightning as he ascended Mt Olympus...Prometheus sentenced to eternal punishment by Zeus for stealing fire and giving it to mortals...while the Cherokee believed that a spider woman built a web and snuck into the land of the light, bringing fire back to humanity deep inside a clay pot.

To ancient Greeks like Aristotle, fire was one of the four elements, fire's opposite perhaps not surprisingly being water.[ccxiii] Fire held special significance because it was the only one of their elements no mortal being could survive in and in fact Aristotle believed water and fire were the two most active of elements. I've often wondered how surprised he'd be to learn that what takes place inside an ordinary candle's flame is chemically what happens inside the bodies of living creatures including people – oxidation – but at a much slower, more measured pace. I creased the single sheet of paper and slid it into my top pocket, definitely traveling lighter these days. It might come as surprise to a lot of folks, but no one ever teaches a newly minted Ph.D. how to be a professor. They just tell you you're qualified to teach and so off you go.

This particular morning had another thing going for it besides being an anniversary. It was also Friday and I'd been planning a two-day hiking trip into the heart of Yellowstone's backcountry, something we'd done as a group at least once a year since I joined the faculty. My department had a special permit for bio-prospecting thermal features – looking for heat-tolerant microbes – a permit that allowed us to venture into places not detailed on any map sold in the gift shops around Old Faithful. I closed the door behind and stepped out into the dungeon, its hallway more welcoming than usual with the warm July sun reflecting off the metal staircase. Just down the hall Amanda was talking with Marjorie, one of my technicians, who was leaning up against a wall listening to her. After seeing the two of them I had the sudden urge to wince, knowing all too well why they were out there.

SUBZERO CFTR

My lab had recently become a source of ridicule within the building. Very few said so to my face, but I was sure now we were the topic of light-hearted banter in every research group.[261] Because our collaborators in London had asked for – no demanded is probably more accurate – additional protein, several times more, I had to call in some favors.

Producing more CFTR meant growing up more yeast, perhaps 20 gallons in additional broth, and as a result of my scrounging around for shakers we now had an interesting assortment of them...all sizes and shapes...speeds and degrees of decomposition...each rattling away to the tune of its own maker. The smaller ones on the countertops were spinning in confined circles, while the larger ones were doing their share of the work in more generous figure-eight-like patterns on the floor, as if daring anyone to get too close. Another was of the roll drum-type, tumbling our yeast like one of those barrels people use to pick bingo numbers out of in church, while another more traditional shaker was wiggling back and forth behind the laminated airflow hood, the same bench with the glass window that slides up and down and we used to streak our Petri dishes behind, back when we had the luxury of keeping things sterile.

But now, because of the lack of space, we were adding more and more antibiotics to the media and streaking our dishes either in the lab next door if it wasn't being used or on top of the centrifuge, risking contamination. Amanda even had a smaller shaker of yeast shuffling away in a spot cleared off next to her desk. Together their springs and belts made such a racket of noises that it was no longer possible to get any thinking done in the lab.[262] "Like a couple of jackasses trying to escape from a tin barn" was how one passing student had described it.

A colleague had even suggested my lab had taken on the appearance of one of those booths at a symposium where you get a free pen and they try to sell you the latest equipment you don't need. Another more kind-hearted soul took the position that my lab resembled a science museum. As for me looking at all the shakers and spinners in various stages of their lifecycle, it reminded me of the way astronomers back in the 20th century were able to gaze out into the nighttime sky at all those stars and conclude that our sun was about half-way through its own lifetime.

One decrepit shaker was pockmarked, with rust making inroads into its metal surface, while another newer one was still wearing its price tag, yet a

---

[261] The average size of a research group in a small department like ours is about 8 including the professor. In some large labs (for example in industry and at NIH) there can be so many workers that the principle investigator in charge may not know all their names.
[262] We even blew a couple fuses.

third had taken on the disturbing appearance of a rickety old Ferris wheel. The most ancient was a spinner that would have been condemned to the dumpster if not for the reprieve I gave it. We kept that one on the floor in the back room because in order to slow it down to a complete stop we had to use the side of our shoe whenever we returned a flask of yeast or took one off just to change the broth and it sent shudders down my spine thinking of what would happen if a safety inspector happened by at the wrong time. Everyone knows they have a 6th sense.

All in all, I was feeling more marginalized than ever, officially existing on the edge of academia now. My group members were easy enough to spot these days, the ones with the headphones in their ears. But if the ancient Egyptians could build the great pyramids without a single wheel, I reassured myself, then the least we should be able to do is grow up several milligrams more CFTR in the next couple of weeks.

The reason the London people wanted more protein had to do with working out all the details for the experiments that were planned. The idea all along had been to use cryo-electron microscopy – or "cryo-EM" for short – which is a relatively new tool for looking at the 3-dimensional structures of large, complex proteins like CFTR. Unbeknownst to me at the time (yet well known in structural biology circles) was the fact that many of the details have yet to be worked out with cyro-EM.

As a relatively new procedure, the London group needed enough "throwaway" protein to figure out all the small but important details like how thin to cut the slices of CFTR-embedded ice with a diamond knife and how strong to make the electron beam without damaging the delicate structure of the protein they would be probing with the free electrons, and so on. Every protein is different, it seems, when it comes to determining its 3-dimensional structure using cryo-EM. As for me, I also had to remind myself to keep enough pure CFTR in reserve for Stanley Beaufort's biochemistry experiments just across campus.[263]

So as in 1804, when no American had yet laid eyes on what would someday be called Montana, in 2008 no scientist had yet seen the full, 3-dimensional structure CFTR. Meaning that no one knew for certain where all the loops, nooks, and crannies would turn out to be. Or the portions of the protein that hold onto small ATP molecules and use this energy to power open the pore or to close the pore, or where the corridor lies within the protein that allows the chloride ions a free passage through the cell's membrane, or exactly how CFTR goes about this job of moving salt around the body – its

---

[263] Stanley's lab was in a different building because when he first came here, there was no room in the microbiology building. He ended up liking it in the physics department so he just stayed on there. I have a hunch he enjoyed the chance to collaborate with more quantitative minds.

mechanism in other words – and perhaps most importantly, how all of this goes wrong in cystic fibrosis.   No one knew what good, structurally sound CFTR looked like, let alone what secrets the mutated structure might reveal.

The kind of experiments others had been doing up until then were more piecemeal, more akin to probing than gazing, biochemistry experiments reminding me of a blindfolded city person trying to figure out the make and model of a tractor just by bumping into it occasionally.   Many had tried to solve the structure of CFTR.   "It's a difficult nut to crack," as Oskar Mosbacher had pointed out during our meeting the previous winter, and as everyone knew by now, I was in need of a Hail Mary.   There was at least one biotech company on the West Coast that had already spent millions trying to catch a glimpse of 3-dimensional CFTR using more traditional methods.   If we could succeed on a meager protein expression budget, our landmark paper would make more than a ripple within the structural biology community, or at least cause a nice enough stir.

I closed my eyes in an effort to picture my interview with the editors of *Nature* explaining how we had accomplished it.   If we succeeded where investment-backed biotech companies and well-funded government labs had so far failed, it would turn more than a few heads and no doubt a nice string of publications would follow our seminal paper, maybe the structure of our CFTR on the cover of *Time Magazine* as "Molecule of the Year" for a bonus.   We'd probably be offered grant money without applying for it, from foundations eager to bring us under their tents.   My tenure, of course, would be a foregone conclusion by then.   Sometimes it pays to get lucky even if it is only once in a while.   The year 121 BC was such a good time for making wine that the Romans were still drinking the vintage some 200 years later.

The problem with seeing the structure of CFTR so far had to do with the requirement for x-rays.   These short, energetic photons of light are the most common way to probe the shape of a protein these days and have been since the 1960's when hemoglobin was revealed for the first time.   X-rays were the very first way anyone ever saw the makeup of a protein (or DNA) close enough to tell where its atoms were.   Unlike cyro-EM, working with x-rays requires that the protein first be *crystallized*, meaning you need millions of them and each of these identical protein molecules have to all line up next to one another, packed together like a…well…like a well-ordered crystal.

Unfortunately, CFTR seems to be one of those proteins that doesn't like close quarters packing.   Perhaps it wants to form a fried egg when it gets too crowded; has to associate with others like itself.[ccxiv]   No one knew for sure.   There is still a lot of trial and error in structural biology these days; large amounts of luck involved, too.   So basically, if you want to do traditional x-ray studies on your protein, you're going to need to coax your protein into forming crystals first.

The promise is that a molecule's shape will reveal clues about what kind of job it performs inside the cell, and perhaps just as importantly, *how* it does this job.[264]   Someday it might be possible to use a computer to figure out the shape of a protein just by tapping on its keyboard the sequence of amino acids making up the protein, use some software to build a model of the protein and figure out what kind of job it does and how the chain folds up into the unique shape it has and so on.   But that's still a Holy Grail in science. No one's yet figured out a way to predict a protein's 3-D structure just from knowing its linear sequence of amino acids, but if they do, it's a sure fire bet there's a Nobel Prize waiting for them.[ccxv]

The little voice inside was reminding me to think ahead as I glanced around the lab trying to figure out where we could possibly squeeze in any more shakers if the need should arise.   In graduate school, I knew a group that simply moved their dishwasher-sized shakers for growing *E. coli* right out in the hallway.

Scaling up to purify more protein can open up a Pandora's box on so many levels, though.   Glass chromatography[ccxvi] tubes that used to flow nice and evenly during purification can become clogged with cellular debris (membranes and tangled chromosomes, for example) not unlike a backed up kitchen sink does after so many potato and carrot peelings during the holidays.   Proteases can contaminate batches of protein during harvest and it only takes a few enzymes to destroy all of your precious protein, and since enzymes are catalysts, a little goes a long way (catalysts get recycled).   Then there's always the fear your protein will end up so concentrated after purification that it will aggregate and form a "fried egg" all by itself.

Drug companies that produce medicinal proteins worry about this kind of thing all the time.[ccxvii]   Antibodies are the new, targeted, "magic bullet" drugs in cancer research – Herceptin comes to mind for breast cancer – and I was sure a company like Genentech, the maker of Herceptin, would rather produce a small drug like aspirin any day than have to deal with large, floppy, vulnerable proteins.

The whole idea for the collaboration with the London people came after I'd read where another group had succeeded in using cryo-EM to determine the 3-D structure of anthrax toxins, but these bacterial proteins are significantly smaller than CFTR is because their only job is to, like guided missiles, seek out mammalian cells and then invade them.   So most bacterial toxins, like those that cause anthrax, cholera and other diseases, tend to be smaller, simpler and easier proteins to work with compared to an ungainly ion channel

---

[264] But not always.   The structure of DNA, for example, told Watson & Crick nothing about how DNA gets *translated* into protein by the cell.   Biochemical work still needed to be done to figure this out (Crick would later do some of this work).

like CFTR. Even in this day and age of molecular biology, the 3-dimensional structures of large ion channels are scarcer than hen's teeth.

The big problem with electron microscopes is that their beam of electrons is so energetic. The electrons used to bombard cells and proteins act more like bullets shot from a gun and can easily rip a protein apart. Subatomic particles just don't know their own strength. What happens is that these energetic electrons released from the "gun" inside the microscope hit and knock the outer electrons zipping around the carbon and phosphorus atoms of the protein sample as if they were billiard balls, and this collision and subsequent kicking out of an important electron makes the protein so unstable that it literally falls apart and reacts (combines) with anything else nearby on its way back down to the ground state (all things in nature want to get back to their lowest energy level, which is also why water runs downhill in case you were wondering).

So being able to see 2,000 times better than Antony van Leeuwenhoek did with even his best microscope brings with it a certain price tag not measured in dollars. There are a few workaround solutions. But the metal stains used to coat and protect a protein or a cell from all these fast-moving electrons can also dehydrate your sample, shrinking it into unnatural shapes. In short, these harsh stains will form artifacts that can make your data practically worthless. Which is what's so promising about cryo-EM compared to traditional electron microscopy: cryo-EM uses extremely low temperatures to first stabilize the protein in what's called "vitreous ice", without the need for any harsh stains coating it, meaning these electrons can then be shot at the naked, unstained, unadulterated protein to probe its natural 3-dimensional structure thanks to the low temperature. All in all, it means a lot less GIGO to worry about here (garbage in, garbage out).[265]

On the way to class I thought about how I probably wouldn't have recognized Amanda these past few days if I'd seen her smiling so much just a few months ago. Even though we'd all been busier than a pack of coyotes in a meat house with the cystic fibrosis project, Amanda seemed to have blossomed under the increased pressure. Maybe it was because Josh had been going through a good period...still it was tempting to give some credit to the CF project. To be sure, the thought of Amanda having to find a new mentor if I lost my bid for tenure found its way in and out of my thoughts these past few weeks. Chris would probably be in medical school, my two technicians had enough experience that they could get jobs again even if it took them a couple months to locate someone with enough grant money, but

---

[265] Water can form at least 15 different types of crystalline ice, each with its own properties, as well as two types of non-crystalline, or amorphous ice (so-called "vitreous ice" which is similar to glass), one of which is found on comets. Most water in the universe is amorphous ice not found naturally here on Earth.

Amanda...her blue eyes sparkled when she talked about the project, dance almost, like they did when she described Josh's new front teeth or his recent discoveries in the backyard. She had a sense of purpose, more so than before, and it gave me energy I didn't know I had too. I hated the thought of my failure taking all of that away...not when we were getting so close to obtaining some real results.

ABILITY TO DO WORK

As I trudged up the hill, the geology department came into view, the place where my bleary-eyed class would be waiting, and I ran through the energy lecture one more time in my mind to fix the highlights that would serve as mile-markers. I would begin by telling them my grandmother's story about how, when she was a little girl she and her sisters would run down to the corner where her ranch's driveway joined the main road just to watch the "horseless carriage" pass by, the one with their neighbors in it on their way to church. Automobiles were still such a novelty in 1911 that few could afford them, and even the ones that could only drove them on Sundays. In fact, for almost the whole of human history, animals were our only real source of energy, the only way to get things done on a ranch as well.[266]

Sailboats in ancient times were built to harness the energy air possesses when it moves and were in fact the first method of reliable transport that didn't depend on animals.[ccxviii] The Egyptians used the current of the Nile as it flowed northward towards the sea, and then on their way back upriver they would unfurl their reed sails to harness the power of the wind, allowing its energy to carry them all the way back to Thebes because, conveniently enough for Egypt, the prevailing winds off the Mediterranean were to the south, in the opposite direction as the river's current...yet another reason their civilization lasted so long...this ability to trade and communicate so reliably.[267]

Progress in history seems to arrive more like steps in a staircase than an inclined plane and so within a couple thousand years, people finally figured out how to use the kinetic energy of moving water to turn waterwheels connected to grindstones. By the time Christ walked the earth, waterwheels were being used to irrigate fields and grind grain for bread. Its possible carpenters in Christ's time even used moving water's energy to saw wood.[268]

---

[266] It's possible up to 30% of the inhabitants of ancient Greek cities were slaves.
[267] The Egyptian hieroglyph for "traveling south" was a picture of a boat under sail, while the hieroglyph for "traveling north" was a boat having oars but no sail.
[268] The Egyptians had waterwheels for irrigating fields thanks to the Ptolemaic dynasty, but they used donkeys to turn them. Their wheels had a series of buckets attached that could lift

The Romans seldom took advantage of water's kinetic energy because slave labor was so cheap. At times, a slave could be bought for as little as a flagron of wine in Rome, still they did build a notable waterwheel complex in southern France that had some 16 wheels for grinding grain.

When people found they could put animals to work on farms, it led to what's been called the *second agricultural revolution* and to the first significant population explosion in human history.[ccxix] Villages now had a food surplus and could even sustain a division of labor. For the first time what we would recognize as cities began taking shape in Mesopotamia with names like Ur and Eridu. The Egyptians built the pyramids with only human muscle, even though they had animals like sheep and cows to grind wheat into flour for making bread. They also used animals to stomp wheat kernels with their hooves into the ground during planting by pushing the seeds into the soft soil after the Nile receded each year.[ccxx]

The most common definition of energy in physics textbooks describes it in terms of what it can do. They usually start off by saying something like *"energy is the ability to do work."* At one time everything that existed in the universe was energy if you go back far enough... all the way to just after the Big Bang... and some of that energy left over still bombards us 13.6 billion years later. In fact, you can see some of it on your TV screen in the form of static.

One of the most concentrated sources of energy in the universe is a Gamma Ray Burst, energy released as a whole star is consumed by a black hole.[269] But the *ability to do work* is a useful definition here on Earth because it applies to so many situations. Most forms of energy, if harnessed, can be made to move objects (do work), and movement is something that can easily be kept track of.

To get an idea of the amount of energy something possesses, one simply measures how heavy the object it moved was, and how far the energy in question moved it. This may sound trivial, but it meant that humanity, starting sometime around the mid-1800's, finally had a way of keeping track of – and perhaps even more importantly comparing – all those different forms of energy on equal footing with one another. Not until then was it truly appreciated that they were interrelated, and that energy is "conserved", meaning that it can never be created out of nothing. Like matter, energy always has to come from somewhere else. And conversely, one cannot eliminate it out of existence either; energy simply changes form, sometimes

---

water higher than the level of a canal. Egypt's population increased to 6 million after this one innovation.
[269] A single GRB can be the equivalent to turning the earth's sun into pure energy all at once. It's possible the energy from GRBs have caused mass extinctions in the past. Our sun is no slouch, though. It turns 4 million tons of its own matter into energy every second.

even taking refuge as matter.[270]   It's as if God has this gigantic ledger up in the sky to keep track of things... not really, but you probably get the idea.

So waterwheels, sailboats, and human muscle were about the only way people had of using energy to do work for thousands of years unless that energy came in the form of animals.   There was an upsurge in the use of waterwheels after the Bubonic plague wiped out perhaps half of Europe during the Middle Ages, in fact most people's daily bread in medieval Europe would have been made from wheat ground by a grindstone connected to a waterwheel.

In England, the Domesday Book lists some 5,000 waterwheels, and by the late 1700's in colonial America there were perhaps 10,000 being used by various mills.   Even George Washington had a waterwheel gristmill for making whiskey at Mount Vernon.[271]   Milling grain into flour also happens to be a good way to store and preserve its chemical energy.   When Lewis & Clark were on their way west, in order to lighten their load in Montana they stored their flour underground in caches and when they dug it up on their return trip they found that it was still edible after almost a year.

I'm not sure, but I think the first lecture where I went off just a single page of notes might have been the first time a student didn't have to get up and leave early.   Since then, I've actually had a waiting list to get into some of my classes.   My attitude towards teaching changed from that day on.   Everyone in the department knew that an ace in my hole for tenure would be my teaching evaluations.   And on this particular day just as I did three years earlier, I also brought up the subject of fire, and how it was not really used to do work either, not until the start of another kind of revolution.

ROME'S PLASTIC

We humans are the only animals to ever overcome our fear of fire, which is also why we're the only ones who have ever put fire to any kind of use. The ancient Egyptians didn't have matches but they did know about rubbing flint stones together to produce a spark, in fact some of these stones were found alongside King Tut in his tomb.[272]

---

[270] Einstein showed that matter is a form of stored energy called *potential energy.* / The bow was the first weapon in human history that worked by storing up potential energy (when the arrow was drawn back).
[271] During their trip up the Missouri, Lewis was on the lookout for promising rapids that might one day be a good location for a waterworks.
[272] Book matches were invented in America in 1892 but didn't become popular until a brewer began advertizing beer on their cover four years later.

It's hard to grow up in Montana without hearing about Lewis & Clark on a regular basis so I naturally enough began by explaining the way Lewis came across agrarian Indians using fire to clear forests intended for growing crops. Prior to about 1700, even Europeans hadn't been able to put fire to much better use than the Mandans Lewis saw that day, at least as far as performing work (also defined as the ability to move things).[ccxxi]   Other than for cooking and keeping warm, ancient people had very few practical uses for fire.   It's true the Egyptians used it for weakening granite stones before cutting them into monoliths, and they also used fire for putting glazes onto jewelry and smelting copper, and it's no coincidence that the first metals obtained by heating rock turned out to be ones with the lowest melting points.

The Copper Age arrived in human history long before the Iron Age because the bluish rocks yielding copper nuggets could be obtained with a normal fire...the same temperatures used for hardening pottery and cooking. Lead and tin also have low melting points and so were obtained early on, too, bronze being an alloy (mixture) of both copper and tin.   Special furnaces would be needed if iron was to be obtained from its ore since iron only purifies at a higher temperature than can be achieved with a simple campfire and as a result just a handful of civilizations throughout history – like the Hittites and the Chinese further east – had the advantages of iron weapons for significant periods.[ccxxii]

In 1991 some hikers in the Italian Alps discovered the naturally mummified body of a Copper Age European nicknamed "Otzi the Iceman".   When he died Otzi had been carrying a nearly pure 99.7% copper ax, smelted and cast using a mold, and Otzi even had traces of copper in his hair, inviting some scientists to speculate that he was the one doing the smelting.[273]   Otzi was a Copper Age European, but if he'd lived in ancient Egypt he would have been present even before the pyramids were around.[ccxxiii]

In case any of my students at this point were wondering why I was talking about fire in a biology class, they wouldn't have wondered long because I was about to bring the lecture around to cooking.   Charles Darwin called humanity's use of fire our greatest achievement other than language and Darwin probably would have been interested to know that our digestive systems have actually evolved to become shorter, it's now believed, because of fire.[274]

Thermal energy from fire alters the 3-dimensional structure of proteins and starches, making their calories and nutrients more digestible, more accessible for our enzymes to cut into smaller fragments of amino acids and glucose

---

[273] The astronomer Copernicus's surname may be due to a family tradition of selling copper.
[274] There are several theories as to why our ancestors took to walking upright (bipedalism); one being to free their hands for carrying tools, or perhaps making it easier to spot danger. Another theory has it that walking upright on two legs instead of four saved energy simply because walking took less effort than crawling.

respectively.   These smaller units can then be absorbed into our blood and delivered all around the body.

"Think of what a little heat does to a hard, uncooked kernel of popcorn," I mentioned to my class. "Makes it a lot easier to chew, doesn't it?"[ccxxiv]

When our ancestor's bodies didn't have to work so hard to extract calories from food, their intestines shortened over time.   Cooking food to extract more energy is also what allowed us to grow larger brains and why our teeth have gotten shorter since then.[275]   For the first time, rather than for performing digestion, our ancestors could put more of their calories to work growing bigger brains.   Indeed, no human culture has yet been discovered that hadn't caught onto cooking and while it has been disputed that only humans have language and can make tools, the same isn't true for fire.   Only we have deliberately used fire to make our lives better.[276]

For Otzi and other pre-bronze age Europeans, one of the drawbacks with copper was that, while heavier than stone, copper is also softer.   The edges of pure copper knives dulled too easily.[ccxxv]   At some point an unknown artisan experimented by heating a chunk of copper ore alongside some tin in a campfire and that person was rewarded with the very first nugget of bronze. The properties of this alloy are truly remarkable compared to copper and tin alone and would have been appreciated because hardened bronze manages to combine the advantages of both copper and stone into one material.[ccxxvi] While copper is soft and heavy, when mixed with tin to make bronze, copper transforms into a rock-like alloy, yet still shiny and still able to be cast into different shapes when heated and most importantly, it will hold a sharp tip.

This simple trick of mixing two metals together changed human history more than any other innovation has since the use of fire for cooking.   The Egyptians had uses for bronze too, in fact it's why the Old, Middle, and New Kingdom periods are sometimes referred to by archeologists simply as the Early, Middle and Late Bronze Ages of Egypt.   But because of Egypt's centuries of isolation, they actually got along without much metal for long periods of time.   It wasn't until they were invaded by the Hyksos and later the Hittites, who made full use of metal weapons, that the Egyptians woke up and became serious about metallurgy.   From then on, Egypt took to it like a horse takes to oats and in fact would use metal along with horses to drive the Hyksos back out of Egypt, an event marking the beginning of the New Kingdom period.   Yet metal remained so valuable in Egypt that even now it is why metallic artifacts from the Old Kingdom are so rare to find: the Egyptians were avid recyclers.   Stone, on the other hand was much more common and

---

[275] Aristotle believed the brain's function was to keep the blood cool.
[276] Neanderthals knew how to use fire, too. There's evidence they employed it to break open bones of monk seals hunted along the coast of Gibraltar, extracting the energy-rich marrow.

monuments made of stone are plentiful and have been left to posterity, fortunately for us.[ccxxvii]

When I was in Saqqara I came across a 4,000 year old relief on the stone wall of a tomb depicting four Egyptian metalworkers stoking a brazier with blowpipes; puffing on them to increase the heat of their flames. But in Egypt, there was little wood; in fact Egypt's running out of trees has been called the first energy crisis in human history. It's even possible the scarcity of wood and Egypt's dependence on it led to the empire's eventual decline.[ccxxviii] Mounting regular military expeditions across deserts into far off lands like Nubia and Lebanon just to acquire wood would have been expensive. Lack of wood is also why Egypt never had a great navy, as Marc Anthony would someday discover. [ccxxix]

In the ancient Mediterranean it was this need for metals more than any other commodity that stimulated trade and exploration. Of the twin ingredients to make bronze, tin is by far rarer. As luck would have it, tin and copper ores seldom occur in the same place so ancient people had to go and search far and wide for tin. Athens, Rome, and Carthage grew into the first major cities of the ancient world because of trade (rather than strictly agriculture). These were what we would recognize today as true cosmopolitan places. In fact, it was the search for tin that stimulated the Greek expansion into the Mediterranean, so it could be said that the need for bronze even helped bring about the Greek alphabet and Greek logic so important to Western civilization today.[ccxxx ccxxxi ccxxxii]

Lead's melting point is low enough that ordinary campfires can set this metal free from its ore, but lead had an important drawback pure copper does as well...its softness. Lead melts easily, which is why dentists in ancient Athens could use it to fill teeth and later on the Romans made use of it in their plumbing.

Lead has even been referred to by historians as *Rome's plastic*.[277] The Romans flattened it into thin sheets and then rolled it to make hollow pipes useful for transporting water. The emperor Commodus sometimes fought as a gladiator, but less impressive was the fact that he first made sure his opponent would be stuck using a sword made of lead, which of course bent more easily.[ccxxxiii] In fact, lead melts at such a low temperature that Lewis & Clark's blacksmith made musket balls just by using the heat from a campfire. He used this fire to melt the lead canisters that carried their gunpowder because, cleverly enough, each canister once melted and molded provided just enough bullets to be used by the same amount of gunpowder inside its

---

[277] The chemical symbol for lead is Pb, which comes from the Latin word plumbum...and is where the English word plumber, or plumbarius, comes from: someone who works with pipes. The ancient Chinese about this time were using bamboo pipes to carry water, held together by lime.

canister. And even earlier, the American colonists melted lead during the Revolutionary War to make musket balls using the heat from lanterns in their homes, pouring this liquid lead into wooden molds.

Yet the most common metal ore (rock) found in nature is of iron. The ancient Chinese invented blast furnaces capable of liberating iron from its rock, and they kept this monopoly in part because of China's forests, which provided fuel. Rather than cast it as weapons they used iron mainly for making agricultural implements like wheelbarrows and plows. Being able to obtain more food from the land using iron tools helps explain why China's population explosion came about during the Han Dynasty. Because molten iron could be poured into molds, the Chinese were able to mass-produce their agricultural implements.[278]

PURE CULTURE

Egypt weakened internally when Egyptians had to cross vast deserts in search of wood, while Europe's forests wouldn't dwindle until the end of the Middle Ages.[279] By then, the chemical energy locked within wood was being put to use by artisans making glass and swords and other fire-hardened goods like tiles. But England stood poised above a source of energy that would spark a revolution never before seen... one that would for the first time change the way the average person went about his or her daily life.

Coal had been mined from beneath England since ancient times; in fact the Romans even exported some of it.[280] And by the 1600's holes had been dug so deeply into the earth that coalmines had to be abandoned because of water infiltrating from underground. Miners placed animal skins against the walls of shafts, hoping to slow the seepage and failing that they used waterwheels to pull the water out of the mines in buckets.[281] Others made use of horses, which turned a large wheel called a *horse gin* to lift their water-filled buckets out on ropes. One mine employed 500 horses and you can take it from someone who grew up on a ranch, this would have been pretty darn expensive just to get rid of water. If someone could figure out a way to use coal to pull this water out it would make available a simpler source of

---

[278] They also used waterwheels to operate bellows, which provided air for the molten metals in their blast furnaces.

[279] Thomas Cook's steamboat excursions up the Nile River in Egypt in the late 1800's were more expensive compared to Cook's other tours because he had to ship all his coal into Egypt. Egypt has always been an "energy-poor" nation.

[280] The Greek word for coal is "anthrakis" and is where the name *Bacillus anthracis* comes from. The cutaneous form of anthrax produces coal-black sores on a victim's skin.

[281] They also used waterwheels to provide ventilation for mines.

energy to move things, in fact it wouldn't even have to be that efficient a method since there was so much coal right there at the mine.

In 1699 Thomas Savery demonstrated what he called his "fire engine" to the Royal Society (picture a giant pressure cooker), which used steam pumped directly into the water of the mine, the pressure of which forced this water up and out onto the surface through pipes. It had no moving parts, so it wasn't really an engine, and the high pressures it produced caused pipes to break, which resulted in injuries and even death on occasion. In addition, they were so inefficient that deeper mines required a *series* of Savery engines, one every 40 feet or so, to move the water up and out onto the surface.[ccxxxiv]

The first true steam engine – one with moving parts – was built in 1710 by the British ironmonger Thomas Newcomen.[282] Newcomen's steam engine came equipped with a moving piston, which was attached to a pump at one end while the other end was connected to a compartment with steam coming from the boiler. Newcomen was able to get his engine to do the work of about five horses. When the steam in the chamber was subjected to cold water, the energetic water molecules slowed down and condensed, forming a vacuum, which in turn sucked the piston inward, pumping water up and out of the mine. It was safer than Savery's engines because it worked at atmospheric pressure, which is of course lower than steam's pressure inside a pressure cooker, but it was still inefficient compared to machines built later by the Scottish inventor James Watt, yet Newcomen's engines were useful enough that by 1775 more than 600 could be found working away throughout the English countryside.

In time people got better at using the chemical energy locked within a piece of coal. Just one of Watt's engines was doing the work of about 200 horses following the next 20 years of improvements, and they weren't confined to coal mines by then, either. As early as 1784 James Watt's coal-fed steam engines were working in two London breweries and within 16 more years another 18 breweries would be using his engines for grinding malt, stirring mash and beer wort. In North America with its abundant rivers and fast-moving streams, waterwheels remained popular and so steam engines were slower to catch on.[ccxxxv]

James Watt was repairing a broken Newcomen when he had his breakthrough idea. For some time he had been interested in the way hot steam seemed to carry more heat than was being harnessed by Newcomen's engines. Watt wanted to conserve this extra "hidden energy", which he was sure could be useful. Among his improvements was a *separate* chamber for condensing the steam using the spray of cold water, which meant the first

---

[282] Europe's newly emerging iron technology allowed him to make his steam engines larger and cheaper than previous brass engines had been.

chamber didn't need to be cooled and then reheated with each stroke the engine made.   With just this one innovation, Watt improved Newcomen's engines by 400%.   Watt's engines could even be made smaller, more reliable, and provide a steadier source of power useful for England's industries, which would soon be engaged in such diverse tasks as weaving, grinding, sawing, printing, pumping air into blast furnaces as well as moving locomotives and steamships.   The thermal energy in steam, if properly harnessed, can be made to move all sorts of useful things.[283]

Mankind had, after thousands of years, finally figured out a way to harness fire to do useful work with machines yet ironically no one had any real understanding of what heat was.   And to make further improvements in steam engines, a deeper understanding into the nature of heat could prove useful, for as good as Watt's steam engines were, they were still only 4% as efficient as was theorized possible (Newcomen's engines were only about 1% efficient.[ccxxxvi]).

While Isaac Newton's laws of motion were useful for engineers, they provided no insight into the *mechanism* for how steam engines were putting to work all this heat from coal.   What was inside the coal, for example, that provided useful heat?   And what, exactly, was heat?   No one really knew.

The ancient Greek *atomists* believed their 4 "elements" had distinct properties because the particles that made them up had special shapes.   For example, fire was said to be composed of atoms small and spherical, which accounted for the way fire rapidly penetrated so many different substances when burned.[ccxxxvii]

Scientists in the late 1700's had settled on the notion that heat from fire was the result of a mysterious substance inside things that burned called *phlogiston*.   This idea grew out of work by alchemists like Becher and Stahl in the 1600's who called phlogiston the "5th element", thereby adding it to Aristotle's four others.[ccxxxviii]   The limitations of this theory eventually became too hard to ignore when it was shown that some substances (like mercury) when burned actually became heavier instead of lighter.

So a new theory about heat emerged in the late 1700's, which held that heat was more like a fluid called *caloric*, and when something burned it was because its caloric, which floated and therefore could have negative weight, had drained from it.   I pointed out to my students that morning that if they had lived in the late 18th century they would have been taught by their professor that – whenever they rubbed their two hands together vigorously – the heat they felt was due to microscopic holes being made in their palms, holes through which some of this mysterious caloric could then leak out.   Few

---

[283] The electric company that provided the first electricity for Munich's Oktoberfest (in 1885) used a steam-powered dynamo that ran on coal. It was owned by Albert Einstein's father.

scientists had considered friction, but they soon would and another piece of the puzzle would be set in place.

So if the caloric theory of heat were true, it seemed reasonable to assume that any object that released caloric could hold only a limited supply before it ran out. In 1798, an eccentric Count in Munich named Rumford called all this into question with a simple experiment. He showed that it was possible to take a hollow brass cannon immersed in water, rub the inside of the cannon with a boring tool, and just this rubbing could produce enough heat to the water surrounding the cannon to boil the water in a tub (if rubbed long enough). There seemed to be no limit to how much water could be made to boil just by rubbing the inside of a brass cannon. The more diligently the cannon was rubbed, the hotter the water got. So there seemed to be an important connection between rubbing objects (which is transferring motion to their atoms) and generating heat from them.[284]

In time, James Joule, the son of a wealthy brewer, would do some experiments similar in scope, but using more carefully controlled conditions.[ccxxxix] Rather than a cannon in a tub of water, Joule used wooden paddles in a bucket of water to generate his heat and he used a thermometer to accurately measure the heat his contraption produced. He kept careful track of details like how many minutes the wooden paddles in the water were turned, how much water was being heated by the paddles, and how much its temperature increased. Rather than trust his own muscles to turn the paddles, he used measured weights suspended in the air, allowing them to fall slowly towards the Earth to do his paddling. He prided himself on his accuracy – I picture him a bit like Stanley in the lab – for Joule's experience in his family's brewery learning techniques for measuring temperatures had apparently rubbed off.

Heat would turn out to be the most disorderly, chaotic form of energy known and because of his work, Joule is credited today with the kinetic theory of heat, the accepted definition we now use, meaning that the hotter something is, the faster its constituent parts move.[ccxl] We would say today that an object's atoms move faster when they're hotter, and slower when they're colder. And at absolute zero all movement, even on a molecular scale, would stop.[285]

---

[284] Rumford also invented the wax candle to replace smokier tallow ones, and in 1804 married the widow of Lavoisier (the same year two other notable events occurred: the world reached its first billion inhabitants and Lewis & Clark set out on their journey to the Pacific).
[285] By proving doing work could produce heat, he disproved Lavoisier's theory of caloric. Joule was once a student of John Dalton, a schoolteacher who believed that, like the atomists of ancient Greece, matter was made up of individual particles.

ANTOINE'S PIGS

In Paris around this time another detailed experimenter – a chemist by the name of Antoine Lavoisier – was doing some interesting work using animals as subjects.   To see what was happening during digestion, Lavoisier carefully went about weighing and keeping track of every morsel of food he fed his guinea pigs and even the waste they gave off afterwards including their "animal heat" (body temperature), which he measured by allowing his subjects to melt ice and then measuring the water the ice produced.   He measured the oxygen the animals took in and the amount of carbon dioxide they gave off in their breath.

All this led Lavoisier to the startling conclusion that what was taking place inside an animal was similar to what happens when everyday objects burn. "Breathing is a combustion, like that of a burning candle," he would write, and I often wonder how impressed Lavoisier would be if he had known that scientists would someday recognize that every chemical reaction known to occur, either in the lab or in the natural world around, always takes in or gives off some amount of heat.   As far as we know going into the 21st century, it's a law. [ccxli]

Nicolas Carnot, a French physicist, chose to concentrate on the source of the heat (i.e. the coal) as well as the temperature of the steam inside the engine when compared to the air temperature outside.   Carnot eventually came to the conclusion that what really mattered was the *difference* in temperatures between the inside and the outside of the steam engine, leading to the conclusion that the greater one could make these differences, the more work the engine could do.   It was reminiscent of the waterwheels the ancient Romans used, and how their engineers knew that the higher the water fell before striking the wheel, the more work could be done by the falling water.

Carnot was just coming to the realization that the caloric theory of heat was wrong and Joule's kinetic theory right when he was struck down by cholera in 1832 at the age of 36.   In the haste to bury him, his scientific papers were laid to rest too.   Carnot's landmark book still survives fortunately for us, and its title is "On the Motive Power of Fire".

Because of work by Carnot and others on understanding the steam engine, engineers in the late 19th century were giving in to the realization that all the various forms of energy, whether its origin was electrical, chemical, light, or mechanical in nature, were somehow related on a fundamental level.[286]   Carnot and the others gave us the study of energy and its interrelatedness, a branch of science today called *thermodynamics*.   I took a

---

[286] Michael Faraday showed in 1839 that electrical and magnetic energy were interchangeable, leading to the possibility that other types of energy might also be incorporated into a new, all-encompassing energy law.

moment for this to sink in as my students turned to the bioenergetics equations while I glanced at the clock on the wall to see how much time was left.

"You see, these equations in your book came from the study of steam engines during the Industrial Revolution," I added for emphasis. And it was true. The same formulas used to figure out the amount of work a bumble bee can produce by flapping its wings to hold itself aloft with a given amount of sugar are derived from the same equations used to improve steam engines during the Industrial Revolution. Energy is interrelated in so many interesting and unexpected ways. [ccxlii]

I glanced at my watch to make sure the clock on the wall was right and couldn't believe how much time had passed. Lecturing can be a form of self-hypnosis at times. For some reason it was running long, leaving me just a few minutes to wrap up the rest of our one-day foray into the strange world of energy and metabolism. I thought about what my end target had been all along that morning. I had planned to finish with glucose, the most important and commonest sugar on the planet…and for good reason. Burning coal may have been what powered Watt's steam engines during the Industrial Revolution, but glucose is the carbon source our own bodies depend on, our own version of a "log on the fire". It's where we, and so many organisms on Earth – from bacteria and plants to whales and brewer's yeast – get the majority of their energy as well.[287]

FROZEN BURN

It's been said that the difference between a biochemist and a biologist is that a biochemist can draw the molecular structure of ATP from memory whereas a biologist is going to need his notes. I'm not sure what they say about glucose, though, since glucose is simpler than ATP, having just three elements – carbon, hydrogen, and oxygen – scattered among its 24 atoms, and so like any microbiologist worth his salt I put the structure of glucose up on the chalkboard from memory, in long chain fashion. Glucose is a small molecule by most biochemist's standards.[288] Then, on the other side of the arrow to the right I drew where the products are formed when glucose is

---

[287] Pet scans can image the brain by detecting glucose taken up by cells there, providing a picture of which parts of the brain are most active. The human brain requires approximately half a cup of this sugar every day.

[288] By comparison, a relatively large protein molecule such as CFTR has about 30,000 atoms in its structure. (Aristotle didn't believe in atoms. He chose to believe one could divide matter in half continuously forever.) The formula for a glucose molecule is 6 carbon, 12 hydrogen, and 6 oxygen atoms ($C_6H_{12}O_6$) bonded together covalently.

burned, whether this burning takes place inside a person, a bacterium, or a bonfire. They are quite simply carbon dioxide, heat, and some water.

Feeling a bit like a guide leading a group of tourists through the tombs of Saqqara, I pointed out some of the more interesting features of glucose along the way that morning. What stands out for me are the five groups hanging off the molecule's sides called -OH groups (pronounced "oh-H groups") and these -OH groups, not coincidentally, look like water (H-O-H, i.e. H_2O). Because of this chemical similarity, glucose mixes (dissolves) extremely well in the water that makes up 65% of our bodies, including our saliva and our bloodstream coursing through our hands, feet and brain. Dissolving in water is what allows glucose to reach every cell in your body. Every shore the tide of blood laps onto, glucose goes along for a free ride.[ccxliii]

And these same -OH groups can also function like handles, or places of attachment, sites where other glucose molecules can latch onto and form long chains called *starch* in plants and *glycogen* in animals. Glucose can also be joined to make *cellulose* in plants (the same cellulose my student's desks, their paper, and their clothing were made of, which I of course pointed out). The only difference between starch (which is a food) and cellulose (which is a building material and only microbes in the gut of termites and ruminants like cows can digest) is the specific way the glucose molecules are attached to one another at the –OH groups.

Cellulose is surprisingly strong while starch is a food, even though both are made up of the same long chains of glucose.[289] The glucoses in cellulose are all joined in such a way that they can form reinforced sheets with other cellulose chains, meaning these chains are oriented so they can snuggle closer together the way iron rods can reinforce concrete. This gives cellulose its legendary strength and is one of the reasons you and I can't digest any of it in our diet.

While on a tour in Egypt a guide inside one tomb pointed out a wooden mallet that was still holding up a three-ton sarcophagus lid. Tomb raiders had placed the mallet there 4,000 years earlier, the same ones who had pried the stone lid up. The wood was still supporting the lid even though its cellulose is made of glucose (the same chemical as in your sugar bowl). The devil is in the details in chemistry sometimes because it all comes down to the way each glucose molecule is joined together with the next glucose at its -OH groups.[ccxliv]

When we eat starch, we have enzymes in our saliva that begin breaking it back down to shorter sugars again, which is why if you chew on a piece of bread or a potato long enough before swallowing, these starchy foods will

---

[289] The main reason cooks heat vegetables is to soften the cellulose, which makes them easier to chew. And the more cellulose a plant has, the longer the cooking time required.

begin to taste sweet after a while. Your enzymes are liberating glucose from the ends of the starch chains.

Glucose is small, so it can travel quickly, too. It gets absorbed right through the lining of your intestine and makes its way into the bloodstream (by diffusion) without your body having to do any work on it (i.e. not expend any energy). Glucose is "available" packaged energy, ready to travel anywhere in the body. At any moment your blood contains only a 15-minute supply of glucose, required mostly by the brain.[ccxlv] Sometimes I like to picture the engineer of a steam locomotive shoveling coal into the boiler to keep the train running. Our liver shovels glucose into our bloodstream to keep energy-hungry organs like the brain and heart running. When glucose in the blood runs too low our body begins to break down long chains of glycogen stored in the liver and release glucose, which can then make its way into the bloodstream to nourish the rest of our body. The brain is so heavily dependent on burning glucose that if our bloodstream's supply was to run too low, we would lapse into unconsciousness after a mere 10 seconds.[ccxlvi]

Now the tour was about to take a detour into a more technical area so I pointed out the oxygen atoms on the glucose molecule, comparing them to the six carbon dioxide molecules I had drawn to the right of the arrow. The carbon atoms making up the carbon dioxide molecules (CO_2) on the right are the same exact atoms that *were* part of glucose on the left, but now separated and with more oxygen joined onto them, and this happens when we breath in oxygen. Our glucose gets chewed up, or burned completely by enzymes inside our cells.

Fat, by comparison, has less oxygen bound to it to begin with, which is why fat can hold twice as much energy as carbohydrate, but an important drawback with fat is that fat can't release its stored energy as quickly as sugar can.[ccxlvii] The glucose molecule is the chemical equivalent to a "deer caught in the headlights" when it comes to combustion. It's as if glucose is suspended, chemically, in a kind of "frozen burn". The glucose has some oxygen atoms added onto it already (by the plant when it made this sugar) so glucose is "partially burned", meaning it can go the rest of the way to carbon dioxide and water without as much trouble as it takes to get fat burning.

Getting fat to ignite (using enzymes) is a bit like trying to start a wet log burning with a match. And while glucose is not only smaller, dissolves more quickly, and therefore travels faster in the blood than fat, it's also ready to be burned up completely, too. All in all, a molecule of glucose is the perfect packet of fuel for when your cells need quick energy and no surprise that most cells on Earth use glucose, from brewer's yeast to anthrax bacteria all the way up to elephants, whales, you and me.[ccxlviii]

I'd almost reached the mile-marker I'd been working towards all morning...the part where I finally got to put the whole thing together. But one last topic remained to be covered, the one I had deliberately skipped while discussing energy forms that were being discovered back in the 19th century. I had skipped batteries and the electricity they produce. Class this day was just going to have to run a little longer.

ENRICHED MEDIA

The true nature of current electricity – electrons moving through a wire – wouldn't be discovered until the very end of the 1800's, but that didn't stop experimentalists like Alessandra Volta from making batteries a full hundred years earlier. Like fermentation and heat, electricity was another of nature's phenomena put to use before it was completely understood.[ccxlix]

Volta found that if he used two different metals – zinc and copper in his case – along with a strong acid, a current of electricity could be made to flow between the two metals. In fact, if you have metal fillings in your teeth, you made your own battery whenever you inadvertently bit down on a piece of aluminum foil and then got a slight stinging or perhaps sour sensation on your tongue. For a brief moment, you made a battery in your mouth and got a quick jolt of electricity (electrons) to flow and the feeling detected was the energetic electrons as they traveled from the metal filling in your tooth out towards the aluminum foil. The circuit, or "wire", the electrons flowed through was your saliva.

People like Volta experimented around using different types of metals until eventually they found just the right combinations that produced the strongest currents. We would say today that the best batteries are the ones where one material is more generous than the other material, and wants to let go of its electron the easiest, while the second material wants to take on the electrons just as eagerly as the first material wants to let them go. In other words, the greater the difference between the two materials in their desire for electrons, the stronger the "push" (i.e. voltage) of the battery.

If you insert a light bulb in the wire circuit joining the two metal ends of the battery, you can see the light bulb light up because, as the electrons rush from one metal to the other of the battery, the electrons are forced to pass through the wire and the narrower filament inside the bulb, and as they all have to crowd through this "bottleneck", they produce incredible amounts of friction, which in turn yield light and heat.[ccl] So the greater the difference between the two metals in their desire for electrons, the more push they can give the electrons as they all rush headlong through the thin filament inside the bulb. Stronger batteries produce a stronger push, which of course produces more light.

You may have noticed that batteries (energetically speaking) resemble the waterwheel and the steam engine, in which case you'd be right. But with the waterwheel, the further the water fell, the more energy it provided and the more work it could do in turning the waterwheel, which in turn milled more grain into flour or chopped more wood and so on. With the steam engine, the greater the temperature difference between the steam inside and the air outside, the more work the engine could do. One of Watt's greatest improvements was being able to operate his machines at higher temperatures, thus capturing more of the steam's energy and converting it into work.

Well what Lavoisier had stumbled onto shortly before he lost his head during the French Revolution was that the food he was feeding his guinea pigs was doing essentially the same thing. What biochemists would come to find out during the 20th century – often while closely examining purified brewer's yeast enzymes – was that when food like glucose is "burned" deep inside cells, what goes on is similar to what happens inside the battery when it liberates electricity or when the log burns and forms carbon dioxide and water.

Electrons leave the glucose molecules inside our cells – electrons that start out at a higher energy level – and then these electrons get stripped away from the sugar and "fall" towards the oxygen molecules that we continually breath in. And as these electrons from glucose are stripped away and "fall" towards the oxygen to ultimately produce carbon dioxide and water (which goes into your blood and then you expel it with your breath), our cells manage to get these energetic electrons (that were originally in glucose) to do some useful work along the way, not so unlike the way electrons do work when they have to travel through the filament in the light bulb to produce light, or the way water from a river does work when it falls down onto and turns the waterwheel.[ccli]

Like the inner workings of a Swiss watch, there are literally a dozen or so different proteins in each mitochondria of our cells that are carefully arranged in perfect order and geared up for doing this energy conversion. We all have a series of these proteins that act as the tiny machine parts, analogous to the gears and crankshafts inside the waterwheel that channel kinetic energy captured from falling water into the mechanical energy useful for grinding grain.

For example, some of the proteins in our cells are ordered into an elaborate bucket brigade called the *electron transport chain*, where they carefully take on and then hand off each energetic electron originally stripped from glucose to the next protein down the line and so on all the way to a waiting oxygen molecule at the very bottom (energetically analogous to the water falling down onto the waterwheel). And as they "fall" towards the oxygen, these electrons from glucose give up some of their energy along the way, energy which can be used to pump hydrogen ions (protons) into the mitochondria and these hydrogen ions build up there and are stored (like a battery stores electrons) and then carefully released to turn little "mills" also made out of protein complexes imbedded in the mitochondria's membrane, proteins that continuously crank out ATP molecules as their final product.

These ATP molecules could be thought of as the flour produced by the mill (each of your cells continuously recycles about a billion of these ATP as long as you keep breathing in oxygen), and this ATP will become the "energy currency" – kind of like money – that the cell can then use to "pay" for all the things it needs to do to stay alive and reproduce, help proteins fold into their 3-dimensional shapes properly, copy DNA, make hemoglobin out of amino acids, open and close the gates of ion channels like CFTR, and so on.[cclii] And just like the log on the fire, our cells break down glucose and create heat too, accounting for our constant body temperature, as well as why brewers need to refrigerate their vats.   Yeast generate heat too when they break glucose apart.

TRAVELING MEDICINE SHOW

The last item on my list for the afternoon was to show a prospective student around my laboratory.   Once or twice a year our department invites the top undergraduates in the country applying for graduate school to an all expenses paid interview.   They're supposed to list ahead of time what members of the faculty they want to visit with, and we're supposed to impress them with our work so they will join us.   The whole process reminds me of Lewis & Clark conducting their "traveling medicine show" for the Indians as they made their way across the continent.   But instead of showing off the benefits of the Industrial Revolution as they had, I would give a brief tour of my lab while stopping along the way to look through the microscope at some of the more interesting bugs I've collected over the years.

Lewis & Clark's men would get dressed up in their best uniforms, march around the campfire, and then demonstrate the use of a magnifying glass and an air gun, one that could shoot lead balls with hardly a sound, the mysterious attractive properties of the magnet, and of course a spyglass to count trees on the sides of a mountain several miles in the distance.   To seal the deal, they would save for last a general store-like display of items the Indians could expect to acquire if they accepted Thomas Jefferson as their new "Father", and agreed to trade exclusively with Americans over the British in the future.

I've enjoyed showing off my collection of magnetotactic bacteria the most because they never fail to impress even the most jaded student.   These curious fellows accumulate magnetite crystals inside their cells, joining them into filaments of iron oxide that all line up and you can actually see as dark rods under the microscope inside the bacteria. The bacteria normally use this iron "antenna" to detect the Earth's magnetic field so they can figure out which way is up as they get buried in the sediments of a lake or bog.[290] We would have them inside a glass jar filled with water and then tape a small magnet to the outside of the jar. After lunch when we came back there would be a dark smudge on the inside where all the bacteria had congregated near the magnet. Then after examining some under the microscope, we'd make them all swim away from the jar simply by flipping the magnet over.

If that wasn't impressive enough, I could always pull out of the deep freeze some cyanobacteria called *Oscillaloria* I'd collected from a horse trough on our ranch and let the students watch as the long thin green colonies glided past one another, riding on a thin layer of slime and protein continuously shed by the bacterium. To this day no one knows the exact mechanism *Oscillaloria* uses to accomplish this. And as a last resort, I always held in reserve the yellowing cardboard box of microscope slides from graduate school of quite dead but still (especially since 9/11) fascinating anthrax bacteria in the process of forming endospores.[291]

But these days, trying to make our way through the lab while at the same time avoiding all the shakers was more like stepping across a mountain stream with only a few rocks spaced far and wide. I also noticed the students seem to be getting younger these days, which is of course when I caught sight of my own waistline, making a mental note to go for a bicycle ride after work. At that moment my jeans felt about as tight on me as bark to a tree, erasing any doubt I had managed to maintain that I was slowly gaining back all the weight I'd lost in Egypt.

On this particular day I had a student more interested in asking the tough questions than gazing at microbes – uncomfortable questions for someone in my position – like whether or not I could support him with a stipend if he chose to work for me. I had to level with him by explaining how he'd probably have to teach undergraduate biochemistry labs and do some grading to earn his keep in the department if he worked in my lab…that is if I was even still here. The expression on his face told the rest of the story and our allotted hour was shortened by half when he excused himself for what was apparently an important text message. It's a shame because this senior was also an honors student back in high school, one who had taken first place in a

---

[290] Iron oxide is the same pigment Roman artists used to write graffiti onto the walls of Pompeii. Iron oxide provided them with the color red, while the iron mineral hematite was used to create a yellow paint.

[291] The bacteria become a bright purple when stained (since they're *gram positive*) and each has an endospore, visible as a clear spot inside the cell since endospores don't hold a gram stain.

statewide science competition.   Losing the anthrax grant was yielding more unexpected consequences every day.

I then spent the rest of the afternoon getting things ready so they could be left unattended over the weekend, topping off the yeast with a fresh offering of broth and refilling the cryogenic tissue culture jugs with about a gallon of liquid nitrogen.

For me, going down to Yellowstone isn't about getting away from life's little annoyances so much as it is rediscovering a connection I used to have...one I seemed to be forgetting about more and more lately.   Once it was so much a part of my life that I can't even recall my first visit there, but I do know it wasn't until I was in high school that it finally dawned on me the entire Yellowstone Plateau is actually on top of a live volcano.

A lot of folks who visit here never realize it either.   The caldera is so enormous (35 by 45 miles) that it isn't what you'd expect of a volcano. Situated mostly in the northwest corner of Wyoming, parts of Yellowstone National Park still manage to spill out into the adjacent states of Montana and Idaho.   It's so vast that geologists refer to the Yellowstone caldera as a "supervolcano".   Geothermal anomalies in the Earth like Yellowstone's are how the Native Americans were able cook fish here without ever taking it out of the water first, and how some Romans in Britain heated their baths around the time of Christ.

The plan was to meet up separately at the trailhead on a ridge, where we would begin our descent along a narrow winding trail created by elk thousands of years ago, later widened by Native Americans and fur trappers, five miles down switchbacks to the edge of a rapidly-flowing river, the place where we would set up camp for the evening.   The park has several wide basins – places where geysers, bison, and hot springs co-exist – and we were due to pass one along the way.   I knew for certain because it's where I had spent a good deal of my time those first two summers.

Earlier I mentioned how microbiologists have our side projects we quietly work on in the lab.   Well mine involved finding new species of thermophiles inhabiting the hot waters of the park.[ccliii]   Today there are a few hundred known thermophilic species of bacteria inhabiting Yellowstone, some able to withstand waters so acidic that a single drop could easily burn a hole right through your clothing if you're not careful.

STATIONARY PHASE

The next morning we got to the trailhead first, so with Barney getting acquainted with previous hikers – some of which had visited without the benefit of a latrine – I had a chance to think things through before the others arrived.   Back when my group was just my two technicians and myself, their families and I took this camping trip every summer.   There's something special about a campfire when darkness descends, maybe it has to do with our shared genetics, but a roaring campfire at night never fails to bring all those present closer somehow.[292]

I had a professor in graduate school that liked to compare scientific research to taking a camping trip and that was how he approached every crisis in his lab all the time, which I thought was pretty practical.   Amanda said she was going to catch a ride with Chris, whose parents had recently bought him a new sports car.   Some things hadn't changed, though.   He still insisted on calling me Dr. Ketchum.[293]

Nature comes equipped with all kinds of cycles...water cycles, decomposition cycles, carbon cycles, lunar cycles and so on.   One of the key features besides heat that makes Yellowstone so interesting is its plumbing. It's why over half the world's geysers can be found in the park.   There's a silica material called *geyserite* in Yellowstone's soil that, when it gets hot enough, dissolves in the water and then precipitates back out again as a solid wherever this water becomes cooler, forming the perfect sealer for patching cracks and crevices inside the Earth, as if it were a plastic liner.

This allows for the concentrated pressure geysers need and for the heating of the tranquil hot springs that constantly percolate up the rainwater and snowmelt that manages to penetrate down into the ground.   In fact, one reason you'll only find geysers like Old Faithful in the basins rather than at the tops of mountains is that there needs to be this reservoir of loose gravel riddled with spaces to catch and store water so it has a chance to seep down to where the hot magma of the volcano lies further below, become superheated, and then rise again up towards the surface.   It moves because of a cycle as old as the Earth itself – one based on changing density – the changing density of water due to this alternate heating and cooling that takes place there.[ccliv]

The basin we planned to descend into from the parking lot at the ridge was a good three miles down the far side of where a glacier had made its way from the mountains, scouring the valley below thousands of years ago, widening it in the process.   I knew about glaciers because the Lewis Valley – and indeed most of Montana – was under a mile-high sheet of ice at one time and no force on Earth is more powerful when it comes to erosion than a glacier is.

---

[292] Thanks to fire, humans have been able to inhabit most of the globe for the last 10,000 years.   /   The Earth is the only planet we know of where fire can exist (fortunately, our atmosphere is 20% oxygen).

[293] The ancient Egyptians liked to show off their wealth too, by wearing layers of jewelry.

In addition to sculpting the valley my family's ranch sits on, the glacier changed roles after it had done its job, becoming a giant conveyer belt, transporting an assortment of rocks, pebbles, and even boulders the size of SUVs caught up inside and scattering them all around the valley, down where the edge of the glacier once was. It's a strange sensation to come across a huge boulder lying quietly on its side so far away from the mountain it used to be a part of, motionless in the middle of a tranquil field, that is until you realize that a glacier was once on top of where you're standing and that's how the rock got there. The central character in the drama has simply melted away long before your arrival, becoming a river during one of the opening acts.

Within an hour the rest had arrived and were busy stuffing last minute items, the kind that tend to be forgotten, like quick-energy bars and mosquito repellent, into their backpacks before setting out single-file down along the narrow switchbacks towards the basin. Watching my younger technician shepherding her children around made me feel a twinge of sadness for Amanda having to leave Joshua at home.

The quiet of the wilderness seems almost overwhelming when you first step off into it and you're used to being in a city for too long. It can feel wrong at first for nature to be so darn quiet. No sharp sounds of any kind, no metal-on-metal grinding of brakes past your window at night, just the sounds of running water and birds and maybe the rustling of trees thanks to the wind; that is until it suddenly dawns on you that nature is just fine and that the real problem is inside of you. The soundtrack of civilization has a way of coming along uninvited just about everywhere we go these days, at least for the first couple of miles, but fortunately it wears off, like a cut that takes a while to heal but in the end doesn't leave a scar. Still, sometimes you've got to invest several miles of walking first.

The hotspot within the Earth that created Yellowstone is what geologists sometimes refer to as a heat anomaly and in fact, by all rights it shouldn't even be here. There are a few dozen such places scattered all around the globe, places where the Earth's crust is thin enough that a plume of magma from down below can get up close to the surface, so close that it makes its presence known like a hotplate can underneath a pot of boiling water. The Galapagos Islands and the Hawaiian Islands are other such places.[294] Normally, the Earth's crust is about 50 miles thick, and if the Earth were the size of an egg, that 50 miles would correspond to about the thickness of the egg's shell, but here for some reason the crust is even thinner, only about a mile or so deep. One theory is that a comet came down 17 million years ago, bringing with it so much kinetic energy that it penetrated the skin of the Earth

---

[294] This idea is sometimes called the "mantle-plume" hypothesis. The other major theory holds that hotspots are due to what's called "lithosphere extension".

all the way down to the mantle, puncturing it and creating the volcano, but no one knows for sure.

I buried my watch down towards the bottom of my pack, determined to tell time by the sun for a change as we set off, carefully picking our way within the narrow confines of the trail down the mountain side, only realizing others were getting tired when I undid my own waist strap and began shifting my pack from side to side every few minutes. We stopped when the occasional frown became too hard to ignore. As a pleasure trip there was no reason to turn it into a hell hike. And besides, the side project was history now thanks to what had happened two years ago.

We dropped our packs and took a seat on an old log decorated with moss, the remnant of what was once a pine tree, one of a group of them, the others still standing vertical and looking rather naked without any needles, huddled together as if someone needed to remind them they were no longer alive and so weren't required to stand upright anymore. We had reached the first of the thermal features – a steaming hot spring about the size of someone's backyard swimming pool – and would be zigzagging around several more just like it for the next half hour or so...then it would be an easy twenty-minute walk down to the river's edge and our campsite.

I still couldn't completely shake the "city" in me, estimating we'd been walking about 3 miles an hour, which isn't bad. I was still thinking about the herd of buffalo we'd passed in the distance so still that their silhouettes could have been mistaken for stone statues. The calves had the same shape as their parents, and many of their faces seemed to be touching.

When Clark and Sacajawea were canoeing to meet up with Lewis further down the Yellowstone River back in 1806, they had to yield the right of way to a herd of buffalo so expansive that it took an hour before they could get going again. It seems that buffalo back then could be so numerous, they made the ground shake like a series of small earthquakes.

"From the fire of '88?" Marjorie's husband asked, pointing up at the scorched branches on the naked trees above.

"No. Probably an earthquake," I noticed a few puzzled faces and so explained further, "Sometimes the plumbing in Yellowstone gets rearranged when the ground shifts." I then explained how, being an active volcano means earthquakes visit Yellowstone a thousand times a year, and how without the right equipment nearly all go unnoticed.[cclv] But every once in a while, as happened in 1959, one hits energetic enough that it rearranges fault lines underground, causing some geysers to go extinct. Like crimping a garden hose to stop the flow of water, all of a sudden the supply just gets shut off. On the other hand, heated water has to go somewhere and as a result new thermal features can open up and scalding water will emerge from a place it hadn't before. Sometimes a geyser or a hot spring that's been dormant for a hundred years will suddenly spring back to life due to all this shifting around underground. Our planet is an ever-changing place and the trees Frank was referring to were dead now because a new thermal feature had formed near their base and killed the trees with its heat. Far more tourists at Yellowstone are done in by hot water or buffalo these days than they are grizzly bears.[295]

No one was talking so I squatted to scoop up a handful of soil, which got their attention, everyone's except for the two smallest children who were too busy playing. To them, a fully-grown man scooping up dirt with his hands must not have seemed all that unusual yet. When Lewis first set eyes on the prairie he had been surprised by how rich the vegetation was. Here he had been expecting a mostly barren land but instead found rich grasses growing on fertile soils two-feet thick.[cclvi]

"There must be 100 billion cells right here between my thumbs," I explained, "Mostly bacteria. Maybe a million yeast too. I'd be willing to bet among them are 10,000 different species of bacteria no one's ever catalogued or named in any microbiology textbook." Since the soil is the most bio-diverse place on Earth, I was fairly certain of what I was saying. Just the hindgut of a termite has 250 kinds of microbes living side-by-side, which I didn't mention, for I was working my way towards something else...gradually letting Amanda and Chris in on what my two technicians and their families knew already, about the project we all had wasted so much time on while looking for a new species of bacteria, and what had happened. It was time my students knew too, if for nothing else than their own experience.

I started by mentioning how the very first permit for collecting microbes at Yellowstone was issued back in 1898. Dr. W. A. Setchell (the Ph.D. kind) was looking for thermal algae in the cooler pools away from the center of the hot springs. Able to get their energy from the sun, algae are photosynthesizers and therefore easier to spot since they're green – perfect subjects for a late 19th century microbiologist's project. I'm still not sure if Setchell found what he came here for, but I'd be willing to bet it never occurred to him that bacteria were living within the boiling hot waters down near his feet.[296] In fact, up until the 1960's most biologists believed water like that was too hot for proteins and DNA – the material of life – to function in.

Normally, heat is a good way to destroy proteins, cook them and turn them from neatly folded 3-dimensional structures whose intricacy could put a snowflake to shame into nondescript lumps of "Swiss cheese" in a matter of

---

[295] Members of the Lewis & Clark Expedition were the first white men to see a grizzly bear while York was the first black man to see one.
[296] Photosynthetic algae haven't been found in water over 60° C, which is hotter than the hottest water a human hand can tolerate. / We sometimes use heat in the lab to inactivate (denature) enzymes when purifying DNA & RNA so we don't lose any of the genetic material to destructive enzymes.

seconds.   So it came as a huge surprise when a self-taught microbiologist from Wisconsin named Thomas Brock, along with his undergraduate student, came here in the 1960's and found bacteria growing on glass microscope slides they'd left lying around in the hot springs.

I emptied my hands and pointed towards the edge of the hot spring we were sitting next to.   Brock called it his "slide immersion technique", and fortunately heat-loving bacteria will stick to glass the same way they do silica rocks on the bottom of a hot spring.   He and his student took some back to the lab where they grew and expanded one species they called *Thermus aquaticus*.   They even found later on that *T. aquaticus* was able to grow on ordinary Petri dishes, paradoxically making it almost as easy to culture in the lab as *E. coli* is.[cclvii]   When he finished characterizing it Brock did what microbiologists still do today...he sent a sample of his new find off to the ATCC.[297]

I explained how, during one of my lab rotations in graduate school, I had found myself working with an older microbiologist for a couple months; one who had been using DNA from microbes to probe their interrelatedness.   It's useful to know which microbe is similar to which and by how much because it can turn out that a lot of work has already been done for you by someone you don't even know, and if you both happen to be working on a closely related bug it can make life a lot easier.[298]   So it's possible you can gain a lot of information about your microbe really quickly just by knowing what you've got growing on your dish.

Nowadays, most researchers simply sequence a portion of the DNA of both organisms and compare their sequences.   It's kind of the way you might notice a family resemblance in two brothers by the shape of their noses, except that you can look directly at the sequence of a microbe's genes these days, the sequence of its A's, G's, C's, and T's at specific locations along each DNA strand (their chromosomes).   Biologists can spot similarities in the sequence of this 4-letter code that makes up DNA and infer as to whether or not the two had a common ancestor at some point in the past; and even estimate when this ancestor diverged and the two daughter cells went their separate ways, each becoming a new species.

But the microbiologist I was working for was intent on doing things like they did them in the 1970's, when researchers couldn't sequence DNA so

---

[297] ATCC is the American Type Culture Collection, a nonprofit repository for animal and plant cells, as well as 20,000 different strains of bacteria and viruses.   They regularly receive microbes from researchers all over the world and send them out, too.   Between 1985 and 1988, some 2-dozen vials of anthrax were sent to Baghdad, Iraq from ATCC and into the hands of Saddam Hussein's weapons scientists, fully licensed.

[298] Scientists often work together using the literature, even if they are not contemporaries.   It was Champollion's deciphering of hieroglyphics 100 years earlier that allowed Howard Carter to know about King Tut and even where to search for his long-forgotten tomb: in the Valley of the Kings.

easily but instead took the purified chromosomes from their two microbes, simply mixed them together in a test tube, and then heated the tube and all its contents.[299]   If you heat DNA to a high enough temperature, the two side-by-side strands of each chromosome (which are what create the double helix) will separate, or "melt", a bit like two halves of a sleeping bag coming unzipped, because the main thing holding the DNA double helix together are these relatively weak *hydrogen bonds* in between.   Hydrogen bonds are the same bonds that liquid water molecules need to break with other water molecules when they boil away to become a gas.   Compared to *covalent bonds* like the kind that hold the oxygen and hydrogen atoms together *within* a single water molecule, hydrogen bonds are much weaker, but these bonds can still add up since there are so many of them.

So early biochemists ever since the structure of DNA was first determined in the 1950's have known that if you apply enough heat to DNA, you can separate its two strands from one another.   And if you give it enough of a push with thermal energy, they will even come completely unzipped and float around like pieces of over-cooked spaghetti, separate but still within the same test tube.   It's the thermal energy that overcomes the strength of the hydrogen bonds and keeps the DNA strands apart in your tube.

And if you cool the solution back down to room temperature again, these two strands will find one another and simply zip right back up to form a new double helix again.   This whole thing is reversible.   So if you deliberately mix the DNA so that there are strands from two different species of microbes in the same tube, the two strands will get together and hybridize with each other, forming a new double helix, one strand from each species.   One DNA strand from one microbe will form a new double helix with the DNA strand from the other microbe.

And it turns out that the temperature this all happens at is a good indication of just how closely related your two species are.   In fact, this was the first biochemical method used to show that humans are more closely related to chimps than we are to gorillas.   The more related any two species are, the more similar the sequence of DNA bases along their chromosomes are, and therefore the higher the temperature this joining together of the two DNA strands can happen at.[cclviii]   So the guy I was working for had been mixing the two species' chromosomes, half from an *E. coli* and half from a *Salmonella* if I recall, heating their chromosomes and then slowly cooling the mixture back down to find out the exact temperature these DNA strands formed new hybrid double helices.   It is a quick and dirty indication of the

---

[299] This method is called *DNA-DNA Hybridization*.   The higher the temperature needed to separate the DNA hybrid strands from each other, the more closely related the two species are taken to be.

degree of interrelatedness between two microbes without having to sequence the DNA and is still sometimes used today.[cclix] [cclx]

I was getting closer to what Marjorie and my other technician already knew, about the previous project. I watched as their heads began to sag, like sunflowers entering a summer drought, and told Amanda and Chris how, in the 1980's Kerry Mullis in California figured out a way to heat DNA, then cool it and add an enzyme to make billions of copies from just a single piece of DNA.

His idea was that he could use a *polymerase enzyme* to copy the strand (after they were separated with that initial blast of heat) and that he could then use the newly made DNA strand as another template to make even more copies from it. Mullis found that a person could simply do an amplification of the original piece of DNA if, that is, one repeated the same steps, over and over again...heating, cooling, and copying...heating, cooling, and copying...just like this about 20 times, instead of simply stopping, as what folks had been doing up until then once they got their first copy made. In other words, he used the copies of the DNA to make even more copies of the DNA.[cclxi]   It's the procedure Tahany was using to detect a single schistosome worm egg from her patient's urine, the same procedure used to recover DNA from ancient mummies and Neanderthal bones. PCR is the reason it's possible to amplify the DNA from just one hair left at a crime scene, or a single cheek cell on the back of a postage stamp where someone has licked it, and do DNA fingerprinting analysis to identify the culprit.[cclxii]

With PCR there's little reason to worry about having enough DNA, even if all you have is a single chromosome. This one procedure revolutionized molecular biology and Mullis got a Nobel Prize for it in 1993. Very few Nobel Prizes have ever been awarded just for inventing a new procedure. Usually they're only given out after a lifetime's worth of work. That should give you some idea of just how valuable PCR has become these days.

I still hadn't told Amanda and Chris anything they didn't already know. PCR is pretty routine, taught in just about every undergraduate biochemistry class, but I was slowly getting to the part my technicians and I were working on three years ago. You see, every time he finished a round of copying, Mullis had to heat the DNA again to separate the 2 strands so he could continue copying each strand with the enzyme (you could think of this enzyme as a tiny Xerox machine, but made out of protein). And each time he came to the step where he heated the DNA to separate the two strands, the heat of course killed the copying enzyme, which was a protein obtained from an *E. coli* bacteria cell. And since *E. coli* normally lives inside our gut, where it's always a comfortable 37° C (98.6° F), its proteins tend to lose their activity, denature in other words, at temperatures hot enough to separate DNA strands from each other (in the separation step this temperature is just below the boiling point of water... about 100° C).[300]

What Mullis and his coworkers did next was to send away for a microbe that prefers living at higher temperatures, figuring that if they obtained a copying enzyme from a heat-loving bacterium, then they might not have to add fresh enzyme each and every time they made another round of copies from the DNA inside their test tube. So it turns out they got back from ATCC the same microbe Brock had deposited back in the 1960's, the *Thermus aquaticus* he'd found at Yellowstone living in the hot spring.

And it turns out their idea worked; in fact it worked spectacularly well. And as a result of using this heat-resistant enzyme,[cclxiii] they only had to add the enzyme to the solution once, at the very beginning of the procedure, then put a cap on the tube and the PCR could go on its way and even be automated with a machine, which is also why PCR is a lot cheaper, easier, and more reliable to do these days than it otherwise would be.[301]

Finding a useful new protein this way signaled the start of a gold rush by microbiologists eager to find other useful enzymes from bacteria capable of living in extreme environments like the hot springs of Yellowstone. Today, we have enzymes (called *extremozymes*) purified from bacteria that live in all sorts of harsh places, including acidic copper mines, soda lakes, and sea ice, in fact you've probably benefited from them from time-to-time, enzymes useful for stonewashing blue jeans, making antibiotics, removing bloodstains from clothing as additives in detergents, and even for cleaning up oil spills.

But there was just one fly in the ointment with Mullis's new version of PCR. It turns out the enzyme makes mistakes; in fact it makes a lot of mistakes (called "mutations"). Taq will insert an A, for example, where it should have inserted a G in the growing strand's DNA sequence and so on. Taq is one tough little enzyme and is great for copying quickly and not becoming unfolded at higher temperatures, but when it comes to accuracy it's never going to win any spelling contests. Even Taq, it turned out, had an Achilles' heel.

Which finally brought me around to the reason we were bioprospecting the hot springs of Yellowstone my first two years at the university. We had been looking for another microbe similar to Brock's *Thermus aquaticus*, one that made an enzyme that could withstand the rigors of PCR, but at the same time didn't make as many mistakes as Taq... kind of like having your cake and eating it too. But as often happens in science, we got scooped.[cclxiv]

---

[300] Louis Pasteur used heat (42°C over 20 days) to weaken (attenuate) the anthrax bacteria's proteins so he could safely inject these cells into cows for use as a vaccine.

[301] We bought a PCR kit from a catalog for about $90 and used it to verify that the yeast took up the CFTR gene and that it was sitting in the right direction. Sometimes genes can get inserted backwards within a chromosome or plasmid and as a result the protein won't get made.

Still squatting on my heels, I looked up at Marjorie and there was little need to explain any further how we all felt when it happened. We had wasted a lot of time and effort while someone else got there ahead, beat us to it by finding an enzyme called Pfu, short for the Archaean microbe that makes it, *Pyrococcus furiosis*, in the near-boiling marine sediments off Vulcano Island in Italy by none other than Dr. Karl Stetter, the same scientist with the aluminum briefcase he carries his microbes around the world in, the one with the special lab in Germany just for growing his extremophiles called his "witch's garden". The man is extremely good at what he does.

As if on cue, Barney returned from wherever it was he had gone off to and, assuming he had to be the reason I was still kneeling, began licking my face. He was like a puppy again, at least for the day, and his spirit helped lighten the mood for the rest of the hike. I knew tomorrow we'd both pay, as he'd lie around camp all morning and I'd even have to carry him part way up the steepest section of the ridge to the truck in the afternoon; but it was worth it to see the dog I used to know. It seemed just then a good idea to pick up and move on down towards the river.

In a way it was fitting we had been doing our experiments in Yellowstone those first two summers, I thought as we picked our way along the trail, for even Yellowstone itself was an experiment, started back in 1871 when it was carved out of wilderness by an act of Congress. The world's first national park must have been an experiment worth repeating, though, because today there are more than 6,000 national parks all around the world, making Yellowstone a pretty successful experiment by anyone's measure.

DOUBLING TIME

By the time we made it down to the river Barney had beaten us, already having come across the remains of a bison carcass from the previous winter, lodged on some sticks at the river's edge, with only square bits of fur stretched over its bleaching skeleton, the eddies of water running over the animal's ribcage as if a series of small rapids. Most likely the animal hadn't had time to put on enough fat for the winter due to the short fall we had the previous year.

Within an hour five tents were evenly spaced 25 yards or so from the campfire, the light from which was taking the place of the sun as it disappeared behind the jagged mountaintops. Just looking at the differences in sizes of the tents it would have been easy for a park ranger happening by to guess that our group consisted of three single hikers and two families.

The next order of business was to gather up firewood, after which the men began spreading hamburgers and hot dogs over the grill while the women set about uncovering side dishes. By tradition, we each brought one to share and I carefully guided mine out of my pack sideways – a saran-wrapped macaroni salad – while wondering if this was to be our last trip together. I didn't have long to wait. In just two short months we'd all know. Barney, who had been keeping an eye on the simmering meat, became distracted only briefly, cocking his head to one side when a coyote began howling off in the distance, managing to sound even lonelier than usual for some reason.[302]

By the time darkness took over completely the fire had formed a welcoming dome of light and warmth. My grandfather would have been proud and we were soon giving in to our appetites, which felt well earned after the long hike. Before our ancestors discovered cooking with high temperatures to soften up their food they probably would have spent half a day or more just sitting around, chewing on raw seeds and stems. Uncooked food doesn't yield up its energy so easily. Our primate ancestors also preferred ripe fruit because it contained the most sugars, which is probably also why we evolved to metabolize alcohol so effectively. Every once in a while the heat from the flames sent into violent motion some water trapped within the wood, producing an explosion in the gut of the fire that rang out like a gunshot and Barney would recoil at my feet. The human world must be a strange and difficult one for dogs to inhabit sometimes.[cclxv]

The best part of the evening came when our appetites had been satisfied and the sky was dark enough to make out the familiar ribbon of white light stretching across the sky like a vaporous cloud. It was our home galaxy. Always ready to use whatever he had at his disposal to impress the Indians of the new nation's strength, Lewis warned them that back east America had "cities as numerous as the stars in the heavens."[303]

When the Earth is on the side of the sun that brings summer to the northern hemisphere and you're far enough away from civilization, it's possible to look towards the heart of the Milky Way galaxy. It's the same place where the ribbon of light bulges out slightly – as if you're looking at a coin edge on – but with a pale marble stuck in the middle.

Amanda pointed towards its center and exclaimed, "There's a super-massive black hole in there somewhere. All galaxies have one in their centers."

---

[302] Lewis was the first American to describe a coyote and also what would someday become the state flower of Montana – the bitterroot.
[303] And if angered, America would "consume you as the fire consumes the grass of the plains." Instilling fear was often necessary since the expedition usually found itself outnumbered.

We listened to the fire noises, the same ones the early the travelers would have heard 20,000 years ago while crossing the Bering land bridge.[304] When the fire had died down enough, the rushing of the river became audible again and everyone became quieter because nature can do that to you when it's on full display...render you speechless. Thinking about the rest of the universe can put everyday problems into perspective.

Here I was almost 40. When light had been sent out from a star near the center of the Milky Way the day I was born, it wouldn't even be a thousandth of the way across by now, which made me feel a bit more youthful as I held my thumb and forefinger up to estimate the distance. And if that same light had been emitted back when the ancient Egyptians were laying out their pyramids, it still wouldn't be a quarter of the way across, even traveling at 186 thousand miles every second for the past several thousand years. And to think there are billions of other galaxies out there, as large or larger than our own.

As with matter, energy comes in different densities too. It can be concentrated, diffuse, or in between, which is why a laser beam can be so effective at burning steel, unlike light from my flashlight, which spreads out so easily that I can barely see more than a few feet in front of me on a night like this.[cclxvi]

The light up in the sky got me to thinking about the lecture I'd given on metabolism the day before. It could be the last time I'd ever give it and I still had never finished it the way I wanted it to end. All four times I'd stopped somewhere along the lines of "All animals burn glucose in their cells using oxygen, which is why both campfires and people give off carbon dioxide, water, and heat." But time always ran out before I got around to telling my students about the other part of the flame, the part that was every bit as interesting. I never got around to explaining about the light.

So I waited for a lull in the conversation wide enough to drive a combine through. They'd seemed up for a story all day so I took my chances at finishing the lecture the way I always thought it should end, beginning conveniently enough with the Milky Way right above us.

I pointed to it, explaining how the ancient Greeks believed the goddess Hera had suckled the infant Heracles in her arms, and how she was careless and that some of her milk spilled, which for them was what formed the Milky Way. To the ancient Egyptians the narrow band of light represented a celestial version of the Nile. They saw the nighttime display as a counterpart to the one that flowed through their lives every day. At other times they explained the Milky Way as the place where the goddess Isis, while fleeing from the monster Typhon, had dropped shafts of wheat, and that the Milky

---

[304] The oldest mummy that's ever been found is *not* from Egypt but from the Western Hemisphere: 9,000 years old (Acha Man) in Chile.

Way was the glittering of those grains. I looked around, pleased to see I still had everyone's attention, even the two small children.

The Cherokee believed the Milky Way was put there by a disobedient dog that had stolen a sack of corn and scattered it as he ran. The Bushmen of Africa saw it as the dying embers of campfires other fellow travelers had lit. Still, it was Democritus – one of the atomists of ancient Greece – who first realized that the ribbon of light was the collective illumination of millions of tiny stars too small to be seen individually. When the telescope was invented 2000 years later, Galileo was able to confirm that the Milky Way was indeed a vast collection of stars invisible to the naked eye."

I got sidetracked briefly when Marjorie's husband, who was more interested in knowing where all the heat that warms the water at Yellowstone comes from, asked about it, so I put my lecture on hold and began telling them about my roommate back in graduate school...Howie. In the interest of time, I skipped going into how, for a country boy living in the big city for the first time it can be a lot like always being at someone else's family reunion, always feeling like you're left out of things, and being on the shy side probably never helped.

I also didn't mention how meeting someone from Wyoming was like having a member of your family drop by when you're far away and how the both of us had had similar experiences. Howie was the youngest and like me had an older brother who would inherit the ranch someday, which was why he was there studying geology. He had plans of becoming a professor and our conversations lasted well into the night, I going on about the incredible diversity of microbes in the gut of a cow one evening, he about the unique geology of the west and the energy it took to create it out of level plains the next.

Energy is a common denominator in so many areas of life. No one's ever seen energy, or ever will and so we usually think of it in terms of what it can do. During those sleepless nights I listened to Howie and learned how, by the end of the 1800's scientists had realized a few important things about energy, like how thermal energy possessed by moving atoms and molecules was what caused the sensation of heat, and that electricity and magnetism were really just two sides of the same coin. Scientists had also come to the conclusion that energy couldn't be created out of nothing, or destroyed for that matter either. Energy merely hid out when it wasn't doing anything.

Then, someone came across a curious phenomenon that didn't fit so neatly into this newly emerging picture. Henri Becquerel in 1896 took a close up look at some crystals that glowed in the dark, exposing them to photographic plates, but because the physicist had taken the trouble of wrapping his crystals with black paper beforehand, the negatives shouldn't have gotten exposed. And yet there it was...an image of the crystals right on the negative.   It seems the crystals had given off some kind of energy similar to light yet more energetic since it could travel straight through paper.[305]

He didn't have an explanation for this, but Becquerel did manage to interest two others in the phenomenon, a married couple in France, who would go on to discover that the purer they got this substance, and the more of it they had, the more energy it gave off.   No one realized this strange energy could be due to the inside of an atom falling apart since the nucleus was just being discovered around this time.[306]

All things in nature given enough of a chance will seek their lowest energy level, a boulder at the top of a mountain, an electron leaving one end of a battery in search of the other end, or a glucose molecule bathed in oxygen; even the nucleus of an atom will try and reach its lowest energy level.   And as the nuclei of the uranium atoms making up Becquerel's crystals spontaneously fell apart, they released some of their mass as energy.   But the amount of the nucleus's matter that turned to energy was so miniscule that its loss couldn't possibly have been detected even with the best balances of the day.

Some immediately took to claiming that Becquerel had found a violation in the brand new conservation of energy law.   That maybe it *was* possible to create energy out of nothing after all.   Eventually, Einstein put things right again by explaining that energy and matter were just two forms of the same thing using his equation $E = mc^2$.   A small part of the mass of the atoms in Becquerel's crystals and the Curie's uranium was simply becoming energy. The opposite is also true, because just as the equal sign in the equation indicates, it's a two-way street and you can also make matter out of energy.[cclxvii]

Neither the Curies nor Becquerel ever found out how useful radiation would someday be or where it would take human machines, how two spacecraft named Voyager would someday use the energy released from radioactive decay to power their electronics as they exited the solar system, nor would they know how geologists would predict that radioactive decay deep inside the Earth is destined to create another supercontinent like Pangaea given another hundred million years or so.[cclxviii]

Most of the heat energy trapped inside the Earth is due to this falling apart of nuclei, the radioactive decay of elements like uranium and nickel.   The rest of Earth's heat trapped within it is mostly thermal energy left over from its birth several billion years ago.   Like most substances, rocks expand when they get hot, so molten rocks in the interior of our planet become less dense and will therefore rise to the surface.

---

[305] Henri's father was also a physicist and demonstrated the first fluorescent light in 1867.
[306] Pierre & Marie Curie.   Marie was one of Becquerel's graduate students and would go on to win two Nobel Prizes, as well as coin the term *radioactivity*.

A lava lamp is a good way to see this principle in action. The "lava" in the lamp gets heated by a light down near the bottom. Then this material expands and rises because it is now, having increased in volume due to the heat, become less dense and therefore more buoyant than the cooler liquid surrounding it, so it travels up. Now further away from the light source the lava begins to cool and contract again, falling back down to start the cycle all over.

I turned towards the panorama of mountains in the distance, visible only as dark patches where they blocked the stars near the horizon. "Evidence of the tremendous heat still trapped inside the Earth," I explained as I pointed in their direction, "An example of thermal energy within the Earth being transformed into gravitational energy of rocks up on the surface. If not for the hot magma underneath Yellowstone holding us up, we would all be almost two thousand feet lower than we are now."[cclxix]

Howie had also explained to me all those nights ago how the geology in western Wyoming and Montana was like trying to decipher a story that's been written on a single sheet of paper, except that the paper has not one or even two, but maybe half a dozen stories all written on top of each other. And of course when one story is put down, it partially destroys some of the previous stories underneath. So geologists out west spend a lot of time piecing things together like this, trying to figure out stories.

Back in the 1960's Francis Boyd, a graduate student from Harvard, came to Yellowstone looking for a subject to do his Ph.D. work on, and in the process became the first to realize that an extraordinary event happened here, and more recently than anyone had thought possible. He noticed Yellowstone's newest rocks were made during a volcanic eruption that was so energetic it actually obliterated part of an ancient mountain range lying just above it. Visualizing the eruption of a supervolcano in his mind allowed him to make sense of how there could be rocks over a billion years old lying right next to youngsters…rocks only a mere million years in age.

Boyd realized the volcano and the mountains surrounding it in the distance were formed by unrelated events separated very far apart in time. The older mountains were formed due to moving continents shifting and colliding with ocean plates, the same way most mountains are formed today, kind of like scuffing up a rug with your feet. But the volcano that created Yellowstone was more recent, powered by a hotspot directly underneath it.

It's because the North American continent has slowly been inching its way over top of this hotspot, kind of the way a sheet of paper can be slid over top of a candle's flame. It's also why the Hawaiian Islands are in the form of a chain, and why people studying satellite images of the Pacific Northwest after Boyd published his field work in 1961 were able to notice this huge 400-mile long swath called the Snake River Plain where the Yellowstone volcano has slowly inched its way across in the past, taking some 17 million years to get where it is now – in the northwest corner of Wyoming – having started out in Oregon towards the Pacific. In fact, it's possible to count at least 140 past eruptions as the volcano burned its way to its present location. The North American plate is still moving over top of this hotspot at about the same rate your fingernails grow, which is of course how fast the entire continent is slowly inching its way to the southwest.[cclxx]

I wrapped up the detour by mentioning how Howie graduated a year before I did, but I left out the part about him not being very fortunate with his choice of a mentor; and how this would eventually come back to haunt him. There's no question about it, your mentor is the single most important factor in your success – or failure – after leaving graduate school. I've heard that most people who have won a Nobel Prize worked for someone at some point that also won one. I'm not sure how true this is anymore, but it is true that your advisor is the one who finds you the right project when you start graduate school and later on the one who sets you up with important contacts in the field.[cclxxi]

The guy Howie chose may simply have seen him as future competition, or maybe they just never got along, or maybe the guy wasn't very talented, I'm not sure which it was. Like me, Howie kept a lot of things to himself. Either way, the last I heard, Howie was doing consulting work for an oil company and in the process taking on a drinking problem, never having gotten into a tenure-track position at a university, and I've often wondered how many other Howies there are out there in science no one ever hears about.

My younger technician guided her dreamy-eyed children back to their tent while the rest of us remained to enjoy what was left of the fire. Being a professor has taken me out of my shell, but not so much that I actually enjoy public speaking. Hearing the sound of my own voice still fills me with a sickish feeling but at least I can stand it more than before. Growing up on the ranch, I was always able to let my older brothers do what talking needed to be done and they didn't seem to mind the arrangement much.

And as had become my pattern lately, Egypt wasn't far from my thoughts, so I got back on track by explaining how ancient people who lived near the Nile wouldn't have stayed up this long after dark. To power lamps each night would have been too expensive for the average Egyptian, at least for the ones who worked in the fields from sunrise to sunset, which is what 90% did. The Egyptians fueled their lamps with vegetable oil when they had it, extracted from crops like the castor bean, or they used wax from honeybees. The very first lamps our ancestors made to stave off the night were likely fashioned out of coconut or eggshells. Others may have been a rock with a natural depression in it, this hole then filled with oil and a piece of papyrus or flax thrown in for use as a wick.

The Native Americans of the Pacific Northwest used a whole fish to illuminate their darkness. Small as a sardine, candlefish are laden with fat during spawning, and so the Indians dried them in the sun; then inserted a long strip of bark into the fish's body as a wick or sometimes they just lit the whole fish on fire. Almost a fifth the weight of a candlefish is fat; chemical energy capable of being turned into light, carbon dioxide, water, and heat.[cclxxii]

Custom made terra cotta lamps eventually replaced rocks and shells so that in ancient Greece, lamps had grown sides on them, which would have helped keep the pressed olive oil inside. The Romans took this a step further, extending the sides of these lamps right over the top and joining them like a roof so that the lamp resembled a hollow "slipper". Some Roman terra cotta lamps accommodated as many as a dozen wicks.[cclxxiii] In northern Europe, away from the milder weather of the Mediterranean, candles became more common.

Ancient people would have valued fire for many of the same reasons we do today, however they also likely worshiped it. The Vestal Virgins in ancient Rome were charged with keeping a flame going in the heart of the city and any citizen of Rome could make their way to the Temple of Vesta and get fire whenever they needed it. If the flame at the temple went out it was a harbinger of destruction, taken as a sign Rome had fallen out of favor with the goddess Vesta.[cclxxiv]

At the Acropolis in ancient Athens a flame burned continuously – as it did during the Olympics every four years – to represent Prometheus's gift of fire to humanity, and of Zeus's willingness to punish him afterwards for stealing it. When I was touring Karnak temple in Egypt, I looked up and saw black soot that had accumulated on the stone ceiling above my head; the result of countless fires kept going by Coptic Christians beginning around AD 100 to cook their food.[cclxxv]

It's tempting to consider that the use of candles in so many religious ceremonies from Buddhism to Christianity is still practiced because candles hearken back to a time when a continuous flame would have represented protection and warmth from an unpredictable, alien world, not so different than what religion holds out to so many today. In light of this, it's not hard to see why ancient Egyptians worshiped the sun. In fact, its light represented for them an escape from death.

In 1976, the Olympic torch in Montreal, Canada was lit by the torch in Athens several thousand miles away. They did it by changing the energy of the flame in Athens (using an ionic sensor) into a radio signal, which was then transmitted by satellite across the Atlantic where it switched on a laser beam, igniting the torch in Canada. The energy from the original flame in Greece traveled all that distance, a dramatic example of how energy can be made to move without itself being destroyed in the process.

Without the light from the moon we relied on our flashlights to go back and forth from the tents to get more mosquito repellent. The whole reason we even have nighttime on Earth is that our own star is so far away from the center of the Milky Way. Closer towards the middle, stars are near enough that it's daylight all the time there, but since the Earth is on an outer arm, we have always had periods of darkness, so it wasn't hard to imagine how difficult life would have been in ancient times without the benefit of illuminated streets at night. In fact the largest city of the ancient world, Rome, was really two cities...a safer one where business was transacted, the baths, temples, colosseum, and Circus Maximus visited during the day...and an extremely dangerous one after the sun went down, when few Romans would have ventured outside without a bodyguard. Ancient Rome at night had been given over to the criminals and other riff raff.[cclxxvi]

When street lighting did arrive towards the end of the Middle Ages in London in 1417 thanks to the extremely dark winters, it would have changed the way people lived. Paris got street lighting in 1524 and Antoine Lavoisier – the one most often credited as the Father of Chemistry – was elected to the French Academy of Sciences at age 25 thanks to an essay he wrote on better street lighting in Paris.[cclxxvii] Whaling became the largest industry in North America in the 1800's and sperm whales, whose bodies can be up to a third oil, eventually grew scarce due to their fat's ability to burn brightly and with little smoke in city street lamps. Whale oil also lubricated the new machinery invented during the Industrial Revolution.

While Europe's city streets were being lit for the first time, another development was taking place that would also factor into my lecture, this time a little further to the east.

SPIRIT OF THE COAL

The word coffee comes from Arabic because Yemen was where coffee beans were first imported from Africa by way of Cairo and roasted, then brewed into a beverage. As with other foods, cooking with high temperatures can cause profound changes in the biological molecules of the coffee bean as the intense heat breaks apart stored starches, making them more accessible to join in so-called *browning reactions* with the bean's other major ingredient: its proteins. This carmelization also produces new flavors while releasing stored up volatile oils from the bean, lending roasted coffee its distinct aroma.[cclxxviii]

Moslem monks soon began growing coffee trees in their gardens – coffee being an alternative to drinking alcohol – and for a time in Europe this stimulant acquired a reputation as being an "Arab drink" so coffee was shunned until 1600 when Pope Clement VIII claimed it as a Christian beverage.

Four years after the fall of Constantinople in 1453, a visitor could have come across coffeehouses located near the city's most important mosques. Cairo acquired its first coffeehouses in the 1530's and Napoleon would later fend off a rebellion while in Egypt that was sparked by talk inside a coffeehouse of Cairo.[cclxxix]   Coffee's journey to Europe passed through Venice as this city-state had established trading routes with Egypt and the Middle East, so the first coffeehouses in Europe were, not surprisingly, in what is now Italy around 1645.   England's first coffeehouse would open not long after in Oxford in 1650 and by 1739 there were more than 500 all over London. [cclxxx]

So in the eighteenth century, Europeans living in cities felt safe enough to walk the streets at night thanks to artificial lighting and they began congregating in Europe's new coffeehouses to discuss issues of the day.[307] These were natural meeting places – outward looking and optimistic rather than inward and withdrawn as neighborhood alehouses had always been – and coffee's stimulating effects would have helped fire the vigorous debates inside.   They were the Internet chat rooms of their day.

Customers could also choose to simply relax and play games or learn subjects as diverse as astronomy and chemistry.   At a time when only the sons of the wealthy could afford an education at a university, no where else could someone with modest means lay down the cost of a cup of coffee and rub elbows with the most learned people of the day including Isaac Newton, Edmund Halley, Robert Hooke and Samuel Pepys, all of whom frequented coffeehouses. [cclxxxi]

One of the most significant events during the 1600's was a questioning of religious authority.   This was the period following the bloody religious wars in Europe and the time seemed right for reflection, which is why it's not surprising coffeehouses would come to be viewed with such suspicion by Europe's ruling families.[cclxxxii]

With the invention of the printing press, information was being disseminated faster than ever before.   The first modern encyclopedia was edited by Diderot[308] in a Paris coffeehouse in the late 1700's and by 1784 Europe had no less than 70 scientific societies, some of the first ones having sprung up inside coffeehouses.   Textbooks, pamphlets, and dictionaries were also widely shared, Diderot's encyclopedia being the most tangible example of this new belief that knowledge should benefit everyone in all walks of life.

---

[307] Not coincidentally, it was the beginning of the scientific revolution. Francis Bacon made an analogy between the lighting of a candle and the formation of a *hypothesis*, which he said, along with performing experiments, would someday allow science to shed its own light.
[308] Among other entries in the year 1750, it had descriptions of some 30 French breads baked by more than 250 Parisian bakeries.

Lloyd's of London and the auction house Christies also began in coffeehouses. John Locke and others met in them to transform the scientific principles lain out by Newton governing the motion of the planets into laws governing societies, eventually giving birth to the notion of *natural rights* and the revolution these ideas would ignite in the New World.[cclxxxiii]

In the 1600's chemists knew that coal could be heated without the benefit of oxygen, producing a volatile gas they referred to as the *spirit of the coal*, and by 1807 this brighter-burning fuel was being piped to London's streetlights and factories which, because of the increased luminosity, allowed people to work additional shifts throughout the night.[cclxxxiv]   These gaslights also increased the popularity of both reading at home and going out to the theater at night in the mid-1800's, useful for producing never before seen effects, heightening the drama of an important scene by reproducing daylight or sunsets more accurately than ever before.   Like Hollywood with its latest special effects, this astonishing new use of gas lighting in theaters drew large and admiring crowds.   Abraham Lincoln was among them; in fact he visited the theater no less than 100 times as president.[cclxxxv]

Better drilling methods in the 1850's allowed for deeper oil wells, which meant that a chemical byproduct of oil – paraffin wax – became cheap enough to replace animal fat as the fuel of choice for candles in a similar way that kerosene had become the more practical choice for lanterns.[cclxxxvi]   By the 1860's there were factories in America that could churn out 1500 wax candles each and every hour.[309]

Marjorie's husband tossed another log onto the fire, shooting flames up just as I was coming full circle though no one knew it, finally wrapping up my lecture on energy, the same lecture I had begun three years ago when my teaching changed.   The heat from Aristotle's flame would someday power steam engines of the Industrial Revolution, which would change forever the way the average person went about their daily life.[cclxxxvii]   It would also give birth to the study of thermodynamics, and the principles that governed the steam engine turned out to be the same as those governing our muscles and the opening and closing of neurons in the brain.

So much for the heat, but as for the light, it made city streets safe enough for people to gather in coffeehouses, which in turn helped to fuel the Enlightenment and the Scientific Revolution this gave birth to.   The divine rights of kings to rule as well as for religious authorities to explain the natural world, was finally being questioned.[310]   Readily available information and the light to read it by in this way helped free people from tyranny, paving the way

---

[309] The 2nd patent in US history (in 1790) was for an improved method for making candles.
[310] Napoleon would use the Enlightenment as a reason for invading Egypt, which explains why he brought so many scientists and artists along with him.

for the American Revolution, the first time in history a nation would be created on paper.

LAG PHASE

At about the same time I finished, the yawns were becoming more frequent, the lively sounds of the fire having been replaced by those of the rushing river as one by one everyone returned to their tents, leaving Amanda and I alone on the log.   Frank's piece of wood turned out to be green so the fire had now become a stream of smoke revealing the direction of the wind.

There were many nights Lewis & Clark's men stood within the thick columns of smoke given off by their campfires to ward off mosquitoes; endless swarms it must have seemed.   And when they didn't have smoke, they used bear lard, smearing it all over their bodies – "voyager's grease" they called it – 300 pounds of which they stocked up on in St. Louis before leaving civilization.   The clouds of insects were so overwhelming that at times Lewis couldn't see well enough to aim his rifle or even open his mouth to eat dinner.

All day Chris had been by her side, offering to carry her backpack when she got tired, helping her set her tent up when she fumbled with the poles. But now he had gone to bed and I couldn't help wondering if maybe he hadn't finally picked up on the possibility that Amanda wasn't interested in him romantically.

"Looks like everyone's gone to bed," Amanda said as she leaned in closer to what was left of the embers that were imparting a warm glow onto her skin, as if the coals were trying to kiss her, and I wondered if she hadn't been one of my students whether she would she still laugh at all my dumb jokes.

There were times in the lab I had to make a conscious effort to keep my distance but I was never really sure if anyone noticed, or maybe I kept my distance because I was afraid of how I might feel, or worse, that these feelings might show.   One thing about working with students is that they are quick to notice if you pay more attention to one than another, even if it is innocent on your part.   Teachers are human too, though, and the more I got to know Amanda, the more I came to respect her and think of her when I was alone.   She scooted closer to me now that Chris had gone to bed.

"I just want you to know," she began, "that no matter what happens…with your tenure I mean," she hesitated, then finished her thought, "that I've done more than I ever would have in the other lab."   She placed her hand on my knee and left it there while I thought about how, if she wasn't one of my students, how easy it would have been to place my hand on top of hers.   In fact, I ached to at that moment.   But it was also true that Marjorie's tent was right between Amanda's and my own. And as everyone in the department knew, Marjorie was by far the biggest gossip in all of Clark Hall.[311]

I pretended not to notice, thinking about how what was between Amanda and I was like the campfire…something that could go either way. I could let the flames die down until they faded out completely, or give them new life by tossing another log on top. I always was a coward when it came to my heart. But if something should happen, if anyone on my committee were to get wind of a teacher being with a student in that way, I could kiss my chances for tenure goodbye…that much was for certain.

"You know," I began hesitantly, "my brothers were only interested in roping cattle and getting married. Sometimes I think maybe I should have stayed on the ranch too. Problem is, I was always too afraid of falling off a horse to be a good rider."

"Me too," Amanda said while pulling her hand away from my leg and placing it on her own lap and my heart diminished a little more along with the flames.

In an effort to cover up I added, "Maybe running a lab isn't such a good thing for someone afraid of falling off a horse."

We sat a while longer, listening to what was left of the fire, which was still generating warmth even though it was on its way out and it had been more than three years since my last relationship. I knew professors who had managed to avoid marriage completely because they were so concerned with "making it" in academia. It takes a lot of commitment to stay in science these days. Research is competitive and the money can be scarcer than hen's teeth. I wondered if Amanda even realized my situation. Finally I broke the silence, which was by then descending into awkwardness.

"It's strange to think how Charlie has more job security than I do. He and the secretaries upstairs both have pension plans. Kind of makes me think ten years of college could have been better spent sometimes. Hell, if grant money was skunk oil a hound dog couldn't find me."

"I suppose," she replied and smiled. If Amanda had been hurt by my decision not to follow up, she didn't let on. "What will you do if…" her voice dissipated into the surrounding wilderness…as if consumed by it. "If…"

"If I don't get my tenure?" I finished it for her.

She nodded and then looked directly at me, perhaps to let me know she was waiting for an answer. I had been asked that question by just about everyone in Clark Hall and around town but usually with the intention on the part of the asker to find out how it might affect them. With Amanda it was clear she wanted to know how it would affect me, which meant something.

---

[311] A standing joke was that her first husband was so henpecked, he molted twice a year.

"To be honest, I haven't thought about it much. Probably go back home. Stay there for a while. They can always use an extra hand during calving season. Maybe it would give me a chance to think things through."

"But you're such a good teacher. They can't take that away from you...can they?" She wasn't one for open displays. When Amanda said something, she usually meant it. We sat a few minutes more until it was clear that the fire had made up its own mind...still I was hesitant to drown out the last of its embers with the pail of dishwater. Not just yet. It still seemed there was something more that needed to be said. Somehow I had to end the evening while at the same time leaving the door open for another chance...at least that was my plan. My only wish at that moment was that I could pick up on some sign that she understood my predicament.

"Josh's father wants to come back to us," she stated with a voice that sounded more like resignation, then she turned and looked off towards the mountains. In all the time I had known her, it was the first time I had ever heard her mention Josh's father and it seemed strange to hear about him now.

"What are you going to do," I asked, a dry, irritating feeling accumulating in my throat.

She didn't say anything because she didn't need to. It was obvious by her tone she hadn't decided. The dark silhouette of her profile made her look delicate against the fire, everything about her seemed fragile at that moment and I wondered what could possibly drive someone to leave her. I couldn't understand and deep down hoped I never would. Sometimes being fortunate enough to be born into a stable family can be a disadvantage when it comes to understanding some things.

I reminded her of the big day tomorrow and stood to use the bucket of dishwater to douse out the embers, which began smoking in protest. We parted ways without our eyes meeting...not until the last possible moment.

Just as she was close to being out of earshot, she turned towards me and asked, "Do you have any more room in your truck tomorrow? I think Chris is getting a little tired of me. And besides, he plays that music so loud."

"Of course," I answered, happy for myself, yet sad for Chris. Amanda then slipped into the darkness, the only thing giving away her presence the diffuse beam from her flashlight...until it too disappeared.

Afterwards, I lay inside my sleeping bag with Barney draped across my legs and couldn't help thinking about her again. I sure was a fine one for missed opportunities. First the anthrax grant and now Amanda. Something told me there would be no second chance this time. There had to be a lot fewer successful resubmissions with women than with grant proposals, I lamented.

It was then that it occurred to me I had missed something. No lecture was ever perfect and the one I'd just finished turned out to be like all the others. But how could I have forgotten Robert Bunsen of all people? And at Yellowstone no less? I scratched Barney a little harder behind the ears, silently disgusted in myself, resisting the urge to cuss out loud, remembering that Marjorie and her family were just two thin sheets of canvas away.

Bunsen was the German chemist in the 1800's, the one who designed the burner that cooked with heat so cleanly and reliably and, of course, hot. And he used this new flame to discover some elements, two of them in fact, inside of rocks... cesium and rubidium. One of the ways chemists identify atoms is by the unique color signature, or spectrum of light, they give off when they tickle them with thermal energy.

You could think of it the way brass bells will each give off a unique, identifying tone whenever they are rung depending on their size. The heat from a hot flame "rings" an atom's bell by energizing (lifting) its outer electrons encircling its nucleus up to a higher energy level, and then when these electrons come falling back down again towards the oppositely charged nucleus (the electron's lowest energy level), this extra bit of energy that previously lifted the electron now has to go somewhere and so it is emitted as a specific color of light. These differences in energy levels are dependent on the identity of the atom and so produce their own unique colors. To be reminded of the usefulness in this, one need look no further than many streetlights, which are probably orange because they contain the element sodium, or the two elements Bunsen discovered with his new burner – cesium and rubidium – and the names he gave them, both in Latin... cesium for blue sky and rubidium, which means deepest red.[cclxxxviii]

The city of Heidelberg, Germany had just gotten gas street lighting in 1852 and as a precondition for his becoming a professor there Bunsen had the university extend gas pipes to his laboratory. Now that's pull for you, I thought. As a professor I'd never been in a position to set preconditions about anything, not even my parking space. Bunsen had been studying gases inside iron blast furnaces in Marburg and had already made a name for himself in Europe. His insight for the new burner included the use of an adjustable screw valve down near the bottom, down where the gas first enters, kind of like a carburetor in an automobile's engine, which allows the fuel and the oxygen to mix in just the right proportions *before* being burned.

Barney was smiling now, appreciative of the extra attention and I was silently regretting not mentioning Bunsen the more I thought of it. And I had the perfect tie in, too, and like Amanda I'd let the whole thing slip away. Bunsen was the same chemist who discovered the mechanism for geysers. After visiting Iceland and making observations, he returned to Germany with his new hypothesis and built an artificial geyser with some spare pipes lying around.

Acting on a hunch that the reason the water came out of the ground in geysers so forcefully was because the water boiled deeper down, in the lower chambers first, down in the ground where the water was hotter, he heated only the middle of his water-filled pipe until just this lower section boiled. When this water suddenly turned to steam, the pressure of the expanding gas launched the column of liquid above it high up into the air, not unlike the way gunpowder turns to a gas and pushes a bullet out of a rifle. Bunsen's reasoning is still one of the accepted explanations we have today for how geysers work.

He could have been rich, but he didn't take out patents, not even for his burner as far as I knew.[cclxxxix] He was also a dedicated teacher and he was blind in one eye because of an experiment; I'd read that somewhere too, and I seemed to recall he'd never gotten married. While teasing out a burr from behind Barney's ear, I silently wondered if I'd ever get the chance to give that lecture again. But at least I'd finished it…such that it was…and without anyone getting too bored. Still, it's hard to tell with Montanans sometimes if they are listening to you out of politeness, or because they're genuinely interested. We tend to live by the old saying that you're never supposed to slap a man on the back while he's chewing his tobacco.

TRAVELING AHEAD

It's officially summer in Montana not when a calendar says so, but when you can drive with your windows all the way down on the highway without needing a cap to cover up your ears. Even with the air blowing throughout the cab, the drive back was still uncomfortable and Amanda held onto Barney's collar ring with her index finger, allowing him just enough room to poke his head out the window until we stopped at one of the roadside monuments that mark a place where Lewis and Clark passed by.

Growing up in Montana means knowing things about the Corps of Discovery without knowing for sure where your information comes from, whether it was one of my grandfather's talks around the dinner table, a play at school, a documentary on TV, or a roadside monument like the one we'd just stopped at. When I got older and went away to college I was surprised to find that even there I didn't have to be far away from them. In some ways I just went deeper into their journey is all. In advanced microbiology courses they teach you all sorts of things about infectious diseases and how there are some that can travel so quickly – even before the age of steam or jet travel – that they moved with a rate that seems astonishing to epidemiologists today.

On their way back to civilization the Lewis & Clark Expedition just missed discovering Yellowstone by a mere 100 miles or so, passing to the north, where Amanda and I had stopped for our break. One member, a guide named John Coulter, later returned and trapped beaver and traded with the Indians and he eventually did make it down to Yellowstone, but when he got back to St. Louis no one believed what he had to say about the place. No one thought it possible that mud in the ground could be hot enough to boil or that steaming water could shoot straight up out of a hole one hundred feet into the air.

To the Native Americans, the place had been sacred...even frightening at times. Clark later interviewed one Indian back in St. Louis, which he recorded in his journal: "There is frequently heard a loud noise, like thunder, which makes the earth tremble, the [Indians] state that they seldom go there because [their] children cannot sleep – and conceive it possessed of spirits, who were adverse that the men should be near them." Not until an 1871 expedition that included scientists, an artist, and a photographer, would Yellowstone become accepted as more than just a figment in a mountain man's wild imagination.[312]

At about the same time Robert Fulton was experimenting with steam to travel up the Seine River in Paris, the Corps of Discovery was using a combination of muscle, wind, and skill to fight their way up the Missouri. In 1804, before steamships plied North America's rivers, people and goods traveled at about the same rate they did 4,000 years ago back in ancient Egypt. The strongest of the expedition tugged on the 55-foot keelboat from shore which, when fully loaded, weighed 12 tons, pulling on ropes along the banks, while others stayed on board and helped out with the rowing. When they had the wind at their backs they could put up a sail and make thirty miles in a day. But mostly it seemed as if the wind came head on so naturally things went slower.

Each member consumed 9 pounds of meat on average[313] per day while fighting the current, and they spent the entire first summer going north so that by the time they made it to the Mandan village along the Missouri River in present-day South Dakota, winter was already settling in. They built a fort on the other side opposite the village, where they would await the arrival of spring.[ccxc] While over-wintering they hired a French fur trapper after learning

---

[312] The artist was Thomas Moran. He rode there with a pillow on his saddle because it was his first time on a horse. Several members of Custer's 7th Calvary at the Little Bighorn were city boys and had only recently learned to ride a horse. Custer was often referred to by his troops as "Iron Butt" or "Hard Ass" because of his ability to remain in the saddle for long periods.
[313] Sitting Bull's warriors each consumed about 6 buffalo a year on average.

he had a Shoshone wife kidnapped as a child by the Hidatsa and brought east near the Mandans. Lewis knew from talking with the Indians that crossing the mountains would require horses and the only way to get them was from the Shoshones. In addition, they'd also would need a guide to help ensure safe passage over the mountains once there, and perhaps the Indian woman could help with that too, they reasoned.[314]

And so on the 7th of April, 1805 the group set out from their winter quarters for what remained of the Missouri River all the way to the high peaks of the Continental Divide and eventually, they hoped, down the other side and to the great "illy tasting" sea beyond. The Indians had warned of a waterfall on the Missouri somewhere close to the mountains, but when they got there Lewis and Clark found not one, but a series of five. The only way around was a long and difficult portage, so they made makeshift wagons using cross sections of tree trunks as wheels, and they pulled their boats for 18 grueling miles. It took a month with human muscle and ingenuity as their only resource. Some of the men fainted while wearing out moccasins at a rate of a pair every two days due to stepping on cactuses.

By the time they re-launched back into the river it was already so late in the season they were running the risk of being trapped on the more inhospitable eastern slopes of the Rockies as winter arrived. If they could somehow make it further west, just over the tall peaks in front of them and down the other side they realized they could for the first time use the river's waters flowing all the way to the Pacific; riding with the current rather than fighting against it as they had been since leaving St. Louis more than a year earlier.

As hoped, Sacagawea recognized her people and helped gain horses and a guide, however it still took longer than anyone including the Indians expected to get to the western side. Along the way one member described their ascent up ancient trails as "nearly as steep as the roof of a house" and it wasn't long before they had to sacrifice packhorses for meat. It took 11 days total...some mornings waking up to snow having fallen on their blankets and bare feet during the night. When they finally emerged on the other side of the mountains, they were nearly starved.

Here they built fires to hollow out Ponderosa pine trees felled for making dugout canoes, and described their journey down the Clearwater River as faster than a horse could run. In time this smaller, swifter river would merge with the Snake, and eventually the Columbia and the expedition knew they were getting closer to the Pacific when they began spotting Indians on shore

---

[314] The Hidatsa Indians that kidnapped Sacagawea several years before did so as members of a raiding party intent on stealing Shoshone horses. They also took Shoshone children as slave labor.

carrying brass teapots and British-made muskets. At least one Indian had on a sailor's jacket and knew how to swear in English.[315]

The expedition reached the mouth of the Columbia on the 18th of November 1805, near where it empties into the Pacific, whereupon the Corps struck up a trading partnership with the local Clatsop Indians who went on to explain how there had once been a great many of them, but that something mysterious happened. They had been visited by a devil for they believed they had angered a god and were being punished. But to Lewis and Clark, the reason was more obvious. The Clatsops had been visited by smallpox.[316]

It wasn't the first time they had seen the effects of the disease while traveling in its wake. Not far from the Mandan village where the corps first over-wintered while going up the Missouri the previous summer they'd come across Arikaras living in three villages where once there had been eighteen. The expedition passed abandoned shelters with crops still in the fields, corn and squash but no one left to harvest it. Smallpox had done the reaping instead, reducing the Arikara's from 30,000 in the late 1700's to about 2,000 by the time the expedition made it to South Dakota.[ccxci] Lewis & Clark found human bones still inside deserted earthen lodges as smallpox had made the journey ahead, carried within the blankets of fur traders. In fact, smallpox had become so well established along the East coast that it helps explain why so few Indians greeted the Pilgrims when they landed at Plymouth Rock in 1620. The virus had already traveled north from the Virginia Colony four years earlier.[ccxcii] In fact, the microbe first hitched a ride to the New World just 10 years after Columbus set foot here.[317]

With a 12-day incubation period, smallpox was able to make the journey across the Atlantic in just a handful of victims. On the 12th day after exposure, characteristic red *herald spots* would be visible on a victim's forehead. Once these sores multiplied, a physician would check the soles of the feet because smallpox knew no bounds and could travel everywhere on the body, whereas chickenpox produces similar sores but never invades the soles of the feet. If the unlucky victim had sores there, he or she was given only a 50% chance of surviving with just some scars and perhaps blindness.[318]

Smallpox had become entrenched in European cities long before Lewis and Clark were born for, along with glassmaking, lemons, rice, and dyes, the returning Crusaders also brought smallpox back from the Middle East in the 1100's.[ccxciii] Queen Elizabeth I was bald thanks to the disease, having had

---

[315] One knew how to say "son of a bitch".
[316] Geographically, smallpox was one of the most widespread diseases in human history; the only disease ancient people devoted gods and goddesses to.
[317] In 1540, the explorer Hernando Desoto also noticed abandoned villages due to smallpox that had traveled up the Mississippi River ahead of him.
[318] Most blindness in 18th century Europe was caused by smallpox.

the same telltale facial scars Joseph Stalin would someday insist on removing from his photographs.

Because smallpox was a childhood disease, by the time someone reached adulthood in England and Europe they were likely immune to it. But in the New World, the disease felled whole segments of society, young and old alike, so quickly that it led to the destabilization of entire tribes.[ccxciv] The explorer Cortez was able to defeat millions of Aztec warriors with just 600 Conquistadores because smallpox did his killing.[ccxcv] The Spanish when they entered the ancient city of Tenetchilon described the bodies of the Aztecs as being so numerous that it was difficult to step on the ground without walking on a body.

Smallpox can travel through the air not unlike the common cold...in respiratory droplets. And the same type of sores that form outside of the body can also develop within the narrow confines of the throat, meaning victims could expect to die from pus oozing out of these inner sores and choking off their air. Either way, smallpox meant a horrible death.[319]

A decade before Lewis & Clark arrived at the Pacific in 1805, the British admiral George Vancouver was finishing up his exploration of the westernmost coast of North America including the Puget Sound area near what is today Seattle.[320] And Vancouver's men also came across abandoned villages just as Lewis & Clark would later describe in their journals. Vancouver had also witnessed the blindness and telltale scars on survivors. And in 1787 as far north as Sitka, Alaska, fur traders described the decreased numbers of Indians there because of smallpox.[321]

There were ways of preventing the disease known since ancient times. It was realized for example that anyone surviving smallpox as a child didn't get the disease a second time; therefore survivors were often caregivers. The Chinese recycled smallpox scabs, grinding them into a powder and then had their children inhale it through a straw.

In 1716 a British diplomat's wife, Lady Mary Montegu, came across a safer method of prevention while living in Istanbul, Turkey. Like so many, she too had survived the disease and had facial scars and lost eyelashes as a reminder whenever she looked in a mirror.[322] During the 1700's smallpox claimed on average 400,000 lives a year in London and the rest of Europe

---

[319] Microbiologists refer to smallpox as a *front end & back end-loaded virus* because early in the course of the disease the virus gets passed to the next victim by coughing, while later, after the victim has become incapacitated (or perhaps has died), the virus still manages to get passed on via scabs falling off the skin. Smallpox virus is one of the most stable viruses known, still viable 10 years after being shed.
[320] It was Britain's need for whale oil in lamps that helped finance of his voyage.
[321] Columbus introduced livestock including pigs to the island of Hispaniola on his second voyage, which also gave rise to the first epidemic in the New World caused by an Old World disease: influenza.
[322] Her brother had died of smallpox.

and the universities of Harvard and Yale were both founded in America due to parent's fears of sending their sons off to Europe to study in cities where smallpox reappeared with regularity.

The Ottomans learned from inhabitants of the Caucus Mountains about variolation, the act of deliberately giving someone a mild case of smallpox through an incision in their arm. The sultan preferred women in his harem unblemished, and so he had adopted variolation as his method of choice. Lady Montegu brought the procedure back to England, where it was tried out for safety and effectiveness on prisoners condemned to death. They not only survived exposure to the pus (or "smallpox venom") inoculated deliberately into their arms, but more importantly had immunity afterwards when exposed to smallpox patients and were thus granted amnesty. Variolation caught on rapidly in England after the Royal Family began using it and the practice then spread to the European continent.

So when George Washington and his Continental Army faced off against the British at the Battle of Quebec in the winter of 1777, it was the British who won in large part because a smallpox epidemic had already swept through Washington's camp, weakening and killing his soldiers, while the British had either survived the disease as a child, or had been variolated; explaining why Canada is not a part of the United States today. Washington learned from this setback and soon had his entire army variolated. His campaign against smallpox was so successful that it would be the only disease that decreased during the Revolutionary War.[ccxcvi]

But variolation was not without hazards. In fact, about 3 out of every 100 who underwent the procedure came down with the disease and died.[323] Variolation could, on occasion, even trigger epidemics where none existed before. Because of this, the practice was relegated to hospitals outside towns and cities, or on remote islands. Variolation was so dangerous that 18th century physicians who performed it in their homes might have reason to fear for their lives. As a lawyer Thomas Jefferson represented a doctor whose house had been burned after performing variolations, while Cotton Mather had a grenade thrown under his house for the same reason. Benjamin Franklin was at first opposed to variolation but after he learned more changed his mind and even wrote a pamphlet describing it in 1759.

This fear surrounding variolation helps explain why – when a safer method to acquire immunity was reported in the late 1700's – it caught on so rapidly. Once again it was local knowledge that made a difference. An English doctor (the MD kind) had been practicing medicine in the dairy counties where he observed milkmaids immune to smallpox as long as they'd first gotten a similar yet milder kind of disease.

---

[323] King George III's son died after being variolated in 1783.

Prior to his years in the countryside, Edward Jenner had worked under the great John Hunter (in 1770), a scientist, teacher, and surgeon who had drummed into the youth the importance of making careful detailed observations when doing science. Thanks to Hunter's influence, Jenner also became a gifted naturalist and helped prepare plant specimens brought back by Captain Cook in 1771 from one of his voyages. The following year when Cook sailed again, Jenner was offered the position of ship's naturalist but declined, choosing instead on what he assumed would be a simpler life practicing medicine in the English countryside.[324] Jenner himself had nearly died after being variolated at age 8 and so he knew first hand the dangers of putting smallpox venom into someone's arm and the need for a safer method.

On May 14th, 1796 he vaccinated an eight-year-old named James Phipps with cowpox pus and later exposed the boy to smallpox (by variolation). But when Phipps failed to contract the disease, Jenner tried to publish his results in the Royal Society's journal. His work was not only rejected but he was reprimanded for having wasted so much time on the procedure.[325] He was even warned that if he continued, he'd only cause further embarrassment for his career. Fortunately, Jenner got his results into print even though he had to use his own savings. Thanks to his determination, word spread of what he called *vaccination*, which is Latin for "from the cow".

By 1799, 1000 patients had received Jenner's new vaccine and as the country doctor's reputation grew, he had to give up much of his practice because people were traveling great distances to him just to be vaccinated. Some cowpox virus was sent to America after drying the pus out on threads; in which condition the vaccine was able to make the 30-day journey across the Atlantic intact. Thomas Jefferson acquired some of Jenner's cowpox and had his family, slaves, and neighbors vaccinated at Monticello in 1801. Jefferson himself had already been variolated in Philadelphia as a young man in 1766.

One of the goals of the Lewis & Clark Expedition was to impress upon the Native Americans the power of the white man's medicine, which is why the expedition carried Jenner's new vaccine and Lewis was likely trained by Jefferson on how to administer it. To bring the Indians fully under the tent of the Americans (who had just acquired the Louisiana Territory from France), it was important to demonstrate the power of their medicine as well as their technology.[326]

---

[324] Jenner also studied the cuckoo bird and published a paper on it in 1787.

[325] The Royal Society, chartered in 1662 by King Charles II, had as its motto: "Take nobody's word for it."

[326] Smallpox virus is stable up to 10 years and has settled upon a tactic similar to anthrax when it forms a spore, the *sit & wait* strategy. During the winter of 1801-02, Jefferson had a visiting delegation of chiefs in Washington vaccinated against smallpox. Protecting the Native Americans from disease was one of his many interests.

Jenner's smallpox vaccine was the first medical treatment that could safely prevent a communicable disease...tangible fruit of the Enlightenment. The contacts made within the scientific community through which news of Jenner's vaccine spread had begun in coffeehouses, which led to the formation of scientific societies. Even the Janitor from Delft – Leeuwenhoek – contributed to science by dictating letters, up until a few hours before he died at age 90.[ccxcvii]

Meriwether Lewis was an Enlightenment thinker yet with almost no formal education. Well-versed in 200 botanical terms he used them in his journals to describe plants found along the way as he walked alongside the river while Clark, the better waterman of the two, piloted the keelboat. Above all, the Enlightenment stood for the idea that knowledge didn't belong in the hands of a few, but should be used for the benefit of everyone regardless of their station in life.[327] In the early 1600's, as in ancient Egypt, the majority of Europeans still believed their kings had divine powers that could never be questioned. Commoners would travel great distances just to be touched by a king if they had an affliction in the expectation of being healed. Yet by the end of the Enlightenment entire nations were being governed by a free people.[ccxcviii]

Jefferson's plan to use Jenner's cowpox vaccine at places the expedition over-wintered to inoculate the Native Americans was farsighted, unfortunately still during an age before preservatives like glycerol and refrigeration were available, ordinary things we take for granted in the most basic laboratory today. The cowpox virus had at some point become inactivated before Lewis had a chance to try it out. As Lewis later put it, the "vaxcine matter" had "lost some of its virtue" and so was never used on the trip.

It's been estimated that as many as 80% of Native Americans would die of diseases brought over from Europe and Africa in 93 separate waves. But with persistence and financial support, smallpox became the first (and still only) human disease completely eradicated from nature, in fact the first disease whose spread across an entire continent was documented, and while President, Jefferson appointed the first vaccine agent for the newly established National Vaccine Institute: Dr Benjamin Waterhouse (the MD kind).[328]

Modern medicine sure has come a long way in a relatively short time, I thought to myself as Amanda trailed behind Barney who was tugging at his leash, not used to such restrictions. When my grandfather was born in the closing years of the 19th century, kinetic energy from the wind was still being

---

[327] The reason we know so little about the average Egyptian is that his life wasn't considered worth recording. Since he wasn't rich or of royal blood, he had no access to an elaborate tomb to share his story on its walls.

[328] Waterhouse got the vaccine from Jenner and was the first to use cowpox in the USA. He administered it to his family.

harnessed to transform the American West from a barren desert into rich farmland; pumping water straight out of the ground with the aid of countless windmills. [ccxcix]   And when he was three in 1899 there were vaccines available to protect against just two diseases...smallpox and rabies.   Now researchers at my own university were working on vaccines that held out the promise of preventing cancer someday.

Armed with its *sit and wait strategy*, the smallpox virus is nothing if not persistent.   In the 1830's the painter George Catlin lived and worked among the same Mandan tribe Lewis & Clark had over-wintered with thirty years earlier, but he left before smallpox returned – this time with a vengeance – during the outbreak of 1837.   Smallpox would go on to wipe out over 90% of the "friendly Mandans", including one of the chiefs Catlin himself had painted, a man by the name of Four Bears.

ACTIVE TRANSPORT

Two hours later we were pulling off the main highway that bisects the Lewis Valley into two nearly equal halves and onto one of its side roads that twist like a tiny capillary branching off its main artery, leading us back to the heart of campus, the dust from a recent dry spell rising up behind us in short-lived puffs, then drifting off to the side above the cornstalks and disintegrating, as if a series of smoke bombs had been tossed behind the truck.   Amanda's husband's truck was still parked out on the lawn and since I never was one to be accused of grazing in another man's pasture, I dropped her off and drove the few blocks more to the microbiology department, the 5 story red brick building looking more out of place than usual for some reason, as if it had been picked up and dropped off there by some time-traveling spaceship.   The campus itself is a strange hodgepodge of different building styles, some recent next to others looking rather dilapidated, each representing the era it was put up, back when no one could settle on a master plan apparently.

Since it was still Sunday, Barney came along with me down to the lab and while pulling into my empty space I couldn't help thinking about my own bit of maneuvering within the department.   In just two short months my tenure committee would meet again for the last time regardless of what happened and it seemed as if my truck wasn't the only thing running on fumes these days.   Then I thought about Howard Carter and how he had entered his final year of digging with his benefactor getting ready to pull the plug on all he had worked towards.   Carter even offered to pay for the digging out of his own pocket but then they uncovered a staircase leading down into an unopened tomb.   So it seemed as Lord Carnarvan's help was for Carter, the London people were my only real chance now.

As soon as we got down to the dungeon, I had the same uneasy sensation I sometimes get after entering a darkened movie theater and the film has already begun.   Out of instinct I reached for the light switch but it ignored me and in what was left of the light filtering in through the window I could just make out a large crate of some sort propping open the door to my lab.   Aware of the noise my boots always make on a newly waxed floor I walked as gently as I could with most of my weight on my heels, noticing several yeast shakers had been unplugged and dragged out into the hallway too.   I was unable to think of anyone who would be working on a Sunday in the middle of summer, and how few burglars would be afraid of an out of shape professor accompanied by his ancient dog.   None of my technicians would have ordered anything expensive; not without asking, I knew.   Even the secretaries upstairs were aware I was on thin ice here.

Closer still and I could make out a humming noise; a few steps more and I could tell it was actually two human voices and I began to wonder how long the electricity had been off and what the lack of shaking had done to our yeast cultures.   My eyes fell upon the address on the crate as I slid past it, my back literally up against the doorway to my lab, just three letters standing out large enough to be seen in the dim light.   It was from some company I'd never ordered from before, a supply house in California named DOO.   Once inside I could see Charlie's shadow against the wall.   He was wearing his baggy jumpsuit and leaning against a wooden ladder, one foot resting on the bottom step and in his hand he was holding something.   Nearer still and I could see it was a screwdriver and the person he was holding it for was the campus electrician.

"What's going on?" I called out before getting so close as to startle them for the electrician was by now balancing on the very top step.   "Our shakers blow a fuse again?"

"Oh, how's it going Big Dog?" the crackle in Charlie's voice betraying a decades old cigarette habit.   He liked to call several of the professors in the department "Big Dog" for some reason.   Taking one foot off the bottom step, he replaced it with his other in an attempt to get more comfortable and then added, "So what brings you down here on a Sunday?"

"Just getting back from Yellowstone.   What's with the crate?"   I pointed towards the doorway.

"Oh Christ!   You mean you didn't know?"

"Know what?" was all I could say, a queasy feeling building inside my stomach by the way he said the words so forcefully that it sent him into a mild coughing fit.

When he caught his breath again he simply added, "The Brainiacs upstairs finally figured out what to do with the new laser."

In one horrible instant it all came together. I'd misread the crate. It wasn't from DOO. It was from the Department of Defense. They were changing the voltage in the outlet to accommodate the new laser.

"It's going in your lab."

~ ~ ~

EPILOGUE

TWO MONTHS & 8 DAYS LATER

My other leg having fallen asleep, I leaned further back in my chair, tempting fate with my left hand still gripping the edge of the desk if my chair should happen to slide out from underneath me, my other hand inching its way out into the hallway, feeling for the door handle, locating it and then guiding my door closed, making my already small office seem tiny. But I had no choice. I was in need of some privacy for – like a flurry of papers carried aloft by a sudden breeze and then tugged on by gravity – everything was finally settling down as the university was by now a second week into fall semester. And as a result, I was finally getting some much-needed time to myself; using the lull to take care of personal business I'd been putting off. What with my tenure committee meeting, the London people demanding more and more protein, my interview, and the added teaching responsibilities, I'd been busier than a mosquito in a nudist colony lately.

Having done it so many times in the past, I logged onto the Internet and checked my e-mail without giving it much thought, clicking first on the spam files, the keyboard resting comfortably in my lap. For some reason it always feels good to delete junk e-mail after you've been away for a while, kind of like pulling up weeds in a garden, as if you're accomplishing something important without giving it much effort. After I'd cleared enough space in my inbox I sat a while longer, gradually gaining a sense of proportion but without losing focus on the screen, the rest of the room dissolving from my vision. When the mood finally settled in, I clicked on the *compose* button and typed out the following letter:

Dear Tahany,

Sorry for the delay, but as you probably know by now, things have been a little hectic around here lately. The good news in your last letter couldn't have come at a better time, about Gamal's book being published, and I'm grateful for the small part I could play in it. Please pass on my congratulations to him, as I'm sure he's quite relieved.

As for me, things didn't go quite as well as I'd hoped. I was turned down for tenure and am now finishing out my contract here, which means this will be my last semester. I was offered the chance to stay on and teach as an adjunct professor, and while this means I could still be near my family, in the end I decided not to go that route. There are some things that don't seem right at this point in a middle aged man's life and settling for consolation prizes is right near the bottom of the heap. I will miss teaching, though, as I've always seen myself as a professor, ever since I knew what one did. But then again, maybe I'll find a way back someday. Both Copernicus and Kepler never earned a dime as scientists and look what they accomplished. Even Charles Darwin somehow avoided holding down a real job his entire adult life (lol).

And I've been thinking too lately about that day in your lab when you said how some days are like dates, others like onions (do you remember?) because it's a funny thing the way a single one can change the course of so many others.

A friend of mine, Dr. Stanley Beaufort, the one I've been collaborating with on cholera, arranged an interview for me at a non-governmental organization, one of those NGOs that secure loans from the World Bank in order to build massive water treatment facilities in developing countries. It seems that this NGO was in need of an experienced microbiologist to do some water testing and to help keep track of algae counts and "organic load", whatever that is (lol). To make a long story short, just a day after my committee met, I flew to Washington for the interview and yesterday, thanks to Stanley's recommendation, I was officially offered the job. My very first assignment will be carrying out tests at a facility already under construction near Dhaka, Bangladesh. The plant is scheduled to take two more years to complete with its sister plant going up in Nepal shortly afterward. It seems as if there are 5 million people projected to get cholera in the developing world this year alone and 100,000 will die of it, most of them children, and Bangladesh, well it's always at the center of the storm for some reason. I'm scheduled to go through orientation at the company's headquarters, which is in Cairo, for two full months starting as soon as the semester ends here so it looks like I will be able to attend your wedding in February after all. I guess it's true what they say about every cloud having its silver lining.

Please don't feel sorry for me in the least, as I've had a great run here at the university and I've learned things I never could have elsewhere and besides, I may never have been cut out to be a full professor anyway. I was raised on a ranch and have always had a hankering for the outdoors. The memories of my recent trip to Egypt, and of you and Gamal, none of you have strayed far from my thoughts either, especially these past few days. When I came back here I was in the best shape I'd been in since working on the ranch as a kid. Traveling seems to suit me and I'm now looking forward to more of it.

As for my lab, I have only one student left now (the other having gotten into medical school) and she will transfer into Stanley's lab for the remainder of her education, after which she plans to return with her husband to the same small town they grew up in and teach high school science. Stanley was awarded a grant from NIH and so now has enough funding to hire a postdoc and continue making CFTR and sending it off to London on a regular basis. He is getting a fine student and has even hooked up with a researcher in California who wants to try splicing the CFTR gene into goat embryonic stem cells with the aim of producing the human protein in goat's milk, oddly enough.

Well, I should close for now. I'll forward you more of the details as soon as the secretary e-mails me my flight information. Keep in touch.

Your friend as always;
Benjamin

P.S. I finally got around to putting that photo you liked of me up on my university's webpage. Kind of ironic timing though, isn't it?

With just the low humming of the computer as a distraction I read and reread the words I'd typed out, each time allowing the enormity of the last two weeks to settle in a little more, a life-altering two weeks, there was no doubt about it now. After I had convinced myself again that it had all really happened – losing the anthrax grant, gaining the yeast project in its place, making CFTR, trying to recover from my first real failure in life, and then the interview and a chance to return to a place I never thought I'd see again – I clicked the send button, closed the browser, and leaned even further back in my chair, staring up at the bare spot on the wall where it had once been covered by wallpaper. I continued looking at it until a few teardrops appeared and began washing away the burning sensation. Damn eyestrain. I guess I could have worse problems, though, I allowed.

And it's funny the way life works sometimes because one of the things I hadn't mentioned to Tahany was what I did that day *after* my job interview. I hadn't even told my family for that matter that, while in D.C. I had been assured the job was mine if I wanted it by the head engineer, so I went ahead and rented a car and drove on up to Philadelphia. It's not that far and besides, Philadelphia is where Thomas Jefferson sent his former secretary, a young Meriwether Lewis, back in the spring of 1803, to help prepare him for the trip that would take he and his friend Clark, a dog, an Indian woman, a slave, and over two dozen other men across an entire continent and back.

By the spring of 1803, Jefferson had already taught Lewis much of what he knew about botany while the two strolled his gardens together at Monticello, and he'd even lectured Lewis on the finer points of mineralogy and astronomy and how to use a sextant. Then Jefferson drew on his knowledge of the Native Americans the Corps would be living with...yet he wanted to know even more about this new land the nation was acquiring so he went further still.[329]

But why send Lewis all the way to Philadelphia? It seems Benjamin Franklin had come up with the idea half a century earlier – back in 1743 – of starting a scientific society similar to what Europe had had for over a hundred years. It was to be the first of its kind not only in America, but in the Western Hemisphere, and Franklin called it the American Philosophical Society.[ccc] Jefferson sent Lewis to Philadelphia because Philadelphia was where the leading scientists of his day were. At the APS hall Lewis could learn about anatomy and fossils[ccci] and how to collect them from Caspar Wistar, and botany from Dr. Benjamin Barton who could show Lewis the proper way to press and dry plant specimens (both were the MD kind, along with Benjamin Rush[cccii], who taught Lewis medicine) so that by the time he met up with Clark along the banks of the Ohio on the 14th of October, 1803, to begin their

---

[329] The author of the Declaration of Independence also wrote the first paper on paleontology in America.

journey into history together, Lewis was not only an officer in the Army, but also a taxidermist able to prepare animal specimens, a doctor of sorts, an amateur botanist and even a geologist to boot.[330]

Along the way Lewis collected and sent back hundreds of specimens to Jefferson and Jefferson in turn sent many of them to the APS hall in Philadelphia, where they remain to this day.[331] The museum has some of the original pressed and dried plants Lewis had prepared around his campfires all those summers ago, and I even got to touch a few with my own hands. I also couldn't help noticing where some had a leaf or two torn away.

Museums are good at reminding me that some things don't change, not if they're important, for it seems science was competitive even during Lewis's time, not so different than it is now I suppose. One botanist, a contemporary of Lewis's named Fredrick Pursh, was so intent on publishing the expedition's finds ahead of anyone that, without asking he simply removed many of Lewis's samples from Philadelphia and took them with him when he sailed off to London in 1811.[332]

It seems that the Lewis & Clark Expedition had generated quite a buzz over in Europe because of all the new plants they'd come across and anticipation had been building and Pursh, well let's just say he wasn't one to let a good thing pass him by. He published the expedition's finds as a book,[333] enabling him to parlay his newfound recognition among Europe's naturalists into another job, this time as a botanist sent to Canada to produce a similar volume on the plants in that region there. The places where I'd noticed some of the leaves torn away were where Pursh had taken Lewis's samples with him when he left Philadelphia. It reminded me of a saying the Native Americans have here…that someday we'll all be known by the tracks we leave behind.

After the museum and with a few review articles on "organic load" beside me on the seat, I drove the short distance to Saint Peter's Episcopal Church. It's a peaceful place in the churchyard and it wasn't hard to find a place to sit, all alone on a wooden bench underneath a tree, finally able to get caught up on some reading.[334]

Later in life when Jefferson had grown tired of politics, he found himself spending more of his free time brewing beer at Monticello. Widespread approval of his purchase of the Louisiana Territory still had to await the development of steamships capable of carrying goods and passengers up the

---

[330] Lewis was elected to the APS in Philadelphia.
[331] Some of the plants Lewis sent back east were chosen because, along the way he had noticed Sacagawea gathering and using them for food.
[332] At least 70 of the preserved plant specimens Lewis sent back east were new to science.
[333] *Flora Americae Septentrionalis* published in England in 1814.
[334] Eight Native American Chiefs that died of smallpox in 1794 after visiting George Washington for a peace council are also buried here.

Missouri River against its ample current.[335]   Things didn't go so well for Lewis after returning to civilization, having gotten malaria and then taking opium for the chronic pain afterward.   Plus he drank heavily and was appointed governor of the new Louisiana Territory, a responsibility that was probably more than he could handle.   Lewis was an adventurer at heart, never a politician like Jefferson.   He committed suicide in 1809 at the young age of 35, but that wasn't the reason I was here either.[ccciii]

It seems that, among the 100's of specimens Lewis had cataloged and sent back east during their 8,000-mile voyage, several made it all the way to the nation's capital while still alive.   One was a black-tailed prairie dog Lewis called a "barking squirrel".   They were so clever at eluding capture that the expedition had to use a bucket of water to force one from its hole just to catch it.   Another was a magpie and both the bird and the prairie dog survived the trip down the Missouri and Mississippi Rivers in cages on a boat, then around the tip of Florida and finally up the coast clear on to Washington D.C. where they lived with Jefferson for a while in the White House.

Lewis also sent back some live cuttings, or slips, from an Osage orange tree.   When he got them, Jefferson forwarded the cuttings on to a botanist friend he knew up in Philadelphia, who eventually nurtured six of them into trees.[336]   The pioneers who would follow in Lewis and Clark's footsteps west during the ensuing decades found they could grow the Osage orange trees as fences and windbreaks when they were planted close together in a line, like a hedge, in part because the younger trees still have thorns on them.   And if their trunks had been straighter the railroads would have used them as ties more often.   The Native Americans didn't mind this imperfection, though, because they valued the wood more for making bows out of, it being so strong and resilient compared to other hardwoods.   Lewis once wrote that an Indian warrior would walk for weeks just to get some wood from an Osage orange tree to make a bow with.[337]

But by the 1880's, the Indians had been subdued and barbed wire had been introduced and the Osage orange trees were no longer needed like they once were.   The one I sat under that afternoon is about 50 feet tall now, its canopy completely disease-free and spreading.   I tilted my head back as far as I could to look up at what had become of one of the very cuttings Lewis sent to Jefferson so long ago.

---

[335] About 200 gallons of ale seasonally, which he bottled, corked and gave away to friends and neighbors.   Jefferson also foresaw the influence the steam engine would have in travel. In 1802 he predicted: "The introduction of so powerful an agent as steam to a carriage on wheels will make a great change in the situation of man."

[336] The botanist in Philadelphia was William Hamilton.

[337] When the Mandan Indians hunted game animals, Lewis observed how the arrow sometimes had so much energy, it would travel completely through a buffalo...and they accomplished this while riding bareback on a horse.

At first I assumed it was the museum with the preserved plants that was the reason I drove all the way to Philadelphia that afternoon, but I have no doubt now it was really the tree. I squinted through my spectacles, tracing each of its branches, thick where they joined the main trunk, then thinning out gradually as they crisscrossed one another to form the protective canopy higher up.

The ancient Greek philosopher Aristotle believed that all of life on Earth was arranged like a ladder, with mankind situated on the very top rung, plants and dogs and even trees themselves somewhere nearer the bottom. But then Darwin came along and because of him we now view the panoply of life more as a tree these days, as if mankind has climbed out onto one of its many slender branches.

It's a strange thing to have a single goal in mind for so long – tenure in my case – and after 10 years of total commitment and preparation, to miss it in the very end. Very few things are "all or nothing" like that in life, I suppose.

But then I considered how, no matter where I found myself, or what ups and downs I would encounter along the way from here on out, the memory of the shade from that tree on that fine fall afternoon would be a reminder for me, something to carry around with me no matter what else happened. No matter what, no committee could ever take *that* away from me. I'm sure now that the real reason I drove all the way to Philadelphia and sat underneath Lewis's tree that afternoon was that I just wanted to be reminded that any living thing – whether that thing should be a tree, a dog, or even a man – can be a long ways from where it started off in life…and still continue to grow.

THE END

###

Connect with me online at http://www.facebook.com/davidwoosterphd

SOURCES & ACKNOWLEGEMENTS

Among the many books and other resources that have been helpful are *Life's Matrix: A Biography of Water*, by Philip Ball; *Undaunted Courage: Meriwether Lewis, Thomas Jefferson, and the Opening of the American West*, by Stephen Ambrose; *A Field Guide to Bacteria* by Betty Dexter Dyer; *The Ancestor's Tale* by Richard Dawkins, *Innocents Abroad, Letters from Hawaii,* and *Life on the Mississippi*, by Mark Twain; *Anthrax: A History*, by Richard M. Swiderski; *Demon in the Freezer*, by Richard Preston; *The Last Stand: Custer, Sitting Bull, and the Battle of the Little Big Horn*, by Nathaniel Philbrick; *DNA: by Watson*, by James Watson; *Milestones in Microbiology* and *Robert Koch: A Life in Medicine and Bacteriology*, by Thomas D. Brock; *Children of the Sun*, by Crosby; *Pompeii: The Living City*, by Alex Butterworth and Ray Laurence; *Ambitious Brew: The Story of American Beer*, by Maureen Ogle; *History of Bread*, by Bernard Dupaigne; *The Biofilm Primer*, by J. W. Costerton; *Thermophilic Microorganisms and Life at High Temperatures*, by T. D. Brock; *A Short History of Nearly Everything*, by Bill Bryson; *Or Perish in the Attempt: The Hardship and Medicine of the Lewis and Clark Expedition*, by David J. Peck and Moria Ambrose; *Galileo's Daughter*, by Dava Sobel; *The Scientific Study of Mummies*, by Arthur C. Aufderheide, M.D.; *The Rise and Fall of the Third Reich: A History of Nazi Germany*, by William L. Shirer; *Cleopatra: A Life*, by Stacy Schiff; *Washington: A Life*, by Ron Chernow.

Dr. G.A. Scarborough (the Ph.D. kind) of the University of North Carolina at Chapel Hill and colleagues, whose work served as inspiration for the protein expression portion of this novel, were the first to produce human CFTR inside the brewer's yeast in 1996.

~ ~ ~

ENDNOTES

---

[i] In graduate school I routinely used a special chamber with radioactive cobalt to *fry* anthrax spores whenever I finished an experiment, just to make sure they were not going anywhere.

[ii] Koch's mentor, Ferdinand Cohn, discovered in 1875 the spore stage as part of the lifecycle of anthrax.

[iii] Koch's work with anthrax eventually led to a method for growing pure cultures of other bacteria. Before Koch's breakthrough, they were often grown as complex mixtures, which would lead to misidentification. Some people employed elaborate explanations for having so many different shapes of microbe rather than accept the notion they might have contamination; for example, that they were looking at the same type of cell, but at different stages in its lifecycle while in reality, they were working with different species in the same dish (called *pleomorphism* this notion was championed by Carl von Nägeli). The ability to work with pure cultures (a single type of microbe), as Koch pioneered, is therefore one of the most important milestones in the history of microbiology.

[iv] Dr. Koch also went on to discover why anthrax could persist season after season: its spores were still viable even after drying out, sometimes for years. Since Koch's day, spores of *Bacillus subtilis* – a relative of *Bacillus anthracis* – have flown to the moon on Apollo 16 and survived the vacuum and radiation of space for 22 months outside the International Space Station.

[v] *Anthracis* probably forms spores as a way of competing with other microbes in the soil, like those that produce antibiotics for defense. Chemical warfare is waged among unseen microbes all the time. *Anthracis* "opts out" of this competition simply by outlasting whatever it finds itself with in the soil, by forming these protective spores and then going dormant as conditions become unfavorable. Spore-making is so important to *anthracis's* lifestyle that it devotes one third of its genes to accomplishing this. / In 1997, researchers revived 80-year-old anthrax spores from a museum specimen.

[vi] A mixture of strong bleach & hydrogen peroxide will kill them on contact after about 2 minutes most of the time.

[vii] The smallpox virus uses a similar strategy. And the spore-forming bacterium that causes tetanus, another close kin to *Bacillus anthracis*, causes lockjaw when allowed to invade deep within oxygen-less wounds.

[viii] Which would help explain how 94-year old Ottilie Lundgren was infected by just a few spores that rubbed off at a mail sorting facility and onto a letter bound for her home in Hamilton, NJ in 2001.

[ix] In 2008 on Ted Turner's ranch near Bozeman, Montana anthrax spores that turned up naturally in the soil felled around 300 bison. Fortunately, there hasn't been a human case of anthrax in Montana since 1961. / The Aum Shinr Kyo cult in Japan tried spraying anthrax spores from blowers at the top of an 8-story building in Tokyo in 1995. They used a weaker strain of the bacillus, one for making vaccines. Unfortunately, they then turned their attention to sarin gas with more lethal results. / *Anthracis* can, on occasion, become coated with silica without the help of scientists and achieve flight. In Alberta, Canada during an unusually dry summer in 2000 at Wood Buffalo National Park, 42 buffalo kicked up sand to cool themselves and inhaled anthrax spores and swiftly died. Contrary to news reports at the time, the anthrax used in the 2001 letter attacks wasn't weaponized. The silica detected in the spores turned out to be a natural part of the spore

[x] Like *Thiobacillus*. While often vulnerable to changing conditions, bacteria as a whole are successful at inhabiting an astonishing variety of habitats on Earth, from deep-sea vents to freezing water, the atmosphere, and most everywhere in between.

[xi] It would no doubt increase it. How much time a scientist spends doing actual research has probably always been important. I sometimes wonder what the monk Gregor Mendel could have gone on to do in addition to discovering genetics if he hadn't become saddled with administrative duties later in his career at the abbey. Galileo made sure he didn't have to teach at the University of Pisa before he took the position of professor. Theoreticians as well as experimentalists need their "alone time". Einstein chose to work at Princeton over Caltech because Caltech wanted him to teach.

[xii] Even Lewis & Clark were flat broke by the time they crossed the continent. They traded nearly everything they could do without to get to the Pacific and had to rely on their wits just

to make it back.   Lewis administered medical treatments to the Indians in exchange for horses to re-cross the Continental Divide.   Their men found the Indians were fond of brass buttons on their jackets, so they cut off and traded them for roots to eat.   The expedition carried with it a letter of credit from Jefferson, but had nowhere to spend it.

[xiii] The technology to *freeze-dry* microbes like the yeast is a 20th century innovation.   During the Civil War, soldiers had to forgo the luxury of yeast when baking "quickbreads" around their campfires.   To leaven bread dough, they relied instead on basic high school chemistry – a combination of baking soda & cream of tarter – which produces an acid-base chemical reaction along the lines of what happens when one drops certain tablets of antacids into a glass of water (plop, plop, fizz, fizz).

[xiv] Flowers, tree bark, and other plant parts often secrete a steady stream of sugary fluids to the outside world as a way of encouraging friendly microbes like the brewer's yeast to take up residence there, space that might otherwise go to pathogenic (disease-causing) microbes (the *better the devil you know* strategy).   Plants also make use of sugar as a reward for seed dispersal.   Grapes contain sugars not for people to make wine with but for birds to spread their seeds.   Some plant seeds won't even germinate unless they've first been through the digestive tract of a bird or an elephant.

[xv] We have a planet in our solar system less dense than liquid water. If you could somehow place Saturn in an ocean large enough, unlike the yeast cell Saturn would float up on the surface. Rock here on Earth can attain a density less than water.   During the early stages of the eruption of Mt. Vesuvius in AD 79, the pumice it ejected into the air landed and floated in Pompeii's fountains, which must have been a curious sight for the volcano's victims.

[xvi] Actually, yeast do find a way to move – up and down in liquids while fermenting by changing their buoyancy – which is why anyone who brews ale will see a white foam develop on the surface towards the end of fermentation. Increasing its buoyancy is the yeast's way of escaping to the next batch, which in nature might be the next piece of rotting vegetation.

[xvii] Galileo, after building the best telescope in Europe, was offered what could be considered tenure today – a lifetime appointment at the University of Padua – in 1610.   He turned it down to accept an offer by the Medici Family of Florence, becoming a professor at the University of Pisa.   No fool, Galileo named the moons of Jupiter (which he discovered in 1609-10 with his improved telescope) the *Medician Stars:* to ingratiate himself with the Medici Family.

[xviii] Karl Stetter, the German microbiologist who figured out a way to grow the majority of the world's extreme, heat-loving microbes in his lab (called the *witch's garden* due to elaborate equipment he had to invent just to culture them) is well known in microbiology circles for bringing his newly discovered extremophiles back to his lab…in an ordinary aluminum brief case.   He also named the genus *Thermotoga* because the outer membrane of this thermophilic bacterium looks like a Roman toga.   I also had a frozen sample of the thermophilic archaean *Thermoplasma acidophilum,* which has the admirable ability of growing & reproducing in either a burning coal pile or a hot spring, whichever it finds itself inhabiting.

[xix] A diary belonging to one of the astronauts from the space shuttle Columbia survived the craft's demise and even the 37-mile plunge through Earth's atmosphere, yet when it was discovered lying in a field in Texas 2 months later, its paper was undergoing digestion by soil microbes.   Many of the space suits worn by the Apollo moon astronauts were stored in humid environments and as a result acquired black mold, a fungus that was consuming the polymer materials as food.

[xx] 17th century chemists knew protein was unusual because fluids rich in them – like blood or egg whites – wouldn't boil but instead formed an opaque solid upon heating.   Proteins must have seemed mysterious to medieval alchemists used to working with inorganic minerals and acids.

[xxi] It's these molecules that fall off bacteria during an infection that can get into the blood and cause fever.

[xxii] Even *E. coli,* the bacterium that lives in our intestines, manages to stuff its DNA chromosome within itself in spite of the fact that, if fully extended, its chromosome would be 1000 times longer than the cell.   The DNA inside one of your own cells would stretch 2 meters if fully unwound and joined end-to-end.   Of course it would be so thin that you'd still need an electron microscope to see it.   In fact, if you could stretch out and line up, end-to-end, all the chromosomes from all the cells in your body, your DNA would cross over 700 million miles; or to the Sun and back about 4 times.

[xxiii] The rise of the budget traveler can be attributed to a guidebook written in 1957 by Arthur Frommer called *Europe on $5 a day*.
[xxiv] On their trip Twain and his fellow travelers formed a club that met onboard the *Quaker City*. Each member took turns reading aloud from guidebooks, learning about things they would see the next day. Even the ancient Egyptians were fond of travel stories. One was *The Shipwrecked Sailor,* a fantasy where a sailor returns to Egypt with exotic gifts having traveled to the Land of Punt.
[xxv] The Art Deco movement in America was influenced by interest in Egyptian art in the 1920's while Napoleon's invasion of Egypt and the subsequent "Egyptomania" it inspired influenced women's fashion in Europe back in the 19th century. (Napoleon also took along a printing press to Egypt that was equipped with Arabic type) / The ancient Egyptians had no word for "art" or "artist". They saw the people who made wall paintings, sculptures, and other works as simply copiers. Egyptian art was never supposed to change, so creativity wasn't valued then like it is today.
[xxvi] An Egyptian would have started his or her day with some beer, yet did without money until the 6th century BC. Besides beer, payment might also have been in the form of sacks of grain. Stamped coins – what we'd recognize as money today – weren't used until around 350 BC, having been invented in Greece. Information was a 2-way street in the ancient world, which is how the Greeks got their brewing knowledge, in turn, from the Egyptians.
[xxvii] Just from 1450 to 1500, some two million books were printed in Europe thanks to Gutenberg's invention. Within a year after Galileo finished *The Starry Messenger* in 1610 for instance, there were already bootleg editions being printed throughout Europe without his permission. Before Gutenberg, books were a luxury item and a typical university student during the Middle Ages would have had to rent them.
[xxviii] Anthrax is surprisingly common and found in soils throughout the world. Researchers comparing the DNA sequences of different strains have discovered the bacterium is actually several related strains still found along what was once an ancient trading route from China to Europe. The oldest strains were in China, intermediate strains towards the midway point, while younger strains evolved closer to Europe. The *Ames strain* used in the 2001 attacks is related to one found today in China, suggesting it arrived in the US via trade in furs sometime during the last 300 years and has now (for some unknown reason) settled in Texas and Oklahoma.
[xxix] One of the original 7 wonders of the ancient world, the Lighthouse of Alexandria, tumbled into the sea after earthquakes in the Middle Ages. But it could be claimed the landmark is still around. Just as the lighthouse had 3 stories, so too do many minarets of Cairo's mosques have them. So in a sense the lighthouse never really left. It simply became incorporated into the City Perpendicular's skyline.
[xxx] Some strains can cause necrotizing fasciitis, or so-called flesh-eating disease. The 2001 Nobel Prize winning physicist Eric Cornell lost an arm and shoulder to this microbe. / One reason doorknobs are often made of copper is the innate antibacterial properties copper ions possess, some of which always come off onto your hands whenever you touch a copper doorknob.
[xxxi] The captain would have been given the copy while the library would have kept the original papyrus. By the time Julius Caesar made it to Egypt there were perhaps 700,000 manuscripts in the library. / The largest collection of ancient papyri ever uncovered (1,785 scrolls) was at the Villa of the Papyri in Herculaneum. Napoleon was given several charred papyri as a gift while Sir Humphrey Davy failed to unroll the damaged papyri using 19th century chemical techniques.
[xxxii] Homer paid tribute to the skill of ancient Egyptian doctors in the *Iliad*, a work in which he also described anthrax as a "burning plague". Homer used the Greek word for bread ("broti") to mean "mortals" since only the gods could drink ambrosia while mortal men and women were stuck eating bread.
[xxxiii] Other languages have taken turns. While well versed in hieroglyphics, an Egyptian scribe would have written a letter to a foreign ruler in Babylonian – the *lingua franca* of the time. Later on (thanks to Alexander the Great), Greek would take the place of Babylonian as the most essential language of the Mediterranean. A merchant, mercenary, or someone attending the theater would have needed to understand Greek during Hellenistic times and is why many Romans had a Greek slave in their home: to teach their children the Greek alphabet. Julius Caesar, for example, spoke Greek when he was with Cleopatra. Before

the first century, Judea was Hellenized, meaning Jesus would have known some Greek though his main language was Aramaic. Later still, Latin would predominate in the West (Newton wrote *Principia* in Latin, for example). When the poet Milton met Galileo in Italy the two were able converse because both men spoke Latin. Galileo was the first scientist to claim that the language nature uses is mathematics (subatomic particles today, for instance, can only be described using mathematical terms since physicists have few or no analogies in everyday life to describe them). Galileo wrote in Italian because he believed it important to reach as many readers as possible with his ideas and is also why his lectures were popular and the church considered him a threat. Galileo was unusual among scientists of his day because he also wrote about his discoveries in his native tongue.

[xxxiv] There are many partnerships in nature, some extending into the microbial realm. Cholera is caused by the bacterium *Vibrio cholerae*, yet this diminutive comma-shaped microbe is completely harmless to humans until it first becomes infected by a specific virus, one that attacks only *Vibrio cholerae* (the bacteriophage is CTX-φ)

[xxxv] The scientist declaring to know the exact species of a microbe while looking through a microscope is a myth invented by Hollywood. It usually takes several tests (for example, to determine what kind of metabolism it has), sometimes stretching over days, to identify with any certainty a particular strain of microbe. Does it grow on one type of media with sugar, or another with only acetate as a food source; does the microbe turn a particular dye blue, or does it leave the dye colorless; does it grow in the presence of oxygen or does oxygen actually inhibit its growth? Just as Koch did in the 1800's, a technician today might need to inject the pathogen into an experimental animal to see if it causes symptoms similar to the disease in humans before it can be said with certainty what the microbe is. Lager yeast and ale yeast can be distinguished in brewery labs by the temperatures they grow at and by their food preference. Ale yeast like a higher temperature while lager yeast strains prefer a lower temperature and can use the sugar mellobiose for energy. When spinal fluid was examined from the first victim of the anthrax attacks in 2001 it was recognized (using a microscope) as a type of *bacillus* by its box-shaped cells, but not more definitively identified as *Bacillus anthracis* until other more elaborate tests were performed at the CDC in Atlanta. An important clue that all the anthrax letters came from the same source was when the spores formed a mixture of different shaped colonies while growing on agar plates in the lab and this "signature" of different shaped colonies was the same from all the letters. It later turned out to be due to different mutations in the genes of the bacteria. The recent outbreak of cholera that killed 6,000 in Haiti was identified as being similar to a strain found in Nepal by using genome-sequencing techniques pioneered from the anthrax investigation of 2001.

[xxxvi] As detailed in the Hearst papyrus (so-named because William Randolph Hearst helped fund the expedition that discovered it).

[xxxvii] Species of mosquito that carry the dengue fever virus have evolved to live with humans in cities and have also changed their behavior, for example the times of day they take a blood meal: from early morning and evening (as most mosquitoes in the wild do) to during the day, when human activity is at its peak.

[xxxviii] It is a phenomenon found throughout nature including the ancient world. Amplification was why the Egyptians had to be extremely careful when laying out the foundations for their pyramids. One small mistake at the pyramid's base could easily be amplified into structural instability towards the top. The 13-acre base of the Great Pyramid is so level that it barely changes in elevation by 2 inches.

[xxxix] Christianity is also why the Roman tradition of public bathing disappeared in Western Europe after the fall of the Roman Empire in the west. Christians being ashamed of their bodies avoided public nudity. Istanbul, Turkey – once the capital of the Byzantine Empire – is the only place where the tradition of public bathing has remained unbroken since ancient times.

[xl] Then again, "Faint heart never won the fair lady, neither did it ever pursue and overtake an Indian village" were some of the last words Custer wrote in a letter before arriving at the Little Big Horn. / The same week Custer's Last Stand happened, Mark Twain's *The Adventures of Tom Sawyer* went on sale in London (June, 1876).

[xli] Cats were domesticated in Egypt around 2000 BC for the same reason pottery was invented: to keep stored grains safe from rats. So many cat mummies were discovered in the AD 1800's that sailors used them as ballast for ships. Some cat mummies were then ground up and used as fertilizer.

[xlii] Hieroglyphs remained essentially unchanged even though they were used for 3,500 years and have a formal look to them because their original use was for monument inscriptions. It sometimes reminds me of the way the overall shapes of microbes like bacteria and certain fungi haven't changed either over millions of years...we know because many have been found locked inside amber 220 million years old.

[xliii] While searching for Tut's tomb, Howard Carter was reassured he was on the right track when fragments of wine jars thrown into a pit after Tut's funeral feast were identified nearby (his royal seal was on them).

[xliv] Clark saw a grizzly bear footprint before he or any American encountered a grizzly bear in the flesh. He described it as the biggest footprint he'd ever seen.

[xlv] Acidic molecules are often negative in charge while clay tends to be positive so they bind each other. Many toxins in nature are negatively charged too, which is why elephants will eat clay from the bottom of watering holes. Clay absorbs plant toxins found naturally in their diet. Some human travelers carry Kaopectate, which used to contain a type of clay with the ability to absorb toxins from bacteria consumed after a bad meal (now it has a similar substance called bismuth subsalicylate).

[xlvi] When Watson & Crick were determining the 3-D structure of DNA, they used data previously collected by Chargaff showing that DNA always came in equal amounts of the bases guanine (G) and cytosine (C) as well as adenine (A) and thymidine (T) (i.e. the amounts of G = C and A = T). To make his discovery, Chargaff used a strip of ordinary filter paper to separate the DNA bases since each base has a different "mobility" while traveling up the paper. Stanley Miller also used paper chromatography in the 1950's to identify the amino acids he created with an electric spark and some gases to recreate the earth's early atmosphere inside a sealed beaker. Paper chromatography can be used to separate (purify) many small molecules like chlorophyll and amino acids. In the undergraduate chemistry labs every year we have the students draw a dot of green ink at one end of a strip of paper, then dip that end in water and watch as the water carries the ink up the paper, separating the green into two new spots, one yellow, the other blue. Usually the blue one travels faster.

[xlvii] Fertile Crescent to the north of Egypt. In western Montana we get one crop a year while in ancient Egypt they got three thanks to warm weather and the generosity of the Nile. In ancient Pompeii, farmers often got two crops a year due to the unusual fertility of the local soils complements of Mt. Vesuvius.

[xlviii] A method that can "weigh" molecules and thereby identify them by the specific atoms making up the molecule. The unknown sample molecules are first given an electric charge and then made to travel through an electric or magnetic field, which exerts a bending force. The heavier the molecule is, the less it wants to curve its path (due to inertia). The amount of curvature the molecule takes is directly related to the weight of the molecule.

[xlix] At some point during their evolution grapevines became self-fertilizing – or *hermaphroditic* – having both male and female parts on the same plant. Recent DNA sequencing of modern and wild grape vines suggests grapes were first domesticated in Anatolia (modern day Turkey) about 9,500 BC.

[l] The tomb of King Scorpion I held jars of resinated wine made around 3150 BC – about 4,000 liters worth when the tomb was stocked – as well as some beer and molds for making bread. The dried figs and grape skins, or lees (dregs), at the bottom of the wine jars were still in good enough condition to be analyzed. It too had become mummified due to the dry climate. Scientists used electron microscopes to see the original yeast cells, and even extracted some of its DNA. Wealthy ancient Romans were fond of wearing makeup and used dregs on their cheeks to color them red. / The Incas of Peru made a beer called *chichi*, using saliva to digest the starch into simpler sugars, rather than performing malting as in the Near East (and in modern breweries today). In Honduras, the Indians were brewing a *chocolate beer* from the fruit of the cacao tree around 1000 BC. The Vikings brewed a barley beer they concocted out of bread soup, a source of much-needed calories during the winter. The Pilgrims carried barley seed to the New World to brew beer. / When the Corps of Discovery set out in 1804 from St. Louis, it was a mere outpost at the edge of civilization, a small town lacking even a commercial brewery. When Clark canoed past the first settlement in Missouri – Ste. Genevieve, 50 miles below St. Louis – its brewery had been producing beer since about 1779. St. Louis wouldn't get its first commercial brewery until 1810. By the time Mark Twain was born in nearby Florida, Missouri in 1835 this small settlement already had 3 whiskey distilleries. / Before the American Revolution, cities in the east including

Philadelphia had streets like *Brewer's Alley* set aside for beer makers. / Benjamin Franklin invented a recipe for spruce beer though he preferred drinking wine.

[li] In 1866, Mark Twain spent 4 months in Hawaii as a correspondent for The Sacramento Union. In one letter to the newspaper he described the Kanakas (native Hawaiian agricultural workers) and their tradition of poi. "The poi looks like common flour paste, and is kept in large bowls formed of a species of gourd and capable of holding from one to three or four gallons. Poi is the chief article of food among the natives, and is prepared from the kalo or taro plant. The taro root looks like a thick, or, if you please, a corpulent sweet potato, in shape, but is of a light purple color when boiled. When boiled it answers as a passable substitute for bread. The buck Kanakas bake it under ground, then mash it up well with a heavy lava pestle, mix water with it until it becomes a paste, set it aside and let it ferment, and then it is poi – and a villainous mixture it is, almost tasteless before it ferments and too sour for a luxury afterward. But nothing in the world is more nutritious. When solely used, however, it produces acrid humors, a fact which sufficiently accounts for the blithe and humorous character of the Kanakas."

[lii] Fermented gruels that were porridge-like breads. The ancient Egyptians could turn ordinary bread into a "cake" for special occasions by adding eggs, fat, and honey to their dough before leavening. / It was likely the necessity to store grain centrally, and distribute it using the Nile as transportation that led to the formation of Egypt's Old Kingdom around 2686 BC.

[liii] One of the discoverers of oxygen – Joseph Priestly – lived near a brewery, which he visited often and is where he became interested in carbon dioxide (since this gas could extinguish candle flames yet paradoxically would allow plants to grow vigorously). The owner of the brewery let him set up a small lab, whereupon Priestly transferred water between two glasses over the tops of brewing vats to collect the carbon dioxide, in the process inventing the first soda water. He sold this recipe to a Mr. Schweppe. The new soda water sold well at health spas and made Priestly famous.

[liv] Moses and his followers fled Egypt so quickly they could take only unleavened bread, which is why unleavened bread is eaten today during Passover.

[lv] It's interesting to consider the contrast in textures between the brown outer crust, which is crunchy, and the hot, moist interior of fresh baked bread, which is chewier. One of the reasons modern bakers use so many preservatives in bread is that bread today is made without as much crust, which would otherwise seal in moisture. Ancient bread in Egypt would have tasted more like sourdough bread today due to the presence of bacterial contamination. These stray microbes produce vinegar (an acid), which is sour. It's often incorrectly stated that yeast provide structure to dough by producing bubbles, because in actuality this gas weakens (tenderizes) the bread dough, making it easier to chew.

[lvi] Libations. There are several theories as to how civilization got started. Gamal also mentioned religion may have helped bring about humanity's first settlements. It's possible when people first began making regular pilgrimages to temples and other holy sites 10,000 years ago, the worshiper's requirement for food is what encouraged the first regular planting of crops.

[lvii] Memorizing ancient songs like the *Hymn to Ninkasi* would have been how illiterate brewers remembered all the ingredients and steps brewing beer calls for. In ancient Sumer there were 16 different types of beer, some made from barley, others wheat, or perhaps mixtures of the two. Brewing was done outside the temples in Mesopotamia in private residences – probably by women – and sold in nearby shops. Many ancient people later on, including Germanic tribes in Europe, had gods similar to Ninkasi, also devoted to brewing.

[lviii] Egyptians brewed a wide variety of beers too including: "beer of the protector", "friend's beer", "beer which does not sour", "iron beer", "sweet beer", and "dark beer." Many were only for special occasions.

[lix] Memphis was the ancient capital of Egypt, not far from present-day Cairo. Its strategic location allowed for control of the Nile's Delta region, yet was still near the river enough to remain in contact with the religious center – Thebes – in Upper Egypt. Building materials from Memphis were later requisitioned to make Cairo in the Middle Ages.

[lx] There are other stories like this that predate Christianity, for example: Imhotep, builder of the first pyramid in Egypt was believed to make barren women as well as fields fertile, and he was also half man, half god. Another half-god was Egypt's Horus. His father was Osiris, his mother the mortal Isis. / Cleopatra's family – the Ptomemy's – claimed they were descended from Dionysus.

[lxi] The audience of perhaps 10,000 drank wine and ate meat while using bread as a napkin. Thespis was a member of the chorus – a group of 12 singers – and he developed a dialogue with the chorus, in the process becoming the world's first stage actor. / Unlike in Greece & Rome later on, ancient Egyptians had almost no public entertainment or spectacles.

[lxii] Participants reclined around a circle (or square) so each would seem equal in importance. The Greeks also drank in rounds so all would reach the same level of intoxication at the same time. And yet the symposium wasn't a place to imbibe so much as it was a social situation where younger citizens were supposed to "display their minds", to show that they understood all the "rules" and that they were fit, upstanding members of Greek society. / The Roman equivalent to the symposium was the *convivium*. The same freedom the Greeks enjoyed to pursue knowledge (due to lack of a central authority) worked against them too, and is also why Greek city-states often fought amongst each other, eventually weakening their empire from within.

[lxiii] Temperature is involved in many interesting mechanisms in biology. One example is the jacky dragon lizard. Temperature determines the sex of its offspring during incubation. The temperature the embryos develop at inside the eggs can also influence the animal's behavior later in life. Another example is yawning. We probably do it to help cool the brain.

[lxiv] There are some amoebae large enough to be seen with the naked eye. *Pelomyxa palustris*, for example, can grow up to half a centimeter across. / People have lost their lives to microbes in unexpected ways. Methane gas is produced by a type of archaea called *methanogens* as a waste product, which has in turn exploded, killing sewer workers in the past. Traces of methane have also been found in the Martian atmosphere, leading some scientists to speculate that it was created by a methanogen since methane is unstable to UV light and so something must be replenishing it.

[lxv] Robert Kock discovered the bacterium that causes TB in 1882. Even today it's estimated 1/3 the world's population is infected with the TB bacillus. It kills more every year than any bacterium on Earth, nearly two million. Abraham & Mary Lincoln had only one child that lived to adulthood; two died of bacterial diseases. Their youngest son Tad fatally contracted TB in 1872 while Willie died of typhoid fever in the White House during the Civil War. / Slave children prior to the Civil War were not auctioned off before the age of 8 because by then it was clear they had survived most serious childhood diseases.

[lxvi] The teeth of victims from Pompeii also show wear due to abrasive sand in their bread from grinding stones. Even bison in Yellowstone National Park can have this problem since grass growing near geysers contains silica due to the heat dissolving these minerals within the thermal waters and then being taken up by the plant's roots and stored in the leaves.

[lxvii] As late as the 1890's wine was given as a medicine in Paris hospitals – some two million liters a year. During Prohibition in the USA almost a million prescriptions for medicinal alcohol were written, one of ways people found to get around the law.

[lxviii] Laudanum (or "tincture of opium") was carried by the Lewis & Clark Expedition – a mixture of opium and alcohol useful for alleviating severe pain and coughing among other ailments. The alchemist Paracelsus first made laudanum in the 16th century but its active ingredient, morphine, wasn't purified by a chemist until 1804. / After the death of her 3rd son, Mary Lincoln became addicted to laudanum, which may have contributed to her eventual mental decline. Laudanum was actually cheaper than alcohol in the mid-19th century.

[lxix] Beer was also common in Rome. The most well-known beer lover in history, Julius Caesar, drank it after crossing the Rubicon. After the introduction of wooden barrels by the Gauls in the 1st century AD, the Romans made and stored beer in them.

[lxx] Lead poisoning may have been why Roman aristocrats tended to have few children, a side effect being infertility. Julius Caesar fathered only one child (if that), by Cleopatra while in Egypt. / The higher lead concentration in the skeletons of men in Pompeii and Herculaneum is evidence that men drank more wine than women in the Roman world. Written on the entrance to a bar in Herculaneum: "...drinks cost an as here, but you get a better drink if you spend two. If you hand over four asses, you can drink fine Falernian wine."

[lxxi] Ancient Rome tended to have more docile mobs, unlike Alexandria, which didn't have a dole, or "social sedative". This meant that Alexandrian aristocrats lived in fear of mob uprisings. Roman citizens eligible for free bread or flour included veterans, widows, and the unemployed. / A severe grain shortage in AD 6 led to nearly all the slaves being sent out of Rome and into the countryside to fend for themselves. The emperor Claudius had the port city of Ostia constructed to keep Rome supplied with grain from Egypt. / Ironically, Egypt

today is the world's largest importer of wheat. Hosni Mubarak was forced from power in 2011 by riots fueled in part by poor grain harvests in Russia, which led to higher bread prices in Cairo. / In 2006, a US military patrol captured thumb drives belonging to al-Qaida detailing the organization's strategy for combating the surge. Among al-Qaida's targets were Baghdad's many bakeries. Their plan was to blow them up to deprive Iraqis of fresh baked bread in the morning.

[lxxii] Some vineyards in Pompeii were recognizable to archeologists because Vesuvius erupted before winemaking season. Wine amphorae were still being stacked, which lends support to the August AD 79 date for the eruption rather than the November date as some others believe. The fertile soil around Vesuvius was so productive that locals believed the god Bacchus preferred the mountain as his home. Containers for ancient wine exported from Pompeii have been found as far away as the Red Sea.

[lxxiii] Some proprietors watered down their wine much to the displeasure of at least one customer, as this nearly 2000 year old inscription found on one tavern's wall suggests: "Would that you pay for all your tricks, innkeeper. You sell us water and keep the good wine for yourself." Grapes were the most important crop grown around Vesuvius and artifacts depicting Dionysus are a common find in Pompeii, including mosaics, statuettes, busts, and paintings of him. Other graffiti on Pompeii's walls included political ads, with some candidates promising "good bread" if elected. / Lewis & Clark watered down their men's whiskey so it would last longer.

[lxxiv] Porters would have brought beer and bread to workers on site each day for lunch. For a worker in a temple or tomb in ancient Thebes, "brewing beer" or "having a hangover" was an acceptable excuse for missing a day's work.

[lxxv] Typhus, typhoid, cholera & dysentery. The Crimean War is also where brucellosis was noticed for the first time...in horses. More recently, the presence of the bacterium that causes brucellosis has been detected in cheese recovered from Pompeii and the vertebra of several victims of the eruption also shows signs of brucellosis infection.

[lxxvi] The idea of guidebooks isn't recent. The *Book of the Dead* was a series of papyri in ancient Egypt meant to help guide the newly deceased into the next world. Some were written on the mummy's bandages and described 200 magical spells.

[lxxvii] Of all the animals on Earth, flying insects pose the greatest threat to humans. On a more benign note, the adult mayfly has the shortest lifespan of any animal...a mere 30 minutes. The fruit fly drosophila's generation time is just 10 days with the female laying 300 eggs each generation, meaning that two flies can produce 27 million offspring in a month, making drosophila an attractive choice for geneticists.

[lxxviii] The sprinklers irrigating the fields are a reminder that Egypt today stores more than just grain. Because of dams, Egypt also stores water in the form of large lakes not only for the water but also the hydroelectricity it generates to run the pumps for irrigation. Lake Nasser alone holds enough water to supply Egypt's needs for about 3 years.

[lxxix] Cook was an English printer credited with inventing the modern tourist industry when it comes to package tours. As a temperance man, Cook viewed his excursions as a religious calling. Hotels his guests stayed in weren't allowed to serve alcohol and he printed the first informational brochures about sites along his tours. His devotion to the Bible is why he expanded his operations into the holy lands of Palestine and Egypt. Cook also saw an opportunity after the Suez Canal opened to give tourists a glimpse of Egypt. He realized that passengers would want to disembark in Cairo and stretch their legs. In the 1840's (before steamboats were common) as in ancient Egypt four thousand years earlier, it would have taken about 20 days to travel by sailboat from Cairo to Thebes. No longer dependent upon the wind, by 1869 Cook's company was offering faster, more reliable steamboat rides up the Nile, cutting the same journey by two thirds the time. By the 1880's he had acquired a monopoly. In 1865 Cook was arranging tours of US Civil War battle sites for curious Europeans. By 1872 his company offered a 222-day round-the-world tour by steamship and stagecoach and introduced the forerunner of today's traveler's check in 1874. With his note a customer could obtain local currency in several countries. Cook invented the package tour, which included not only transport but also his guidebooks. Customers could expect food and lodging included in the price. In the late 1800's it was often said that the British had two empires in Egypt – a military one as well as a tourist one run by Thomas Cook.

[lxxx] Similar to the way Arab scholars and monasteries held onto classical knowledge during Europe's Dark Ages, ancient Egypt safeguarded Greek knowledge during Greece's own Dark Ages, which began around 1200 BC and lasted 300 years.

[lxxxi] Even before the fall of Rome, religious travelers like St. Martin of Tours (AD 316-397) had gone on pilgrimages, planting vineyards and spreading viticulture. St. Martin visited the Touraine region of France and introduced the Chenin Blanc variety of grape. The ancient Romans expanded winemaking, creating some of the most productive vineyards in Western Europe today (in Spain, Germany, and France).

[lxxxii] The English word *yeast* can be traced to the Germanic word *gishen*, a term used by brewers to describe the roiling and foaming that accompanied a good fermentation. Today's beers usually have carbon dioxide or nitrogen gas bubbles injected after fermentation because the yeast has already been removed by centrifugation.

[lxxxiii] The Paulanerkeller was originated by Paulaner monks who migrated to Bavaria from Italy, while the Augustinians laid the foundation for the famous Augustiner brewery in Munich in AD 1328. Another nearby brewery is in a Benedictine monastery and may be the world's oldest brewery, founded in 1040. The Benedictines were founded by Saint Benedict, who believed monasteries should be self-supporting, which included the need for growing crops. One wonders if not for him, whether a later Benedictine monk would have been experimenting with pea plants in the 1800's.

[lxxxiv] The Domesday Book is the result of an extensive census compiled for William the Conqueror in AD 1086 for the purposes of taxation, and listed 42 vineyards throughout England. / One of the signers of the Magna Carta in AD 1215 was also a member of a winemaking guild. / The English word for "lord" originates from "loaf-giver".

[lxxxv] The word *bridal* reflects this relationship, created by joining the words *bride* and *ale*. The word *honeymoon* may refer to the tradition of newlywed couples selling fermented honey to finance a trip. Another explanation has it that in ancient Babylonia the father of the bride was expected to supply the newly married couple with mead for one month.

[lxxxvi] The ancient Romans put their stamp on Western cities even earlier; London is an example, founded around AD 50 and called Londinium. Six of London's city gates (including Newgate) date back to Roman times. / A European visitor to London during Shakespeare's time described English beer as being dark and cloudy, like horse urine.

[lxxxvii] As a rancher, I was surprised to learn that, aside from brewing, even today there are no additional uses for hops. The ancient Romans did find one, though – they put them in their salads – while Londoners in the 1660's burned hops in an attempt to ward off plague as they gave off such a strong smell. When early settlers to North America didn't have hops for brewing they often turned to spruce as a substitute.

[lxxxviii] Anyone who wanted to become a baker in London circa 1300 would first have been apprenticed to a baker before being allowed to join a guild. / The first modern chemist – Robert Boyle – had an interest in fermentation and in the 1600's predicted that whoever came to understand the cause of fermentation would someday understand what caused disease. / The guild system of the Middle Ages, with its apprentice-mentor relationship, is what today's graduate schools in universities evolved out of and how I got my Ph.D.

[lxxxix] In 1789, a journalist in Paris reported: "The bread is black, gritty, and sour. Flour is of the vilest quality. Yellow in color and foul smelling. Clotting hard masses that could only be broken up by repeated blows of an axe." / The variety of French pastries today has its origins in government price controls of the 1780's. Because Parisian bakers could no longer charge as much for ordinary bread, they made pastries for wealthy clients containing luxury ingredients like sugar & chocolate, effectively allowing them to sell bread at higher prices. / A bread riot occurred in Richmond, Virginia during the Civil War in 1862 when hundreds of Confederate women took to the streets, breaking into stores in search of food and clothing. In 1917, bread riots in Russia would grow into the Russian revolution.

[xc] Einstein tried and failed to get a job as a professor (or even a full time job as a high school teacher) after finishing college, due to his having rubbed influential professors the wrong way as a student. Einstein didn't have patience for those who held onto old views in physics without challenging them. Some have speculated that if Einstein had managed to land a job in academia sooner, he might never have come up with relativity. Working at the patent office gave him the freedom to pursue his own ideas. Isaac Newton also did the majority of his important work as a young man when he was forced to leave Cambridge in 1665 due to

an outbreak of the Bubonic Plague. As Francis Crick once said, "by the time most scientists have reached age 30 they are trapped by their own expertise."
[xci] Harvard University had its own brewery. / Beer brewed in monasteries is called *klosterbrau*. The world's oldest brewery – Weihenstephan – was built in 1040 and is near Munich. / Einstein was just 6 when his father's company provided the first electricity for Oktoberfest in 1885, lighting it up and refrigerating Munich's beer.
[xcii] It's estimated that at one time the entire human race probably numbered only about 10,000 - making us as endangered then as the great apes are today. / The beer hall I was in could hold 1,300 in the main room, the beer garden an additional 400, which is still a drop in the bucket compared to the Circus Maximus of ancient Rome, which could hold up to a quarter of a million. / A true beer culture, Germany's drinking age (for beer) is just 16. It's also not unusual to find labor contracts that stipulate the right of workers to drink beer during lunch.
[xciii] Or "bottom-fermenting yeast". Rather than mule, most biologists would use the term "proto-lager" yeast. This lager yeast is also called *Saccharomyces pastorianis* (a hybrid of *S. cerevisiae and S. bayanus*) even though it was originally called *Saccharomyces carlsbergensis* by Hansen, who first described it in 1883. / A 2011 study by Diego et. al (published in PNAS) after searching around much of the world, identified *S. eubayanus* from Patagonia as the likely yeast that mated with the ale yeast *S. cervisiae* 500 years ago in Bavaria. / Bananas are a hybrid from the plant kingdom. Those tiny black spots are actually seeds that never formed completely because the modern banana is really the product of two different banana tree species that mated and produced sterile offspring. The only way banana trees can reproduce today is asexually (by producing "runners").
[xciv] The word lager comes from the German word *lagern*: to store. Unlike ale yeast, which produce a sweeter beer and float to the top at the end of fermentation, lager yeast tend to sink to the bottom, and is perhaps how the hybrid was selected for: by staying in the bottom of the vats while the beer was poured out, therefore giving it a head start on reproducing when the fresh wort was put into the vat; another example of early brewers being microbiologists and selecting for specific traits in a microbe, yet without ever knowing about microbes directly. The colder temperatures of the ice caves and bierkellers also limit the number of side-reactions inside fermenting yeast, and helps to make for a clearer, cleaner-tasting beer appreciated in lagers.
[xcv] DNA can add up. In the mucus of cystic fibrosis patients a common problem is too much chromosomal material accumulating in the mucus of the lungs. This DNA is shed by immune cells. One treatment involves inhaling an enzyme to break up this DNA, helping the mucus to flow more easily. Whenever we break open, or "lyse", a pellet containing millions of bacteria in the lab, so much DNA can spill out of the cells that the water turns to a gelatin that has to be repeatedly pipetted through a small hole to break up the DNA and turn the gel back to a liquid. Proteins, on the other hand, can be both a blessing and a curse in brewing. Brewers usually try to get rid of leftover protein because it can cause cloudiness during cold storage. It's possible to add enzymes to chop this protein up (digest it) yet some proteins are needed to stabilize the bubbles in beer foam (foam being a desirable trait).
[xcvi] I could have mentioned that it was these same "gene sequencers" he seemed to have so much contempt for who had recently determined the lager yeast was formed as a hybrid between two different yeast species hundreds of years ago in the bottom of a barrel in an ice cave (still others have sequenced ancient DNA of the yeast from the lees, or dregs, of Egyptian amphorae 4,000 years old). A sure sign of a hybrid microbe like the lager yeast is when the cell has an unusual ratio of chromosomes in its nucleus. This result of acquiring unusual numbers of chromosomes is called *alloploidy*. The lager yeast has 60% more DNA in its nucleus than the more ancient ale yeast has in its nucleus. Since they have been around much longer than lager yeast, ale yeast are, not surprisingly, the more genetically diverse.
[xcvii] Long voyages in warm waters were always hazardous because of drinking water's tendency to become contaminated over time, which is why they added wine. Convict ships to Australia, for example, had higher survival rates when they carried wine. It's been theorized that water becoming undrinkable on ships is what held back discovery of the New World until the Renaissance. Magellan's ship was the first to sail around the world, but it's often forgotten that he began with over 200 men and ended up with just 18 of the original. Even Magellan died on the voyage (but not of bad water. He was killed by natives in the

Philippines).   Pasteurization of milk is a way to kill the *brucellosis* bacterium that infects cattle, a bacterium that can also infect the consumer.
[xcviii] Before government agencies like NIH began funding basic research in the 20th century, science was done by wealthy "gentlemen scientists". Scientific societies also awarded prizes for making discoveries. For example, the French Institute offered a one kilogram gold medal in 1779 to anyone who could explain fermentation.   By the end of the 19th century and until WWII, businesses would take the lead by investing in science, bringing about an end to the solitary scientist working in his lab. / Nationalism was alive and well in the 1800's. Pasteur, for example, was interested in beer not because he enjoyed it but so France could compete with Germany in brewing.   Pasteur made another significant contribution to microbiology, becoming the first to prove using his custom-made glass beakers that microbes were, after all, like visible (macroscopic) life and didn't arise out of nothing.   Microbes were everywhere, floating around in the air. He proved therefore that "even microbes have parents".   Pasteur had put the final nail in the coffin of the *spontaneous generation theory*.   He even showed certain bacteria could, like the yeast, thrive without oxygen. / Details have always been important in science. When Pasteur was developing the first anthrax vaccine, he found he needed to subject the anthrax bacteria to temperatures between 42 and 43 degrees Celsius. This narrow band of temperature weakened the bacteria properly. At other temperatures, they would form spores.
[xcix] The Italian doctor Girolamo Fracastoro (the MD kind) surmised that the Bubonic Plague might be carried by a kind of "seed" in 1546, a full one hundred years before Leeuwenhoek would observe bacteria for the first time. I've often wondered if Fracastoro hadn't seen a burr catching on someone's clothing while walking in the woods and that gave him the inspiration for his "germs". / In Paris as early as 1514, a brewing guild law forbade the presence of pigs, cows, and birds near brewing kettles because it was said to ruin the beer. / It was one of Pasteur's students – Yersin – who first discovered the bacillus that causes plague, during the third pandemic.
[c] In AD 542. Plague killed 10,000 a day at its peak in Constantinople between AD 541-2 (in 2014, DNA from *Y. pestis* was recovered from the tooth of a victim there). / Large cities like London during the Middle Ages still had no underground sewage system, so waste was dumped right into the streets and the rats subsequently thrived there in great numbers. / Archeologists have never uncovered any communal dumps in Pompeii.   It appears the ancient town's residents simply let refuse degrade naturally.
[ci] London's Great Fire of 1666 ended the plague outbreak in that city, but not until the disease killed 70,000.   More than 80% of the city burned from the fire that began inside a bakery. This same outbreak of plague also closed Cambridge University, sending a young Isaac Newton home, where he would have what has since been called the most productive 20 months in the history of science.   While away he laid the groundwork for calculus, the theory of gravitation, and began his work on light and optics. / In 2012, researchers tranced several known strains of *Y. pestis* found in nature today (as well as the *Y. pestis* strain that caused the Black Death during the Middle Ages) back to a common ancestor strain that somehow diversified around AD 600 during the reign of Justinian.
[cii] The horse collar was valuable because it allowed horses to breath while pulling heavy loads (by not cutting off their airway). This allowed horses to replace oxen for plowing and, since oxen plow 50% slower, a food surplus accumulated in most towns and villages.
[ciii] Plague usually begins as a tingling sensation all over the body. Before antibiotics were discovered, the unfortunate victim would likely have died a few days after symptoms appeared.   In fact, plague victims were often still alive as their bodies began to decompose, some describing the smell of their own skin as it rotted, a result of the bacteria having traveled throughout the bloodstream and emerging on the limbs and torso. A common treatment for plague was bloodletting, which would have only increased spread of the disease. A saying in 1300's Europe had it that plague killed so quickly, it was possible to have lunch with your friends and dinner with you ancestors all in one day. / In 14th century Munich as in many large cities of Europe, entire Jewish sections were purged when Jews became scapegoats for plague.   Mark Twain once said, "History may not repeat itself, but it does rhyme a lot."
[civ] Today there are over 700 complete microbial genomes sequenced and available at the NCBI website (NIH) to anyone with access to the Internet, in fact the entire DNA sequence of the anthrax strain in 2001 was available to researchers within a month after the first death in

Florida. Wolf's concerns about protein chemists being relegated to obsoleteness weren't completely unfounded. A Holy Grail in genome research today is to someday use just the sequence of a gene to work forwards and determine the 3-dimensional structure of the protein it codes for. We already can get the sequence of amino acids of the protein simply by reading its gene sequence (looking at the DNA level), but knowing how the protein folds up into the unique and complex shape it has is an entirely different problem. Still, progress is being made and there is no doubt a Nobel Prize is waiting for the first person to do it reliably.

[cv] The prolific Winge – also known as the "Father of Yeast Genetics" – published more than 150 scientific articles, most on yeast, while at the Carlsberg Brewery. Carlsberg also helped pay for the study of young scientists, including Niels Bohr when he traveled to England to work with J.J. Thompson. In fact, the Niels Bohr Institute – which the brewery helped fund – eventually gave birth to quantum mechanics (Carl Jacobsen left his fortune to the Danish Academy of Sciences and Niels Bohr moved into Jacobsen's mansion).

[cvi] This is a handy ability for a microorganism to have because microbes may find themselves needing to colonize a new location alone. In the case of the yeast, it would reproduce by budding. But if it couldn't change sex, all the offspring would be the same sex since budding only produces clones (i.e. copies). Fortunately for the yeast, it is able to change its sex. A popular saying among evolutionary biologists is that "sex increases randomness."

[cvii] The protein the yeast cell uses to detect the presence of another yeast of the opposite mating type is related to the same protein we have in our eyes called *rhodopsin*, which changes its 3-D structure in response to light (energetically, rhodopsin is like a mousetrap ready to be sprung). Rather than light, the yeast's protein receptor changes its shape in response to the presence of a hormone produced by a yeast of the opposite mating type, generating a signal that then triggers fusion of the two yeast cells.

[cviii] Damage to DNA often causes unwanted changes, which is why breweries keep their yeast strains in the dark, thus avoiding UV light that can damage (mutate) DNA. Brewers may also layer mineral oil over the top of yeast stocks growing on agar slants to keep out oxygen, or store the yeast in liquid nitrogen at -196° C, in suspended animation. / During WWII Fleishmann's Yeast (founded in 1868) developed a special strain of baker's yeast that didn't require refrigeration and could raise dough twice as fast as conventional strains, giving American GI's far from home the chance to eat fresh-baked bread, just by adding warm water to activate the yeast.

[cix] In Australia in 1923 the margarine-like spread *Vegemite* was added to grocer's shelves, having been invented by food technologists looking for uses for leftover yeast from breweries, the yeast being a rich source of B-vitamins and protein. Yeast used in brewing and baking are aneuploid, meaning they have extra copies of genes – backup versions so to speak – making it more difficult for them to change due to a single mutation (good for brewers but not so good for geneticists who like to change genes in order to study their effects).

[cx] Base is sometimes called the *mother liquor* by modern chemists, a throwback to the days when all chemistry was alchemy. Alchemy was invented in Alexandria, Egypt by the Greeks. Modern chemists still use several alchemical terms including alembic, crucible & alkaline. The word *alcohol* is Arabic. / Newton wrote one million words on alchemy, in fact it's been suggested he got his idea of gravity by pondering alchemy and its mysterious forces used to explain natural phenomena. / One of the reasons Arabs were the beneficiaries of advanced technology, like blowing glass and making lenses for example, was their close proximity to the ancient Library of Alexandria, the world's original "think tank". Arab scholars kept and transcribed texts that emerged from this melting pot that once existed in Egypt and where learned people from all over gathered for centuries to share, record, and develop new ideas. / Cleopatra lived and was educated at the Library of Alexandria, which is why she could speak several different languages.

[cxi] As Gamal pointed out, due to the high alcohol content, brandy was an effective solution for dissolving herbs as well as honey, and is how Benedictine came about. The medicine was invented by Benedictine monks in Normandy. Alcohol also doesn't freeze as easily as water, making it a more convenient trade item for early settlers contending with harsh winters. Making brandy was also a way of putting to use poorer tasting wines that would otherwise have been discarded. Since alcohol boils at a lower temperature than water, brewers can take advantage of this when making so-called "near beer" (beer without alcohol in it) by simply making regular beer and then heating it to boil off the alcohol before bottling or canning.

cxii They also used whiskey for barter and believed the tax unfair because it was levied on smaller distilleries more than larger ones. The government ended the rebellion but the tax was repealed in 1803. Distilleries located in Kentucky & Tennessee could make bourbon from corn without interference by the government during this period, which is why there are so many distilleries in those states today. The IRS was created in 1862 to enforce laws against making moonshine while taxes collected from liquor sales would help finance the Civil War. The first to *lobby* congress in Washington D.C. worked for brewers who of course wanted lower taxes on beer. In the early 1870's (before they were sent north to Dakota Territory and eventually fought under Custer at the Little Bighorn), the 7th Cavalry was stationed in southern states to enforce federal taxes on distilleries. The modern sport of NASCAR grew out of moonshine drivers testing their fastest vehicles against each other.

cxiii Unlike most whiskeys today, which are dark having been aged in barrels, Washington's whiskey would have been clear (since it wasn't aged). / Thomas Jefferson had two vineyards at Monticello and was known to experiment, growing different grape varieties. The third President also spent $7,500 on wine his first term in office (more than $100,000 in today's money). / Three decades before he became the 16th President, Abraham Lincoln held a state license to sell whiskey by the barrel out of his general store in Illinois. When Lincoln was just 5 years old in Kentucky, he used to carry lunch to his father Thomas, who worked part time in a local distillery. As for himself, Lincoln abstained from liquor all his life.

cxiv Each Pilgrim was allotted a gallon of beer per day during the voyage. The Mayflower was called a "sweet ship" because, before the Pilgrims chartered it, it carried wine & spices in the Mediterranean. The ship's hold gave off a pleasing aroma. The beer, bread, and other goods including tools, gunpowder and salted meats were kept in wooden barrels on the Mayflower. John Alden was therefore hired in England as the ship's cooper, charged with repairing the barrels that would inevitably be damaged during the voyage. The Mayflower was classed as a 180 tun ship because it had the capacity to hold 180 barrels of wine (265 gallons per barrel). During the 1st Thanksgiving there likely wouldn't have been any pumpkin pie, but the pilgrims did have beer, brandy, gin, and wine. In fact, one of the first crops planted by the Pilgrims their first spring at Plymouth was barley so they could begin making beer again. A decade before the Pilgrim's arrival, the first ship to supply a colony in North America brought with it beer (in 1607). The Pilgrims also stocked the Mayflower with 20,000 "ship's biscuits" before leaving England. Ship's biscuits were a flatbread made from wheat flour and water, a key advantage being that they were more easily stacked. They also became infested with maggots by the end of the voyage and one pilgrim later recalled that the best flatbread for a trip of that sort would be so hard & dry that someone would need an axe to cut it. Once in the New World, the Pilgrims began making corn bread. Regiments stationed on the frontier during the Indian Wars (including the 7th Calvary) ate hardtack, a biscuit that could be up to 6 years old by the time they got it and sometimes required a hammer (or some beer) to break it apart. / North America does have native species of grapes, which is why the Viking explorer Leif Erickson called it Vinland ("wine-land") in AD 1000. These species (they may have been cranberries) tended to be more vigorous than Old World grapevines but sourer, as early colonists found out after trying to make wine from them.

cxv Wolf claimed back in the beer hall that the only real contribution America ever made to brewing was being the first to put beer in cans. According to him, American scientists developed a moldable plastic in 1932 that could line cans and thereby prevent beer from spoiling by keeping it out of contact with metal. At the time I thought he was joking. I'm still not sure. It is true that America was virgin territory for Bavaria's lager beers, not having had a previous tradition of drinking ales, like people in England had for centuries. So lagers caught on quickly in the USA. Milwaukee was also endowed with natural underground caves and the Miller Brewing Company has since turned one into a museum you can visit. I decided not to tell Wolf about one of my brothers winning a contest at the county fair for coming up with a way to deep-fry beer inside a pretzel. / St Louis has natural caves underground, which provided suitable conditions for lagering beer as was done in Bavaria (they used ice cut from the Mississippi to cool the caves throughout the summer). Some beer gardens in St Louis were located inside caves and by the 1850's one in three residents was a German immigrant. During the Civil War, soldiers smuggled guns in beer wagons and used the brewing caves as arsenals.

cxvi Welch's innovation was the beginning of the processed fruit industry as well as Welch's Grape Juice.   When Prohibition was first enacted, some towns sold their jails believing abstinence from alcohol would result in the end of crime.   Binge drinking on college campuses today is a relic of Prohibition, back when youth crowded into speakeasies and drank quickly for fear of being discovered by authorities.
cxvii They had reached the limits of diffraction.   The same developments in optics also took place with telescope lenses, paving the way for new planets and planetoids to be discovered. The scientist that discovered an optics law allowing for the correction of chromatic aberration in lenses was Joseph Jackson Lister, father of Joseph Lister. / Ground lenses have even been found at Pompeii, perhaps used by one of the town's engravers.
cxviii By the 1880's most trained physicians had moved beyond the view held by ancient Roman physicians like Galen and others throughout the Middle Ages – that tumors arose because of an internal imbalance of fluids – to believing that they were groups of normal body cells that had somehow turned "rebellious" and that this civil war could occur almost anywhere in the body. / In the 1880's, Wilhelm Roux suggested (correctly we know now) that the reason the cell division "dance" (mitosis) was so elaborate is so each new daughter cell can get the exact number of chromosomes it is supposed to have.
cxix Some early microscopists believed in the "theory of preformation" and convinced themselves they saw complete humans, called a *homunculi*, curled up inside sperm cells, complete with features like a head and torso.   Others using telescopes in the 1870's convinced themselves there were elaborate systems of global irrigation canals built by extraterrestrials on Mars.
cxx Warts are tumor-like growths caused by viruses that carry special genes to cause the skin cells they infect to pass their "start site" and continue to divide when they shouldn't.   Unlike most cancers, fortunately warts don't break up and spread to other organs.   The host cell still maintains some level of control.   Calluses are interesting because they develop when cells under the skin pass their start site and divide again and again, driven by mechanical stimulation.   They continue to grow when they otherwise wouldn't, thus producing a thickening of the skin useful for protection.
cxxi In "molecular biology-speak", Nurse *rescued* the mutant cell using the gene from a normal cell.   Nurse rescued Hartwell's damaged *cdc28 yeast mutant* with the similar gene from his own yeast, a cousin of brewer's yeast called *S. pombe*, and it returned Hartwell's mutant brewer's yeast to dividing again like normal.
cxxii cdc28 would turn out to be so important that this master regulator is regulated itself in at least three different ways, like a child that needs permission to go out and play from not only both parents, but must also be accompanied by a chaperone.   In the case of cdc28, the chaperone that joins with it is a protein called a *cyclin*. Important processes in human cells are often strictly controlled like this.   At each phase of the cell cycle, a different cyclin is produced, and each cyclin guides cdc28 to a new set of proteins, thus the reason for the perfectly choreographed dance of mitosis. There are also special "inhibitor proteins" the cell can make that will shut down cdc28. This is a remarkable level of control exerted over a single protein.
cxxiii As another analogy for cdc28: it's possible to picture it as the owner of a large department store, one who, before the store opens each morning, goes around telling each of his managers to begin giving out orders to their workers.   The manager in charge of produce tells the workers in fruits and vegetables to start stocking grapefruit and other items found there, the manager of the finances tells his cashiers to start counting the money and loading blank register paper to get ready to take in the day's receipts, the manager in charge of maintaining the building tells his workers to start sweeping the parking lot and repairing any damage done the previous day, etc.   cdc28 is like the owner of the store, and each of his managers would be the proteins that cdc28 binds to and modifies further down the line, for example enzymes on the surfaces of chromosomes which start copying each chromosome or line the chromosomes up in the middle of the nucleus afterwards, or break down the nuclear membrane at just the right moment so the chromosomes can separate and begin to form two new nuclei inside of the daughter cells, etc.   Hartwell also found that the proteins acting later during mitosis can go back and put a halt to it further upstream if something were to go wrong.   In the department store analogy, it's as if one of the store managers sees the cashiers have run out of register tape and tells the owner not to open the store until the situation is taken care of.   Normal cells – rather than becoming cancerous when they're

damaged – stop dividing in two and if this damage can't be fixed the cells will commit suicide as a safeguard against becoming cancerous. Biologists have a word for this called *apoptosis*.

[cxxiv] Taxol, for example. / The first human cancer-causing gene (oncogene) scientists discovered – *ras* – is found in not only all normal human cells, but also the brewer's yeast. And *ras* in both yeast and human cells can replace one another because they do much the same job.

[cxxv] The seeds Johnny Appleseed distributed to farmers in Ohio, Illinois, and Indiana in the early 19th century were for apple trees useful for making cider only (he got the seeds from cider mills). / Some breweries in western states got their start in mining towns where German immigrants had given up searching for gold to brew beer near natural mountain springs. Some skipped mining altogether. Adolphus Coors stowed away on a ship to get to America and began gardening in Denver before starting a brewery in Golden, Colorado.

[cxxvi] There are many notable examples, among them the Frenchman who deciphered the Rosetta stone and became the first to understand hieroglyphics in almost 2000 years. Champollion became fascinated by the problem two decades earlier at age 14. And when James Watson needed to escape the cold weather in Denmark he traveled to a meeting in Naples, Italy where he became interested in studying DNA after hearing a talk by Maurice Wilkins who was already using x-rays to look at the structure of DNA. Watson's future co-discoverer, Francis Crick, was a physicist who had changed fields to biology only after reading a book by Erwin Schrodinger called *What is Life?* in 1944.

[cxxvii] Lewis & Clark met a French fur trapper while on their way up the Missouri who was heading in the opposite direction back to St Louis. They hired him on the spot after learning he could speak fluent Sioux. Jefferson had instructed Lewis to invite several Sioux chiefs to Washington for a visit. Later, they hired another fur trapper, this time because his wife spoke Shoshone and could help trade for horses to make it over the continental divide when the time came. This was important since there would be no game in the mountains and the expedition could have easily gotten lost, then starve to death without a guide. / When the 7th Calvary was pursuing the Sioux into Wyoming and Montana in 1876, the map George Custer carried of the Yellowstone and other nearby rivers was drawn up before the Civil War and was filled with inaccuracies. It used dotted lines, for example, to represent the Little Big Horn River because the surveyors had never actually seen it. There were too many hostile Indians in the area. / The discoverer of King Tut's tomb, Howard Carter, had local knowledge that gave him an edge over other archeologists. Carter had already installed the first lighting and doors in tombs of the Valley of the Kings and so had a clearer idea of their layout. Most archeologists believed the important tombs in the valley had all been discovered. Champollion was also able to locate a Coptic church in Paris that still performed mass using the Coptic language, which gave him an edge in deciphering the Rosetta stone.

[cxxviii] It's one thing for a biologist to purify a protein, another to show that what they've isolated in a test tube is still working properly. For example if the protein purified is hemoglobin from blood then it's important to know if it still binds oxygen. Proteins can be damaged during purification and storage so it's a good idea to test them once in a while for activity.

[cxxix] Like most Greeks, Aristotle believed the Earth was like the yolk inside an egg, and so surrounded by water. He also boiled water, noticing sand afterwards, mistakenly believing that the water had changed to earth. The ancient Greek's definition of an element was therefore different than our own. Today we would say an element is *irreducible* because we can't change it (chemically) into another element. Aristotle was the first philosopher to see the universe as an interconnected system, while Democritus before him claimed that atoms had inherent shapes, smells, colors, and that even the gods could be made of atoms. / The Egyptians believed that humans were created from tears and sweat of the sun god Ra.

[cxxx] Earth does lose about a small lake's worth of water to outer space every year, some of which is replaced by water riding in on meteorites while the rest originates from under the Earth's surface.

[cxxxi] No one knows for sure how the Earth came by its water. It used to be held that water was imported mostly on the backs of comets that hit a cooled Earth. It's easier to see how asteroids could have delivered oceans when one considers that all the water on the Earth's surface makes up only about 0.02% of its total mass. Recent experiments suggest that Earth's water may have been here since its formation, and so was able to withstand the high temperatures of the early solar system without boiling away, by grabbing tightly onto minerals

inside the planet. / 450 million years ago, all life would have been found in water. Today there are more viruses in the oceans than stars in the universe and each day these viruses kill half the bacteria in the ocean. Still not impressed? How about: there are more viruses in a cup of ordinary seawater than there are people on earth.

[cxxxii] A Saguaro cactus can store 8 tons of water and survive without a rainstorm for two years. / The Native Americans on the Great Plains often used animals to find water. For instance, if a herd of elk was seen walking in single file, they were likely going towards water, but if spread out, they were coming from water.

[cxxxiii] The lungs are especially susceptible to infections due to their inherent moisture, which is why when authorities unseal an ancient tomb in Egypt these days they wear protective masks. Folks have died after breathing in long-dormant viruses and bacteria from recently unsealed tombs. The microbes settle and become re-activated in the lungs. In fact, there's so much water vapor in our breath that I saw them taking special precautions in tombs because exhaling all that water over time promotes bacterial growth on wall paintings and is also why Egypt protects the royal mummies in the museum in Cairo with glass cases. Studies have shown that the presence of tourists in an Egyptian tomb can increase the air's humidity from 20% to 80%.

[cxxxiv] Dried, or "crisp pretzels" were invented in America (as was pre-sliced bread, in 1928) and have a longer shelf life than traditional (soft) pretzels because of a lower water content. Fermentation – a *partial spoilage* done by fungi or bacteria – and smoking food are the other two methods of preservation. Interestingly, one of the chemicals found in wood smoke (formaldehyde) helps preserve meat and is the same chemical used to embalm bodies. Formaldehyde cross-links proteins in muscle and connective tissue, forming a mesh, so bacteria can't degrade them as easily. The astringent taste of some wines is caused by tannins from the grape skins. Like formaldehyde, tannins also have the ability to cross-link proteins (like mucin in our saliva), which is also why they can be used to preserve leather. The first building Lewis & Clark constructed inside their new fort near the Pacific was a smokehouse for preserving elk meat.

[cxxxv] A trick many cold-climate plants use to achieve frost-hardiness is to gradually remove the water from inside their cells late in the summer. They accomplish this by making the sap slightly higher in salt and sugar concentration, so the water will leave their cells (by osmosis) and flow out into the sap. Less water inside the cells means less ice crystals can form there. Sharp ice damages delicate cell membranes. When Armand Ruffer sliced Egyptian mummy tissue to examine it under the microscope in the 1910's, he first had to develop a technique to rehydrate these desiccated cells using water.

[cxxxvi] Our bodies have about 4 ounces of salt each and Roman emperors sometimes distributed salt to the poor. / Salt's well-known preservative effect in the ancient world is why Jesus told his followers to be "like the salt of the earth" and newborn babies in ancient times often had salt placed in their eyes to ward off infections. Salt's ability to keep food from rotting explains why salt in so many religions was thought to keep evil spirits away. The root of the word "salvation" means salt. / George Washington ran a fishery from his home at Mount Vernon and salted fish his slaves caught from the Potomac River. Thanks to salt, he was able to ship his fish all the way to the British West Indies. Washington preferred using salt from Lisbon, but because of British law he was forced to import a lower quality salt from Liverpool, England.

[cxxxvii] Marathon runners and other athletes in the past have consumed too much water without replenishing electrolytes and lapsed into a coma and died. Water poisoning happens surprisingly often. In fact, water could be considered a drug since too much can alter normal physiology, the very definition of a drug. Alexander the Great's soldiers died after drinking water too quickly upon reaching the river Oxus in Afghanistan having just completed a 50-mile desert crossing. While at the other extreme, the ancient Greek philosopher Thales was said to have succumbed to dehydration (too little water) while watching the Olympics in 546 BC. The drug ecstasy may encourage users to drink too much water while at the same time cause their kidneys to produce too little urine. This creates a dangerous water buildup within the body, causing the brain to swell (while the skull cannot), leading to death.

[cxxxviii] The ancient historian Pliny the Elder described houses made of salt in Egypt. Thomas Jefferson was convinced Lewis & Clark would encounter entire mountains made of salt on their way to the Pacific. / When the Erie Canal opened in 1825, the most common cargo carried on barges (the barges were driven by mules on the shore) was salt. / It's possible

Florida's most significant contribution to the Confederacy during the Civil War was providing salt by boiling seawater. / The Sarno River, which helped make Pompeii an important town in ancient times, was so-named because its delta region was used to evaporate and recover salt.
[cxxxix] When he got to the sea, Gandhi told reporters: "With this handful of salt, I'm shaking the foundations of the British Empire." / The most expensive item Lewis purchased in St Louis before heading west was "portable soup", a mixture of 193 lbs of dried vegetables, which they saved for when there was no fresh game. / Just days before reaching the Little Big Horn, Custer got into an argument with one of his junior officers concerning the already burdened pack mules. Custer wanted the mules to carry more supplies and convinced his officers to at least bring along some extra salt because, as he put it, they'd all be eating horsemeat before the campaign was over. / When Mark Twain traveled back to the Mississippi in 1882 to obtain material for his book *Life on the Mississippi* he wrote of New Orleans "The old brick salt warehouses clustered at the upper end of the city looked as they had always looked; warehouses which had had a kind of Aladdin's lamp experience, however, since I had seen them; for when the [Civil] war broke out the proprietor went to bed one night leaving them packed with thousands of sacks of vulgar salt, worth a couple of dollars a sack, and got up in the morning and found his mountain of salt turned into a mountain of gold, so to speak, so suddenly and to so dizzy a height had the war news sent up the price of the article."
[cxl] In Cairo there is a queen whose skin bears witness to the power of osmosis. The accident occurred while her mummy was being transported to the museum and water leaked into the boat. Because the mummy's skin still had natron salt left over from her embalming thousands of years before, river water rushing in to meet the salt caused her skin to swell and crack. / While crossing the Atlantic during their 9-week voyage, the Pilgrim's food had so much salt added as a preservative they likely suffered from dehydration and high blood pressure. It's been estimated that each consumed on average 6,000 milligrams of salt a day (almost 3-times today's 2300 mg daily recommended allowance) during the voyage. / The human brain is about 77% water.
[cxli] Osmosis is also the reason you can eliminate garden slugs by pouring a teaspoon of salt on them. It literally sucks the water right out their bodies. Their relatives, land snails, are interesting animals, the ancestors of which may have been the first to move onto land from water. Having already evolved a shell for protection against predators in the seas, their shells would have given mollusks a head start in colonizing the land too, since this would have helped keep water in. / In the middle of the Panama Canal is a freshwater lake. Whenever ships pass from one ocean into the other ocean using the canal, the lake's fresh water kills all the barnacles on the ship's hull.
[cxlii] George Custer's father-in-law died of cholera, as did Mary Todd Lincoln's father, in 1849, perhaps contributing to her mental instability. Sons of Harriet Beecher Stowe, Robert Frost, and Daniel Boone died of cholera. Due to fear of this disease in drinking water, the Union Army allowed soldiers to consume lager beer, which was lower in alcohol than other beers. After the Civil War, the popularity of lager beers increased due to soldiers having developed a taste for it during the campaigns. The 7th Calvary (which included veterans of the Civil War) prior to the Battle of the Little Bighorn were stationed at Fort Riley, Kansas, where a cholera epidemic swept through claiming 51 members in two years. Theodore Roosevelt went west to Dakota for the first time in 1885 to recover from a bout of cholera morbus.
[cxliii] Descartes saw the human body as a machine, while the first written reference to the universe as being machine-like dates back much earlier, to the AD 1100's…when the word *machina* was first applied. During the late Middle Ages common thinking held that the universe was like a giant clock the Creator wound up at the beginning, or perhaps had appointed invisible angels to push the planets around in circles. The ancient Egyptians called divine order in the universe *maat*, and their pharaohs were expected to maintain maat. One of the important differences between Aristotle and many philosophers around this time (including Plato) was that Aristotle concerned himself with the *processes* by which things took place; others weren't thinking about processes.
[cxliv] Wegner wasn't a trained geologist but a meteorologist, which is another reason his continental drift idea wasn't taken seriously. There are no shortages of mechanisms for explaining organisms in biology. Complex eyes, for example, evolved no less than 50 times in the animal kingdom, but using the same genetic "toolkit". No one knows how. In the past it's been assumed mechanisms were always a prerequisite for explaining a phenomenon, yet

317

Newton was successful at describing the motion of the heavenly bodies because he avoided what other scientists of his day thought was a necessary first step: to come up with an exact mechanism for *how* the planets moved. Newton simply used *gravity* as his explanation for the "glue" that held the planets in their orbits and didn't try to explain gravity any further than that. In fact, physicists still don't know why gravity works or what it is exactly. Modern physics tries to explain gravity using general relativity (that gravitation arises as a result of the curvature of space-time).

[cxlv] The main reason bacteria stayed so tiny even after billions of years of evolution is that they need to get their food by a process called diffusion…and diffusion works most efficiently across smaller distances. But in biology as in life, many things are a trade-off. Being small also makes bacteria vulnerable to being consumed by larger creatures including amoeba and white blood cells.

[cxlvi] Over 300 Olympic-sized swimming pools worth of water every day. One aqueduct flowing into ancient Rome supplied not only the public baths of Trajan, but also water to turn a grinding wheel in a flourmill for Rome's bakers. The more aqueducts the Romans built, the more people wanted to live in Rome, causing the city to grow. A study published in the journal *Science* in 2011 using tree ring measurements concludes that the fall of the Roman Empire in the west coincided with a period of less humidity and cooler temperatures.

[cxlvii] Even in the late 1700's, Paris was purifying drinking water. When Benjamin Franklin visited the city in 1767 he later wrote that their river water was "…as pure as the best spring water by filtering it through cisterns filled with sand."

[cxlviii] Pacini named the bacterium and published his results, but leading physicians of his day believed so strongly that cholera was spread through the air (the so-called miasma theory) that they ignored him. Since cholera victims often appeared as if they'd been asphyxiated (they turned blue), it made sense, though wrong, to assume cholera was being transmitted through the air. Koch was the first to isolate the germ, calling it "kommabacillus".

[cxlix] Survivors had been washing the clothes of cholera victims in the cistern. / Because *Vibrio cholerae* only infects humans, Koch was never able to accomplish what he had succeeded in doing with anthrax…reproduce the disease in animals. Other animals don't normally get human cholera. Because of his discoveries, Koch was greeted as a national hero after returning from India by the German Emperor.

[cl] The strain of *E. coli* called O157H7 that makes news whenever it causes food poisoning became dangerous after it picked up genes from another bacterium when the two shared the intestines of a cow. Many bacteria are good at swapping genes. Even people get infected with viruses that can insert their DNA inside our chromosomes. Each of us carries the vestiges of 100,000 viruses in our DNA from past infections our ancestors had and then passed onto us in their chromosomes.

[cli] Rosalind Franklin and Maurice Wilkins worked together in the same lab to determine the structure of DNA however they didn't get along. This likely affected the outcome of the race. Credit for the discovery went instead to Watson and Crick. Of the four, Franklin was the only one with a degree in chemistry; Watson was a biologist while Crick and Wilkins were both physicists.

[clii] The lungs are one of the places in the body that, like the cardiovascular system, can be made worse when the body attempts a repair. Exchange of oxygen needs to take place across extremely thin membranes in the lung separating the outside world from the bloodstream, and any thickening of it caused by scar tissue, or fibrosis, can result in too little oxygen entering the blood, leading to suffocation.

[cliii] As with paper towel, water can sometimes travel up stone by capillary action. When I was in Egypt I saw the effects of rising water tables caused by newly constructed dams along the Nile. This underground water tends to climb up stone columns of temples and tombs and other monuments, not unlike wax climbing up the wick of a candle. The limestone of the sphinx has salt crystals depositing and chipping away at the limestone surface due to water bringing so much salt up with it. The underground water carries the dissolved salt and when the water comes out of the stone it simply evaporates, leaving behind the salt, which slowly splits the stone, causing it to crumble at the slightest touch. The walls of Nefertari's central chamber in the Valley of the Queens in Thebes has salt crystals dislodging 3,000 year old paint off its walls. / Albert Einstein's first scientific paper, submitted in 1900, was an attempt to explain the capillary effect and the way it causes water to cling to a straw.

[cliv] Galileo measured time by weighing it (literally) using a water clock, which leaked water into cups he used to time the descent of his falling objects at different intervals, to uncover what would come to be known as the *Law of Uniformly Accelerated Motion*. He simply weighed his cups to record the time interval. As a biology undergraduate, I sometimes wondered why physics textbooks always seem preoccupied with motion until it was explained that motion is what underlies nearly all phenomena in nature, things we take for granted whether it's the traveling of electrons to produce electricity and magnetism for your appliances, or the motion of gas molecules to produce atmospheric pressure and wind, or the motion of hemoglobin when it binds an oxygen molecule in the lung. In fact, all atoms and molecules in the universe including the ones making up this book are in constant motion unless the temperature were at absolute zero. It was Aristotle 2300 years earlier who claimed that to be ignorant of motion is to be ignorant of nature.

[clv] Prior to the fall of the Old Kingdom, Egyptians believed their pharaohs were divine and even had the power to prevent starvation. After the Nile's level fell dangerously low for too long, Egypt descended into chaos. Never again did Egyptians place so much faith in the monarchy. This is probably why no large pyramids were built during the ensuing 2,000 years of Egyptian history. / It was a river that enabled the Egyptians to construct a civilization where it should have been impossible: in the middle of the world's largest desert. Irrigation using the Nile freed them from the need for local rainfall. It's not surprising that Archimedes – the inventor of the *hollow screw* and who used it to move water up to higher ground – got the idea for one while visiting Egypt. / At its lowest times of the year, Egyptians sometimes described being able to wade across the Nile (it froze over twice: in AD 829 & 1010). An Egyptian hieroglyph for water is two wavy lines. Too much of anything is seldom good and flooding may have spelled an end to the Middle Kingdom.

[clvi] Water was carried for miles and built up so much momentum that ancient Roman engineers couldn't pipe it directly into baths and homes without first relieving the pressure; otherwise the water would have broken their lead pipes. So they built fountains at the ends of their aqueducts. The most famous in Rome today is the Trevi fountain, originally built in 19 BC by Agrippa at the terminus of the 13 mile Aqueduct Virgo to supply the thermal baths. This aqueduct is the only one that never ceased bringing water to Rome (even after it was sacked by the Goths). / It was the building of an aqueduct in the AD 1500's that would lead to the rediscovery of the buried city of Pompeii. Recent excavations of Pompeii reveal that just before the eruption a new aqueduct had been built to supply the town with water (during the reign of Augustus Caesar). It had a dramatic effect on the character of Pompeii's gardens in private homes. Flush with extra water, Pompeii's wealthier homeowners went from relying on drought-resistant trees to lush gardens with flowers and even spent lavishly on pools and fountains. About 10% of the buildings in Pompeii were plumbed. / Fountains in Roman cities in Mediterranean climates were often situated in marketplaces to keep shoppers cool due to the water's evaporation (i.e. they functioned like sweat). The Romans were proud of their aqueducts, pointing out that they had practical value, unlike the "useless pyramids of Egypt".

[clvii] Bonds holding atoms together as molecules can store chemical energy. The energy locked within a water molecule's bonds can be retrieved and used, which is what plants do during photosynthesis. Plants use chlorophyll, a pigment that captures light from the sun and converts this energy into enough pull, electrically, that it can actually strip the electrons holding the hydrogen and oxygen atoms together in a water molecule. Plants therefore literally tear water molecules apart using light in an effort to extract, or mine, these energetic electrons making up the bonds in water. In fact, plants breaking apart water molecules into its constituent elements is where we get much of the oxygen we breath. Plants will then use these energetic electrons stripped from the bonds of water and put them to work inside the factory called the chloroplast to help create new energetic bonds within glucose molecules animals will later consume as food.

[clviii] Water was confirmed on the surface of Mars when the Phoenix spacecraft took in some frozen Martian soil, which then melted inside the craft's "oven" at the expected melting temperature of water: 32° F (0° C). This "bake-and-sniff test" used a mass spectrometer to detect water in vapor form (gas). It's been calculated that Mars's ice cap alone has enough water to cover the planet's surface with a 20' deep ocean. Phoenix's laser also confirmed that it snows on Mars, meaning that the Red Planet has its own water cycle.

[clix] Ordinary water inside a cell is not only necessary for life, but can also operate at the other extreme: as a mutagen (cause changes in DNA). Each cell in our body sustains damage to

its DNA about 100 times a day due to water causing *spontaneous deamination*, which alters a C into a U within the DNA sequence. Fortunately, our cells also have repair enzymes that constantly go around and reverse this damage.

[clx] Several members of the Lewis & Clark Expedition suffered from malaria, which they assumed was carried by bad air (*mal* = *bad* in Latin) rather than mosquitoes. While passing through Montana, Lewis wrote in his journal: "Our trio of pests still invade and obstruct us on all occasion. These are the Musquetoes eye knats and prickley pear, equal to any three curses that ever poor Egypt laiboured under." Studies of ancient DNA have recently revealed that King Tut suffered from malaria, as did others in Egypt. The same study also confirmed that Tut's parents were brother and sister, which may have made Tut more vulnerable to malaria's effects because of royal inbreeding (common among Egypt's royalty). Only the female mosquito bites humans because she needs the protein for her offspring. / Malaria is considered an "heirloom infection" because, unlike several diseases that jumped from animals to humans (such as HIV), the protozoan that causes malaria was passed down to us from our primate ancestors millions of years ago. Other heirloom infections include herpes, hepatitis, and yellow fever.

[clxi] Washing a victim's clothing in drinking water is another way it gets spread. In 2010-11 more than 6,000 Haitians died from cholera following an earthquake. Leeuwenhoek was the first to examine diarrhea (his own when he was sick) using the microscope and he found bacteria in it but didn't make the connection that they might have been what caused his illness.

[clxii] The TB bacterium's strategy is to spread from person to person by close contact of respiratory fluids, coughing up of droplets, etc. Since this usually requires prolonged contact to spread to a new host, it's not to the advantage of the bacterium to "lay waste" to its host so quickly (unlike, say anthrax, which uses the smash and grab strategy). / Cholera was far from the romanticized sufferer of 19th century consumption (TB). TB was thought to confer creative gifts like poetry. Some Victorians tried looking like TB sufferers using cosmetics to give them pallor or eating sand to damage their stomachs in hopes of losing enough weight to mimic the chronic wasting TB brings on. When a Victorian playwright needed an attractive corpse, they often had him or her die of TB. Edgar Allen Poe called it "the terrible beauty of consumption".

[clxiii] In India in 1953 S.N. Dee finally proved Koch's guess correct: a toxin was to blame. He took water that cholera bacteria were growing in, removed the bacteria, and found that the water alone (and whatever was still dissolved within it) was enough to cause release of fluid from a rabbit's intestines and therefore mimic the disease. The actual toxin proteins were not chemically isolated (purified) for another 17 years by an American named Richard Finkelstein. Once pure it was demonstrated that a mere 5 millionths of a gram of the protein, fed to a human volunteer, could cause the release of 6 liters of stool; which is pretty powerful even for a bacterial toxin.

[clxiv] For example if you look at, say position 241 (of the 1480 amino acid positions in CFTR) and the amino acid in human CFTR at that position happens to be a glycine, then there's a 60% chance it's also a glycine in the fish's CFTR at that same location along the protein (random chance would put it at 5% since there are 20 different amino acids possible for each location). / According to Richard Dawkins, our 185 millionth great-grandparents were both fish. In fact, all vertebrates were fish at one time.

[clxv] Some questions in biology can only be answered by looking at the *3-D structure* of the biomolecule, not just the sequence of its building blocks, which was, after all, part of Wolf's argument that genetics gets an unfair amount of publicity these days compared to proteins. You still can't get the 3-D structure of a protein just by reading its gene sequence on a computer monitor, for example. You still have to go out and do the experiment, then build a model of the protein's 3D structure to get to know its shape. There is an important enzyme we have in our cells called *glyceraldehyde-3-phosphate dehydrogenase*, for example, whose shape is so similar to a protein found in bacteria and plants that it's basically indistinguishable. While the DNA sequence of this gene has changed considerably over billions of years, the 3D protein structure coded for by this gene hasn't changed much at all.

[clxvi] Perutz supervised the graduate work of Francis Crick, who would go on to discover the structure of DNA. The structure of the DNA molecule, being much simpler and more repetitious (DNA is a fiber) than that of most protein molecules, was solved in 1953, before any protein structure was known. It took Perutz about 23 years to determine the structure

of hemoglobin's 146 amino acids. / As of February 2013, there are about 80,000 different proteins whose 3D structures have been determined.
[clxvii] An interesting example of one of the many potential benefits of knowing the 3D structure of a protein is the anthrax toxin, which normally binds to a human blood cell to bring about disease. I'd read somewhere that researchers had not only determined the 3D structure of the anthrax toxin itself, but also the structure of it *while bound* to its human receptor (which is also a protein), normally displayed on the outside of a human blood cell. Since they now knew the exact location of the atoms making up the toxin that bound to the human protein, researchers could then begin to tinker with, or genetically engineer, the surface of the anthrax toxin in order to *design and direct* it to bind to some other cell rather than the human blood cell, for example a cancer cell, with the aim of destroying the cancer (i.e. turning the anthrax toxin into a magic bullet drug).
[clxviii] Plasmids are circular pieces of DNA, perhaps 100-times smaller than a bacterial chromosome, making them easier to work with in the lab. *Vibrio cholerae*, for example, has 2 circular chromosomes that total 4 million bases, while its 2 plasmids (one carries the toxin genes) total 47,000 bases. If you can get a bacteria cell to take up your plasmid, the bacteria can then make copies of it. These bacteria will all be "clones" of one another. Plasmids are convenient places to carry your gene of interest (the gene is, of course, also DNA), useful for transferring the gene between different kinds of cells...kind of like a shuttle.
[clxix] Along with the gene to make CFTR, we also had placed next to it a bacterial gene (or "marker") that produces an enzyme that makes a colorless dye turn blue on the Petri dish.
[clxx] So can agitation, which is the reason it's possible to beat egg whites into a stable foam. The delicate egg proteins easily unfold and coat the surfaces of air bubbles, which keeps these bubbles from drying and popping. French chefs often use copper bowls to beat eggs in because the egg proteins (called albumin) bind to copper atoms, a metal that helps the albumin unwind, or denature easier. Binding copper is also why the egg whites will take on an orange tint during beating. The copper binds to sulfur atoms in the egg protein (mercury, another metal, binds to sulfur in bacterial enzymes and may be the mechanism of the salves Lewis administered to his men that helped reduce their symptoms of venereal disease). Protein in muscle can denature too, which is the reason victims from Vesuvius are often found curled up into what's called a *pugilistic attitude*, as if shielding themselves from a body blow after death. Forensic scientists recognize this same posture in victims of a hot fire. The denaturing of muscle protein causes the body's limbs to curl inward in a contorted manner, similar to the way a boxer shields himself from an opponent. West of Pompeii, many of the victims of the same eruption at the ancient seaside town of Herculaneum had managed to take refuge in the boathouses, only to succumb to *thermal shock*, their bodies having been heated so quickly that their brains boiled, causing skulls to explode.
[clxxi] Alcohol can denature proteins too, which is how it kills the yeast in wine when the alcohol concentration reaches 20% or so, and also why mouthwashes with alcohol tend to turn clear saliva opaque (like fried egg white) just before it goes down the drain. It's the protein mucin in saliva that aggregates and becomes visible; otherwise mucin is clear and soluble in saliva, much like the *uncooked* egg whites.
[clxxii] Another example of protein turning into a "fried egg" is when a cataract forms within the lens of the eye. Over time, the normally clear lens proteins become cloudy due to the energy of UV light unwinding them, causing these damaged proteins to accumulate. The lens eventually becomes opaque enough to cause blindness. Galileo became blind his last two years due to a combination of cataracts and glaucoma (but probably not because of his previous observations of the sun). Like clear egg whites, protein in fish muscle is also transparent before it's cooked, which is why it's usually a good idea to buy fish at the market with translucent rather than opaque meat.
[clxxiii] Another common example of acid's ability to denature proteins is adding lemon juice to tea and milk. The acid from the lemon curdles, or denatures, the milk proteins, as when milk turns sour. Even a T-bone steak undergoes denaturation of the muscle protein when it cooks...and is the reason the meat changes from flexible and red (when raw) to stiff and brown after cooking. Proteins don't usually tolerate heat, so they unravel and bind together, making the meat easier to digest.
[clxxiv] Pasteurization of milk doesn't usually kill all the microbes in food; just reduce their numbers to a safer level. Most beer is pasteurized today. In fact, Adolphus Busch was using pasteurization to increase the shelf life of his bottled beer by 1873. Busch had a keen

interest in science and kept up with the scientific literature and the latest innovations. Not only was Busch the first to pasteurize beer in America, but the changes he introduced also helped make possible the export of his beer over greater distances. A consumer in the late 1800's could buy bottles of his American-brewed beer in places as far away as Japan & India (I was also surprised to learn recently that as early as 1833, someone in Bombay, India could have ice cut from 16,000 miles away to cool their beer thanks to improvements in transportation, shipping & insulation. The ice was cut from the surfaces of lakes in New England). There are other alternatives to heat pasteurization for beer. The Coors Brewery uses a cold filtering process to remove microbes, one that relies on keeping the beer continuously cold from the time it is bottled until sold. This need to keep the beer cold at all times is why Coors remained a regional beer for so long. / Adolphus Busch was also an innovative promoter. In the 1890's he had lithographs of Custer's Last Stand printed to sell his Budweiser beer. Thousands of saloons in America had the image of Custer being killed by the Indians hanging on their walls.

[clxxv] Usually with electron microscopy (EM), heavy metal stains containing uranium and silver need to be used on cells because the electrons from the microscope are too high in energy and damage the cell's more delicate carbon and hydrogen atoms. EM has been used to look at red blood cells thousands of years old in Egyptian mummies that had sickle cell disease.

[clxxvi] Worrying about someone stealing ideas isn't new in science. In the late 1400's, Leonardo da Vinci wrote his notebooks in code. In fact, it's one of the reasons he was never given adequate credit for his ideas until long after his death. He was the only one who knew the code.

[clxxvii] Thales – often credited as the first philosopher – traveled to Egypt from Greece where he learned geometry (the ancient Egyptians had to re-survey their fields each time the floods receded) and used the length of his shadow and the Great Pyramid's shadow to calculate the height of the pyramid. Upon returning to Greece, he used this same method to estimate the distance ships were from shore. In addition, Thales predicted ahead of time a solar eclipse on May 28, 585 BC using Egyptian math (an event which, so we are told, ended a battle between the Medes and Lydians). In 1609 Galileo would use geometry and the shadows cast by the moon's mountains while looking through his homemade telescope to estimate the height of lunar mountains a quarter million miles away.

[clxxviii] On their way back east from the Pacific (and just before re-crossing the Continental Divide), Lewis wanted to geld some horses they'd gotten from the Nez Perce because they were acting wild. The Indians told him they had a better way, so Lewis had 2 horses gelded his way, and 2 the Indian's way. After 2 weeks, Lewis concluded their method was better even though it didn't appear so initially because it was bloodier. Bleeding probably helped flush out the horse's wound. / The father of genetics, Gregor Mendel, spent 2 years getting the controls right for his pea plants before doing the actual experiments, which would take him an additional 6 years to complete.

[clxxix] He also had several foreign postdocs in the lab, perhaps because they needed a visa, which he held over their heads and could refuse to renew anytime he decided he wasn't getting the proper "results" out of them.

[clxxx] To give you some idea of Amanda and Chris's job: after growing up about 5 liters of yeast they would end up with about a thousandth of a gram of pure CFTR, which is about as much as a grain of salt weighs. When you consider how much else there is in 5 liters of yeast broth (we grew them to a density of about 100 grams yeast for every liter of broth) and all but this tiny amount of CFTR had to be removed, it makes looking for a needle in a haystack seem easy by comparison.

[clxxxi] Proteins can stick to many surfaces during purification – dishes, glass tubes, plastic pipettes, even to one another. They like to stick to metals, too, which is why it's advised not to suck snake venom out of a snakebite especially if you have metal fillings. Snake venom is a protein. The stringy proteins called glutens in wheat flour have sulfur atoms on their ends, which help them form a mesh with each other and thereby strengthen the dough. One of the things I was considering was increasing the concentration of a chemical called EDTA in the CFTR sample before sending it off to London. You may have heard of EDTA. It's a preservative found in some foods because it keeps bacterial contamination down as well as proteins from oxidizing (aging).

clxxxii The importance of publishing in science isn't new. One of the reasons Robert Boyle is considered the first true chemist (rather than an alchemist like his contemporaries) was his willingness to publish results back in the 1600's. Even before reaching civilization, Meriwether Lewis began writing up a summary of their discoveries in a letter to Jefferson while lying in his canoe on the Missouri River (he had been shot in the butt by one of his men while hunting). Much to Jefferson's dismay, there was a delay in publishing the journals of Lewis and Clark for many years, which is why most of the names the pair gave to rivers are not used today. Trappers who came along after Lewis were unaware of the previous names and simply gave the rivers new ones. The chemist Henry Cavendish discovered hydrogen in 1766, yet was so shy that he seldom published. He also discovered that the electrical force of attraction is similar to gravity in the way it gets weaker with distance, but only Coulomb published this finding (both discovered it at the same time) and so only Coulomb got the credit for it. It seems Cavendish even had a separate staircase put in his house so he could avoid coming face to face with his housekeeper. After his death when scientists went through his notebooks they found the shy man had discovered all sorts of useful things he had never let on about, including several laws governing the behavior of gases.

clxxxiii In ancient Egypt, pharaohs chiseled out the names engraved on temples and tombs of former pharaohs and had their own chiseled in its place. It's ironic that today the most recognized name of ancient Egypt – King Tut – is also the pharaoh we know the least about. We know so little because Tut's successors erased him from their monuments. / Egyptian artisans also made mistakes on tombs, covering them up with plaster and then simply working over top of them. I had a guide point out one spot where the plaster had fallen off of a mistake. / It's interesting that the ancient works from the Library of Alexandria that have been handed down throughout history were simply the easiest to understand for the scribes doing the copying. Some experts think the ancient world was even more advanced than we'll know because scribes along the way chose to record only what they themselves understood and left the more complicated works that were over their heads uncopied and so lost forever. The 2000 year old Antikytheri Mechanism may be just the tip of the iceberg.

clxxxiv Curveball had drinking problems, which German Intelligence knew and claims to have told American Military Intelligence. He was a chemical engineer and used his knowledge to lead his interrogators to what they expected. It now seems Curveball was motivated by his desire for a green card. President George W. Bush also used allegations provided by Curveball in his 2003 State of the Union Address. It's interesting to consider that Curveball was never questioned directly by any American before Powell's speech. Curveball admitted lying during a 2011 interview on 60 Minutes.

clxxxv Even the architects of the pyramids made mistakes when they first started out. Early pyramids were built on sand and either fell apart or had to be modified (bent) to keep from collapsing. Paradigm shifts have resulted in new branches of science. Before Karl Webber (and Fiorelli, the site's director) began excavating Pompeii systematically in the 1700's, treasure-seekers would simply smash a hole in the wall of an ancient house. Not surprisingly, miners, rather than scientists, did the first excavations at Herculaneum. Webber began the practice of modern archeology when he started tunneling horizontally, down the length of Pompeii's streets first and then going in through the front doors. People were by then digging for information, not just riches as they had been. Even the idea of a species going extinct was a paradigm shift at the time. Thomas Jefferson, for example, didn't believe the animals responsible for the fossils he'd collected in the east could have disappeared and therefore was quite certain Lewis would come across wooly mammoths on his trip. Another example of a paradigm shift: in the first half of the 20th century, people didn't think the bodies of mummies could yield useful information, which is why they were allowed to deteriorate for so many decades in the Egyptian Museum. When Napoleon's scientists explored Egypt in the early 1800's, they mistook many of the temples for palaces because these structures had wall paintings inside them depicting warfare. It didn't occur to them that religion in ancient Egypt could actually condone war at all times.

clxxxvi Physicians would treat only symptoms prior to this paradigm shift. It wasn't until they recognized distinct diseases that they could begin concentrating on treating a particular disease rather than just relieving its symptoms. More than 60 generations failed to question Galen's teachings from the 2nd century AD on: that humors being out of balance within the body was what caused disease. The monk Gregor Mendel had a paradigm shift when he went further than simply noticing that if red flowering pea plants were mated with white flowering ones they produced either red or white flowers and not a blending of the two (i.e. pink). He actually *counted* the numbers of each color offspring, coming up with a fixed ratio, which led him to the idea that living organisms had specific genes that governed specific traits.

[clxxxvii] As an example of it being too early for a paradigm shift to take place: both Pasteur and Koch witnessed bacterial colonies avoiding each other on the surfaces of dishes, but biologists in the 1800's didn't know about small molecule inhibitors (like antibiotics) so this information was ignored, that is until Fleming rediscovered the phenomenon in the 1920's. By Fleming's time scientists knew about enzymes like lysozyme in human tears that could inhibit bacteria. Fleming's discovery of penicillin eventually gave birth to the branch of biology today called soil microbiology. Even now, the vast majority of new antibiotics still come from the soil. The ancient Egyptians almost never thought to record their own history until the Greeks came along (i.e. the Ptolemies). The Egyptians tended to be more concerned with contemplating the supernatural world than they were the physical one. Virtually everything we know about ancient Greek tragedy comes from the work of just 3 poets, which leads one to wonder what paradigm shifts would take place if a different set of works by 3 other poets had been saved instead. We might have an entirely different view of ancient Greece today.

[clxxxviii] A good scientific theory should also make a prediction. Copernicus, who entered college the year before Columbus sailed to the New World, did use his new theory to make predictions about future celestial events in the sky, powerful evidence his heliocentric system was indeed correct. When Mendeleev wrote down the periodic table of elements he left question marks in places where he predicted new elements would be discovered; and they were. When Maurice Wilkins showed Watson & Crick photo 51 of DNA, Crick realized immediately that it had to be a double helix because the physicist had already used mathematics to predict what the x-ray pattern *should* look like if DNA turned out to be a helix. His prior calculations matched what he saw in Franklin's photograph perfectly.

[clxxxix] While on a sea voyage, Galileo observed fresh water in barrels on the ship were sloshing back and forth and believed he could compare this to the Earth's tides and that the tides could therefore provide evidence that the Earth was in motion. He was wrong, though. The tides are mostly due to the moon's gravitational pull. Fungi were misclassified as plants for many years because they usually grow in the soil (like plants do) and is why today they are still covered in botany classes, even though they consume other organisms for energy (as we do). Thanks largely to DNA analysis, we know now that fungi have their own kingdom and are more closely related to animals than to plants. When Leeuwenhoek saw single cells with his microscopes he assumed they had to be animals…only smaller…with complex tissues like muscles and skin such as we have. Which is why he referred to them as his little "animacules".

[cxc] A version of SDS-PAGE is used by brewers to identify a particular strain of barley, making it possible to recognize the specific proteins inside the barley seed. Different strains of barley will produce different patterns, or protein "fingerprints", on the gel. Gel electrophoresis was used to analyze in 1972 the different proteins making up cholera toxin for the first time.

[cxci] Standards are what their name implies: for standardizing things, creating a kind of "yardstick", and there are many interesting examples in science. In ancient Egypt it was the distance from a person's elbow to their thumb that was taken as the basic unit of length (called a cubit). Of course not everyone's distance was the same since people's arms differed even in ancient times. The problem was solved when the pharaoh decreed that only his arm would be used as a standard and thanks to this the Egyptians were able to build the pyramids. In ancient Pompeii, to check that the right volume of wine was sold in markets, a magistrate would pour the wine into predetermined holes sunk into a stone table. Since Leeuwenhoek was observing things never before seen with his microscopes, he had to invent his own standards of measurement. He used things like a grain of sand or the eye of a louse for standards. As late as 1788, France still had 200 different units for measuring items sold in towns and villages. In England, Richard II set wine barrels at 31.5 gallons while in 1531 Henry VIII standardized beer by fixing the number of gallons coopers needed to make their barrels at 36 gallons. A problem to this day is that there is surprisingly little standardization in forensic laboratories across the USA, which has resulted in the convictions of innocent defendants. It was believed until recently that a skilled analyst could identify the box of

bullets a single bullet at a crime scene came from by slight differences in lead composition between boxes. This turned out to be junk science. There wasn't even a standard definition for what a planet was until recently (which is how Pluto got demoted), in fact this standard may change again as new objects in the solar system are discovered. When Viking 1 landed on Mars in 1976 it carried color cameras as well as a reference chart mounted on its side, but this chart didn't appear in the first images beamed back to Earth. As a result, the Martian sky appeared blue to scientists back on Earth. Photographs taken later included the chart and so could be corrected for, revealing that the Martian sky was, to the disappointment of many, more of a yellow-brown. / Radiocarbon dating was shown to be accurate in the 1940's when Willard Libby used carbon-14 to date wood from a royal barge found in an Egyptian pyramid (1850 BC). He won a Nobel Prize for this in 1960. Religion in ancient Egypt was never standardized and the number of gods differed depending on which part of the country an Egyptian worshiped in.

[cxcii] Collagen is a surprisingly strong protein and about 90% of all the protein in our body is collagen. Lewis & Clark's men often took advantage of collagen's strength in elk skin as ropes to pull their boats up the Missouri River from along the banks when the wind was too strong to paddle against. Collagen also protected Lewis' journal because its cover was made of elk skin, while Clark's shirt and moccasins were buffalo hide, also largely collagen. Different animals have slightly different varieties depending on the temperatures they live at. Fish usually inhabit cool waters, explaining why it's easier to overcook fish than red meat...their collagen loses its structure even in mild heat, resulting in flakiness. Proteins called *glutens* found in wheat give dough its strength and are the reason dough can hold leavening gases created by the yeast. The Chinese knew gluten as the "muscle of the flour". Glutens are, not surprisingly, long chains of proteins making up 10% of flour's weight and can bond to one another and be stretched out like coils, which then contract back again giving dough its springiness. Glutens in dough start off randomly arranged but then become orientated in the same direction during kneading of the dough. When bakers tenderize dough for making piecrusts, cookies and cakes, they add fat, which weakens this interaction glutens have for one another, thus tenderizing the crust, making it softer to chew.

[cxciii] Lewis and Clark encountered so many animals on the prairies because this type of grass is higher in protein. The extra protein can sustain a larger and more diverse animal population and is the reason why, when they finally met up with the Shoshone in Montana, just this one relatively small band had over 400 horses. / Being made out of 20 different building blocks rather than just 4 makes proteins inherently much more complex than DNA, so much so that most biologists were still convinced even into the 1950's that proteins would turn out to be the hereditary material our genes were made of.

[cxciv] DNA codes for the construction of all the different proteins in a cell. Here's an overview of how it works. Most genes can be thought of as a separate, linear blueprint within the chromosome that codes for the construction of one particular protein. One gene sequence in DNA corresponds (on a 3:1 basis) to the protein's amino acid sequence, instructing the cell as to which amino acid is inserted at each place along that protein's growing chain. For example, if the DNA sequence of the CFTR gene at a certain place along its gene sequence happens to be A-G-T, then the cell puts the corresponding amino acid (a serine in this case) in that particular spot within the growing chain of the CFTR protein. This is why DNA is often called "a template for making proteins". Its sequence directly corresponds to the amino acid at that particular point along the gene. The vast majority of your genes code for proteins like this, while the rest still have an unknown function, or perhaps code for RNA to make ribosomes and for the regulation (turning on and off) of other genes, or are the remnants of extinct viruses, etc.

[cxcv] The longest proteins so far discovered are the *titans*, weighing in at over 27,000 amino acids strung together. At the other extreme, some bacteria toxins are just a dozen or so amino acids joined together. The protein Hartwell discovered in brewer's yeast that controls mitosis (made by the cdc28 gene) is an enzyme whose job includes having to move around inside the cell. It is thus 5-times smaller than CFTR. / The exact order of amino acids in a protein can be surprisingly important. For example, by far the most common mutation in cystic fibrosis results in a single amino acid missing in the final structure of CFTR...just one missing amino acid out of a total of 1,480 is enough to result in CF. Another example is sickle cell anemia, which happens when a single amino acid is out of place in the hemoglobin protein (its β-globin subunit). / A typical bacterium like E. coli has an inventory of about

3,000 different proteins, the brewer's yeast makes due with 6,000, while human cells all together can produce some 25,000 varieties (not including antibodies, which are almost limitless in variety).

[cxcvi] When CFTR is made inside the cells lining the lung, if it isn't folded properly, or it takes too long to fold, the protein gets sent to the cell's default pathway – a cellular "trashcan" – where it is chopped up and its amino acids recycled to make another protein. / When the brewer's yeast cell ages, it accumulates clumps of misfolded protein.

[cxcvii] Heavy metal ions can also function as enzyme inhibitors, which is why arsenic was used in embalming fluid in the 1800's: to inhibit proteases in a decomposing body. The corpse of one outlaw from the American West, Elmer McCurdy, was displayed for over a century in carnival side shows after being embalmed with arsenic.

[cxcviii] One of the ways people in the Middle Ages purified table salt was to remove the impurities by dissolving the raw, impure salt in water first. Impurities (like rocks and sand) remain as solids in the water, and so could be removed at this step (simply by gravity). The now pure but still dissolved salt was then recrystallized by letting the water evaporate in the sun.

[cxcix] Costerton and colleagues had previously discovered that bacteria in the *fluid* of a cow's rumen appeared different under a microscope from the same bacteria attached to the *solid food* inside the rumen. / Louis Pasteur invented the use of "selective culture medium" for growing only certain kinds of bacteria in. For example, anthrax bacteria grow better in sheep's blood (Pasteur actually preferred the use of "netural urine") than *E. coli* bacteria do, but *E. coli* from our gut grow better in bile salts. Pasteur worked in a time before colorful dyes were invented to help microbiologists stain and identify bacteria.

[cc] Agar on Petri dishes look solid but is mostly water and some nutrients, meaning bacteria that look like they are growing on solid may be mostly in water.

[cci] Biofilms can also be a *multi-species consortia* (in microbiology lingo). It's even possible that all the different species of bacteria, due to their enzymes that can modify food molecules, have turned our mouths into a kind of "reaction chamber", influencing the way we taste our food. / Our mouths have many different niches inside. It's been found recently that certain species of bacteria will live exclusively on only one side of your tooth but not the other side, for example. And obesity is associated with a less diverse microbiota within our digestive systems.

[ccii] Recent studies indicate that the more diverse group of microbes that are present in a CF patient's lung, the less likely they are to become colonized later on by *Pseudomonas*.

[cciii] Even though cholera is rare in developed countries, *Vibrio cholerae* may still turn up during the warmest months along the Gulf Coast of the USA in undercooked crabs, shellfish and other filter feeders, possibly the same strain from Mark Twain's day (1832 - 49).

[cciv] As a schoolboy, I'd been taught that George Washington had wooden teeth. Turns out, they were actually ivory. The "wooden teeth myth" apparently began because, when Washington took his dentures out at night, he would soak them in a glass of port wine. Over time, the wine stained his dentures, making them appear as wood.

[ccv] Because of the trillions of microbes inhabiting a cow's gut, these ruminants can digest cellulose (which is glucose). And with enough grass growing in between rocks, a herd of dairy cows or goats can turn the worst upland into milk, cheese, and butter, which is pretty amazing when you think about it. / The astronomer William Herschel sometimes used horse manure to cast the mirrors for his powerful telescopes. He even got his wife to help out with this chore.

[ccvi] Another surprise that came about in biology – this time in the 1990's due to the ability to sequence entire genomes – is that humans have only about double the number of genes that go into making up an ordinary fruit fly. Even the single-celled brewer's yeast on your supermarket's shelf has a fifth the genes we do. One reason humans get by with so few genes (~25,000) is that our cells can do more with what they have. It's done largely by a process called "splicing". As an analogy, it's possible to take a sentence and clip out some of the words in the middle, then squeeze the rest of the sentence back together again and get a totally different meaning. "The diver dove into the pool and water flew onto the roof." Remove diver, into, the, pool, and, water, then splice rest together and you get "The dove flew onto the roof". Our cells do this kind of splicing with genes a lot. Spliced one way, a gene's product (mRNA) directs for a certain protein to be made, spliced a different way, that same gene can direct the construction of a protein with very different properties. Still

another reason we humans get by with just 25,000 genes is that some of our proteins can take on more than one task.   Cell biologists call this type of multitasking "moonlighting". For example, the tumor-suppressing protein *p53* can join with more than 100 different proteins inside its cell as well as to DNA and RNA.   One reason p53 can bind to so many different molecules is that its structure is partially disordered.   Perhaps 1/3 of our cell's proteins are partially or completely disordered this way.   This flexibility in structure helps give them the ability to take on different jobs.

[ccvii] There are also lapses for months at a time in Lewis's diaries that have never been explained.   Some historians speculate he was suffering from depression, others that the journal entries were lost. / Modern scholars believe Aristotle wrote an entire book on animal dissections no one has seen for two thousand years. / In 2011, literary scholars were still discovering 3 new letters a week on average written by Mark Twain.   It's estimated he penned some 50,000 letters during his lifetime.

[ccviii] Unlike religion, science is based on change and new branches of science still emerge from time to time.   One example is forensic science.   In the 1920's, Howard Carter wasn't interested in Tut's body to the extent he was his jewelry, so he cut Tut's mummy into pieces, destroying forensic evidence because it wasn't generally accepted in his day that even dead bodies could yield clues.   And most experts now believe only 30% of Egypt's ancient monuments have been discovered so far.   The same year King Tut's tomb was found, the ancient Indus civilization (Harrapa) in India was discovered to be over 5,000 years old, older than the pyramids of Egypt.

[ccix] "So remarkable Shye and watchfull," is how William Clark described them in his journal after trying to track one. / The mummy of Ramsess the Great had 89 different fungi living on it before being sterilized in Paris during its restoration.

[ccx] The most successful textbook ever written was by Euclid 2300 years ago.   While working in Alexandria's library, Euclid laid out the basis for mathematical proofs in geometry still used today.   A fragment of one copy of his textbook has survived, dated to around AD 100.

[ccxi] "It is important to realize that in physics today, we have no knowledge of what energy 'is'. We do not have a picture that energy comes in little blobs of a definite amount.   It is not that way.   It is an abstract thing in that it does not tell us the mechanism or the reason for the various formulas." – Feynman.

[ccxii] Galileo once described wine as "light held together by moisture" (even though drinking too much likely worsened his gout in later years) and Ole Romer later built a crude thermometer out of wine in 1701.   Only later were *enclosed* glass thermometers made based on the expansion of a liquid and therefore independent of the surrounding air pressure (which of course changes).   Daniel Fahrenheit's thermometers around 1724 used mercury and gave the first reproducible, accurate results (finely quantified, measured).   It was inventions like the modern thermometer during the Industrial Revolution that allowed for what has been called the "Age of Scientific Medicine" to come about.

[ccxiii] The Greek philosopher Heraclitus thought fire was the most fundamental element and could give birth to the other three Greek "elements".   He also claimed the human soul was composed of both fire & water and that the pursuit of earthly pleasures "moistened" one's soul (i.e. made the soul less pure) since fire was for him the most "noble" of the two. / If I'd been a professor during the Middle Ages, I probably would have instructed my students to "weigh fire"…by calculating the difference in heaviness between an unburned and a burned log, never realizing that the solid in the log had largely escaped as carbon dioxide. / As was commonly believed at the time, Newton thought the world would end in fire and that comets were made of fire and one would inevitably strike the Earth, which is why he read up on Egypt and even wrote a chronology.   What Newton wanted was to fix exact dates in the Bible so he could then work forwards and come up with the date the world could be expected to end. / It was fire's ability to turn clay into ceramic that allowed civilization to develop.   Ceramic was the first brand new material humanity learned to create.

[ccxiv] The first protein crystallized in a lab, ovalbumin, was obtained from chicken eggs in 1890. Some proteins crystallize so easily they do it without any help from scientists.   19th century sailors on whaling ships noticed red crystals forming on the decks of their ships.   It was the red oxygen-binding protein *myoglobin* from whale's blood forming crystals as the blood dried. The red color in meat is myoglobin, a protein cousin of hemoglobin, and when cooked turns brown as it unfolds.   Modern packaging in supermarkets allows oxygen to penetrate the clear plastic wrap and keep the meat's myoglobin red, which is more appetizing.   If you dig

further down into fresh hamburger, you will see a dark purple color, almost brown; which is the color of myoglobin without any oxygen bound to it.

[ccxv] We are *information rich* in biology these days when it comes to DNA. With all the sequencing of genomes it's easy to determine the amino acid sequence of the proteins the genes code for (it's a simple 3-to-1 relationship). You can read it off the computer monitor. But since we can't predict structure just by knowing the sequence of amino acids making up the protein's chain, biologists still have to go out and determine a protein's overall 3D structure by experiment using x-rays, NMR, or cryo-EM (as we were trying to do). One example of using protein structure to improve a protein is currently being done with the disease hemophilia. Researchers have been using known structures of a protein called *factor VIII* to improve its ability to withstand freezing. This would make factor VIII easier to ship and store since its natural 3D structure is unstable and has a tendency to denature.

[ccxvi] An everyday example of chromatography occurs spontaneously if you've ever spilled a dollop of Catsup on a shirt. After several minutes you probably noticed a light ring, or halo, encircling the drop yet the darker red in the center barely moved. You just performed a molecular separation because the water, salts, and sugars in the Catsup are smaller than the tomato pulp and therefore more mobile in the fabric and simply moved ahead. Protein separation (purification) in the lab is often based on these same simple principles. Size is a common way to separate a lot of things. In London in 1829, sand was used to filter out solid particles from water of the polluted Thames River while in the 1960's cholera toxin would be purified by Richard Finkelstein based on differences in their size from other proteins.

[ccxvii] Many successful microbreweries during the 1970's and 80's went out of business when trying to scale up for a national market. Several made fine beers in small batches, but one of the most important elements in commercial brewing, it turns out, is consistency. Achieving consistency proved extremely difficult for smaller brewers because scaling up using homemade equipment meant an increased chance for bacterial contamination. As Gamal pointed out, commercial brewers making beer on a large scale spend significant amounts of time and money to keep bacterial contamination out.

[ccxviii] The earliest depiction of a sailboat is on a clay vase I got a chance to see in Egypt. It shows two sailors unfurling a square sail around 3500 BC, a predynastic time (i.e. before Egypt had pharaohs). / As a boy, Benjamin Franklin used the wind and a kite to pull himself across a frozen pond. / When Clark & Sacagawea had the Yellowstone River's current to their advantage on their way back to meet up with Lewis in Montana, they made 80 miles in their boat in just one day.

[ccxix] The horse's introduction into North America by the Spanish changed the lives of Native Americans, no more so than the Sioux in the 1700's. Before horses, the Sioux lived as farmers in the southeastern states of Georgia & South Carolina. Horses freed them to migrate to the Great Plains where they became skilled hunters. Before the arrival of the Spanish, the Native Americans had no horses, which is why New World empires were so much smaller in area compared to that of the Romans, Mongols, or Alexander the Great.

[ccxx] Egyptians also had goats and pigs but no horses until the Hyksos invaded in the mid-1600's BC. Camels were another late arrival to Egypt, obtained from the Persians in the early 500's BC. / A bakery in Pompeii used a donkey to turn a millstone for grinding flour, but most bakeries in the ancient town used slaves.

[ccxxi] The first steam engines came about in the late 1600's, but were inefficient. In 1804 Lewis had heard of steam engines, but they were uncommon in post-Revolutionary War America. Thomas Jefferson lived to be 83 yet the founding father that wanted the United States to convert to the metric system in 1790 never rode on a railroad.

[ccxxii] Iron's higher melting point helps explain why the Iron Age followed the Copper and Bronze Ages even though iron ore is more plentiful in the earth's crust. / Heat is a good way to separate things. Physicists believe that the universe was so hot for 300,000 years after the Big Bang that protons and electrons stayed separated from one another. In other words, the universe had no atoms until it could expand and cool enough for them to exist. / The Great London Fire of 1666 was hotter than an ordinary fire; in fact it melted the iron chains on the city's gates as well as the imported steel near the waterfront (the melting point of steel is 2,300°F). / Iron in the ancient world could be more valuable than silver and sometimes traded at a ratio of 1:40. When King Tut's bandages were unwrapped by Carter, an iron dagger fell out onto the floor by his feet. The dagger had been placed next to the pharaoh's thigh by his embalmers 3,000 years before to guarantee Tut's protection in the afterlife. The

dagger was likely made of iron obtained from a meteorite since smelted iron was rare in Egypt due to the monopoly on iron technology by the Hittites. Records exist of Egyptian kings literally begging Hittites for more iron. The Philistines gave the Israelites such a difficult time in battle (as related by the Bible) because only the Philistines had iron weapons.

[ccxxiii] Otzi's ax helped push the accepted date for the beginning of the Copper Age in Europe back 1000 years. His chromosomes have also been sequenced and it appears Otzi was predisposed to heart disease, lactose intolerance, and his ancestors probably migrated from the Middle East. / The Egyptians built the pyramids without benefit of bronze; in fact the only metal they used for tools during this period was copper.

[ccxxiv] Water trapped inside the popcorn kernel turns to steam and does the work of bursting open the tough outer seed coat. Water trapped inside fresh clay was also the reason the ancient Egyptians needed to allow new pottery to dry first before being fire-hardened. Otherwise water left inside the green clay would explode their pots. / The ability to cook is like creating a "second stomach" because cooked food is partially broken down by the fire's thermal energy before it even enters our mouths.

[ccxxv] Metal tools allowed the Egyptians to build stone monuments. / Copper coins found on victims at Pompeii were worn down from having been in circulation. / It was the replacement of clay brewing kettles with copper kettles that allowed medieval English brewers in the 13th century to produce 1000-liter batches of beer and perhaps as much as 4,000-liter batches by the beginning of the 15th century.

[ccxxvi] One of the most common items found in Egyptian tombs are highly polished bronze mirrors. / The swivel cannon Lewis & Clark had mounted on the front of their keelboat was solid bronze and could shoot a 1-lb lead ball or 16 musket balls at once. / Copper melts at a lower temperature than bronze and some historians believe this helped lead to Custer's defeat at the Little Big Horn. His soldiers carried shell casings made of pure copper, which became soft inside the rifles after continuous firing. The casings expanded and jammed the breach, requiring a soldier to stop shooting and pry the empty casing out with a knife. Crazy Horse knew of this problem and is why he urged his warriors to be patient and wait until Reno's soldiers had been firing for a while. One year after the massacre, the US Army switched to bronze casings. Some of the Indians who fought Custer had semi-automatic rifles and could fire 17 shots without reloading.

[ccxxvii] Stone can hold onto thermal energy for relatively long periods and is why the Egyptians baked bread in clay pots with domed lids. These pots were preheated in a fire before the dough was added, the residual heat of the pots being sufficient to bake the bread without the pots having to be put them back into the fire.

[ccxxviii] It also drew Egypt into Mediterranean politics and out of its two thousand plus years of isolationism. Mesopotamian cities also suffered a lack of wood, which is why so little is known about their buildings. As there wasn't enough fuel to make sufficient baked bricks, they had to use mud bricks, which were softer and so didn't stand up to the elements as well as fire-hardened bricks would have. The pharaohs had strict laws against cutting down hardwood trees, the deforested wood already having gone for heating pottery kilns, which happened before the Old Kingdom began.

[ccxxix] Athens, in contrast, would someday depend on a strong navy to defend its grain shipments. / Native Americans had plenty of wood yet never used fire to obtain metals (like copper and iron) from their ores, a reason trade with Europeans caught on so quickly. The Europeans had plenty of reduced metals like brass and iron (Leeuwenhoek made his microscope frames out of brass). / Lewis & Clark took along 288 small knives inlaid with brass to trade. While over-wintering at Fort Mandan the expedition's blacksmith cut an iron stove into 4" squares and traded these scraps with the Indians for 8 gallons of corn apiece. The Indians sharpened and used them as buffalo hide scrapers and arrow tips. Corn from the Mandan got the expedition through their first winter. / Some Indians at the Battle of the Little Big Horn had mounted triangular, iron tips onto their arrow shafts. The skull of one soldier was found with an iron arrow tip still embedded. Two arrows recovered from the nearby Battle of the Rosebud had iron nails fashioned into arrowheads. The Native Americans made use of whatever was available. / Our galaxy, the Milky Way, is classified as a "high metal" galaxy because it contains more metals than "low metal" galaxies.

[ccxxx] *Periplus of the Erythraean Sea* is an ancient guidebook written by a Greek sea captain who seems to have lived in Egypt around the year AD 60. In it, he details more than 20 ancient ports of call around the Indian Ocean from Egypt to India including the best places to

buy cotton, circus animals, Chinese iron, and various dyes.   He also describes places to anchor off of, where to get good wine, the way one could expect to be treated by the locals upon arrival, what month to travel in, places to avoid due to disease, and where glass, copper and tin could be used to pay for things.
[ccxxxi] The blacksmiths on the Lewis & Clark Expedition made battleaxes for the Mandan Indians out of scrap iron while they over-wintered at Ft. Mandan in North Dakota (to trade for corn).   When the expedition was on the other side of the Continental Divide 14 months later, the same blacksmith recognized one of his axes in the possession of a Nez Perce warrior. The Mandan's had already traded some of his axes with other tribes and due to this preexisting 2000-mile trade route all across the Rocky Mountains, the Nez Perce already had them.
[ccxxxii] During its heyday, an ancient Greek could have walked around Athens while enjoying the aroma of fresh baked bread made from grain harvested near the Black Sea in the morning, had fresh fruit from the island of Rhodes for lunch, taken in the smell of incense made in Syria, and snacked on salted mackerel from the Hellespont in the evening.
[ccxxxiii] The Egyptians used lead weights to sink their fishing nets. / Six of the Seven Wonders of the Ancient World have been destroyed.   The last to vanish was the Lighthouse at Alexandria in Egypt.   It survived into the Middle Ages because its stones were held together by lead (as were the Hanging Gardens of Babylon). / The Roman forum fell apart due to an earthquake after the Byzantines took the original iron & bronze clamps holding the masonry together to recycle it.   And it was the use of Greek fire (the recipe of which the Byzantines kept such a good secret no one knows what it was to this day) that helped keep Constantinople part of the Eastern Roman Empire for 1000 years after Rome fell in the west. Only one family in Constantinople knew how to produce Greek fire, which was used to burn ships since this substance could ignite on the surface of water (perhaps because it contained naphtha). / In the 18th century during the initial excavations of Pompeii, gladiator helmets made out of bronze and iron plates were discovered, some decorated with elaborate silver inlay.
[ccxxxiv] Savery was also the first to use the term "horsepower".   A dramatic everyday example of steam's ability to do work can be seen when popcorn is popped. It's the liquid water inside the kernel violently expanding and turning to steam that causes the hard seed coat to rupture and make the kernel edible.
[ccxxxv] Newcomen steam engines weren't confined to coal mines.   Some were also used to supply English towns with drinking water by pumping it from wells.   Other Newcomens were in copper and tin mines (the first steam engine in America was put to use in a copper mine in 1753).   English merchants tried to keep steam engines and how they operated mills a secret, forcing American merchants to offer bribes to anyone with the know-how to build a spinning mill in the United States.   Eventually, a cotton spinner agreed to make one, traveling to America with the layout in his head. / When Mark Twain returned to the Mississippi River in 1882 looking for material for his book *Life on the Mississippi* he saw steam plows on a sugar plantation.   He wrote, "We saw steam plows at work, here, for the first time.   The traction engine travels about on its own wheels, till it reaches the required spot; then it stands still and by means of a wire rope pulls the huge plow toward itself two or three hundred yards across the field, between the rows of cane.   The thing cuts down into the black mold a foot and a half deep.   The plow looks like a fore-and-aft brace of a Hudson River steamer, inverted.   When the Negro steersman sits on one end of it, that end tilts down near the ground, while the other end sticks up high in the air.   This great seesaw goes rolling and pitching like a ship at sea, and it is not every circus rider that could stay on it."
[ccxxxvi] This means Watt only obtained about 4% of the work theoretically possible from the coal to do actual mechanical work with his machines, the other 96% of the chemical energy being wasted as heat.   A typical car engine today is only about 20% efficient…meaning it wastes 80% of the chemical energy stored in gasoline molecules as heat.   The main reason car engines are more efficient than steam engines is that the chemical reaction takes place internally, inside the cylinder, whereas with the steam engine, the fire and its thermal energy lie outside the cylinder.   Some energy will always be lost (wasted) as heat whenever it changes from one form of energy to another.   When our cells harness energy by burning glucose using oxygen, they capture only about 40% of glucose's energy to make ATP.   The rest of our food's energy becomes heat (just like the steam engine) and keeps us at a comfortable 98.6 degrees F (37C).

ccxxxvii Aristotle, however, had rejected the atomists' views, choosing to believe instead that matter could be divided endlessly. The Greeks knew at least 8 of what we would today call true elements even though they grouped them together as "earth". They were aware of lead, gold, silver, copper, tin, mercury, sulfur, and carbon. It wasn't until the 1600's that Boyle gave science its first *modern definition* of an element as something irreducible. Interestingly, it wasn't until the 1930's that most biochemists came around to believing proteins could exist as individual molecules and could therefore be purified. Prior to this, all proteins were thought to exist as complex aggregates.

ccxxxviii Alchemists saw fire as the agent of change, for example when converting metal ores into pure ("reduced") metal. Moslem alchemists discovered how to use fire to separate alcohol from wine in stills. And when alchemists in Europe unsuccessfully tried to turn base metals into gold, they used fire to speed up processes they believed took place naturally within the "womb" of the Earth. Unlike our current view of an element, the ancient Greek philosopher Aristotle believed the 4 primary elements could change, or transform, into one another, which is where alchemists probably got the notion that it was possible to turn base metals like lead into gold. Alchemists also believed only fire could cause changes in metals...until the first concentrated acids from vinegar were obtained by distillation (Geber did it first). / These days, physicists can achieve temperatures over 7 trillion degrees F inside supercolliders.

ccxxxix Joule worked in the brewery (the largest in Manchester) as a boy and later in his career managed the brewery and became interested in replacing their steam engine with an electric motor. As a boy his parents allowed him a chemistry lab in their home. It's not surprising he would take an interest in measuring temperature when one recalls how small differences in temperature can affect the quality of a beer greatly. It was the use of cooler temperatures in Bavaria's ice caves that helped give birth to the first lager beers (i.e. the selection of lager yeast). Brewers in the 1800's could use thermometers to monitor the beer's temperature during fermentation and storage. They also kept track of its density, or "specific gravity". Joule claimed he could measure temperature differences to within 1/200 of a degree Fahrenheit (and he never gave up trying to brew a better beer).

ccxl Scientists during the Industrial Revolution were in search of mechanisms to explain phenomena like heat. It was about this same time a German doctor named van Mayer (the MD kind) traveled to the tropics where he noticed sailors had blood a brighter shade of red than in northern climates. He concluded it was because less oxygen was needed in the tropics to keep the human body warm. He even came up with a theory that all energy was *conserved* because instead of disappearing and reappearing, it simply became transformed. But because he wasn't as well known as Joule, and didn't do Joule's kind of careful, detailed experiments, van Mayer got no credit for anticipating the theory of the conservation of energy before anyone else.

ccxli Lavoisier – by showing that when something burns, it requires oxygen from the air – eventually laid the phlogiston theory to rest. Joseph Priestly would later say that if he (Priestly) hadn't been holding a lit candle after heating some mercuric oxide about this same time, he would never have discovered oxygen. / The ancient Greek physician Hippocrates in 460 BC claimed that the human body had an *internal fire* that animated it and that this fire was located precisely in the left ventricle.

ccxlii With CFTR, for example, previous biochemical work had revealed the exact number of ATP molecules consumed by the protein just to open and close its gate: one ATP to open it and one to close it. In other words, the cell has to spend 2 ATP molecules each time it opens and closes CFTR, which is an insignificant amount to the cell. CFTR is unusual among ion channels because it also functions as an enzyme. / It's been estimated that a honeybee could theoretically make ten trips to the moon and back on the chemical energy equivalent to what's stored inside a single gallon of gasoline. / Sigmund Freud's model of the subconscious mind in the 1800's was apparently based on the inner workings of a steam engine.

ccxliii Chemists have a saying that "like dissolves in like". Fat (oil) & water, for example, are nothing alike, chemically, so they don't dissolve within each other but separate out instead. That's why the Egyptians put a layer of grease over top of a layer of honey to prevent infection on someone's cut. The grease kept water out, similar to a band-aid. "Like dissolves in like" also explains why contaminants like PCBs stay in the body so long once ingested and will actually build up over time. PCBs are "oily" molecules so they dissolve in the body's fat and don't get washed away in the urine as easily as more water-soluble contaminants. / One of the main differences between RNA and DNA is that DNA has one less -OH group than RNA (hence the prefix "deoxy" in deoxyribonucleic acid). This lack of an extra -OH group makes DNA more stable, chemically, than RNA and is the reason DNA is used by the cell rather than RNA for the long-term storage of the genetic information.

[ccxliv] In their ships, Egyptians didn't use nails, but instead relied on ropes made of cellulose to hold the planks together. Cellulose is surprisingly strong. In fact, cellulose in marsh plants was fashioned by the Egyptians into papyri up to half a football field in length (40 meters). Many papyri have survived into the 21st century. It's been said that even as late as the Middle Kingdom period, the most useful weapon an Egyptian soldier had at his disposal was a defensive one: the wooden shield.

[ccxlv] About 1 gram per liter. The amount of glucose in a packet of sugar is similar to the total amount in your bloodstream at any moment. / Ancient physicians knew that blood tastes sweet to the tongue.

[ccxlvi] Diabetics can go into a coma if they take too much insulin. Insulin is a hormone that tells cells to take glucose out of the blood and store it in the tissues (muscle & liver). But if capturing and storing glucose is done too enthusiastically, there can be no glucose left in the blood for the brain.

[ccxlvii] It's no coincidence life became larger and joined together to become multicellular (plants, fungi, and animals) only after oxygen became plentiful in the Earth's atmosphere about 600 million years ago thanks to photosynthetic bacteria. More oxygen meant more energy could be extracted from food, and this extra energy in turn meant larger, more complex organisms could "afford" to come about.

[ccxlviii] Living cells are factories in many senses of the word. Another use cells have for glucose is as a starting material, for example when plants make *vitamin C* or when our own cells use glucose to manufacture one of the twenty amino acids called *serine*. As a dog, Barney makes vitamin C, but I can't. About half the Pilgrims would die of scurvy their first winter at Plymouth (the Pilgrims also had 2 dogs onboard which would have been safe from scurvy).

[ccxlix] Contrary to popular belief, Thomas Edison didn't invent the first electric light. Alessandro Volta made one in 1800 by causing a wire to glow using electrical energy. The first practical use for electric current produced by a battery was by railroads in the 1830's to keep track of trains.

[ccl] You may be wondering why the electrons don't simply flow from one metal to the other metal inside the battery instead of having to first travel through the wire to get there. It's because, inside the battery, there is an insulator that keeps the two metals out of direct contact. The electrons therefore have to traverse the wire circuit to get from one end of the battery to the other.

[ccli] Bacteria – not plants – invented photosynthesis. It was a bacterium a couple billion years ago that first began using water for photosynthesis, producing molecular oxygen as a byproduct, that allowed complex life like our own to exist. A waste byproduct of photosynthesis, this oxygen allowed animals to extract more energy from their food, and become multicellular. The crew of Apollo 12 while walking on the moon discovered how much energy sunlight can have when they accidentally pointed a TV camera at the sun and destroyed its "eye". Sunlight's energy is also why the Hubble Space Telescope was never able to take a photo of the planet closest to the Sun...Mercury.

[cclii] Fats are higher in energy than glucose because the electrons making up the bonds between their atoms can "fall" further than the ones from glucose, and therefore they can do more work. It's as if the roller coaster starts off higher for fats than sugars. About 80% of a human cell's proteins need ATP – along with special helper proteins called *chaperones* – to fold properly into their 3D structure (biochemists would call this an "energy-dependent step"). Cells often use ATP for actions they'd prefer were not easily reversible...i.e. can't be easily undone, for example making glucose or fat for energy storage).

[ccliii] In fact, the only paper I had been a first author on was for a species we'd discovered living on sulfur in a mud pot on one of those weekends. It's apparent from DNA analysis that the very first life forms on Earth were likely thermophilic organisms similar to those found at Yellowstone, having evolved back when the planet was much hotter than today.

[ccliv] Most substances become less dense when heated, and thus more likely to rise, which is what happens to rock and water when heated by thermal energy inside the Earth. Air also follows this rule. When I was in Cairo a guide explained how the tall pointed minarets in

mosques were built to allow warm air inside the building to rise as it gets heated during the day, creating ventilation. This steady stream of warm air flowing upward to an open window at the top of the minaret encourages cooler air to enter through windows down below.

[cclv] Weight from above can also cause tremors. There's some concern that the artificially created Lake Nasser in Egypt today has so much water, it creates small earthquakes. Aswan High Dam was the most ambitious building project in Egypt since the pyramids and took 17 times more material to build than the pyramids did.

[cclvi] The grass on the prairie is especially high in protein, which helped account for the rich variety of wildlife they observed along the way. Plants with more protein are able to sustain a higher diversity of animals than grasses without as much protein to offer the food chain.

[cclvii] It's interesting to contemplate that, from a bacterium that lives without oxygen in boiling water's point of view, it is we humans that are the extreme life forms. These same bacteria that easily survive scalding heat and other extreme conditions often aren't able to tolerate ordinary air like we do all the time. Our cells have special enzymes that detoxify oxygen's dangerous byproducts. We also come equipped with special enzymes that help repair DNA from the mutating effects of ultraviolet radiation of the sun. Some microbes don't have this advantage and so are easily sterilized by solar radiation. It's been speculated that heat-loving bacteria were among the first life forms to come about since they would have been better able to hide out in the hot, oxygen-less conditions of an early Earth. Yellowstone may be among the last vestiges on the planet these microbes have found to survive in, meaning they could be holdouts from a now-alien world that was once widespread across the surface of the Earth four billion years ago.

[cclviii] Because the sequences are so similar, they have more of a tendency to zip back together. A particular base on one strand of DNA is always attracted to its opposite base on the other strand. The two strands can then come together, like a zipper, forming a double helix. An 'A' on one DNA strand attracts only a 'T' on the other, and a 'C' on one strand only pairs with a 'G' on the opposite strand across from it. You could think of an A and a T as two halves of a rung on a ladder that fit together like a round peg into a round hole, while a C and a G are a square peg in a square hole, chemically. If two different species are closely related, they're more likely to share the same sequence of bases (A, T, G, C) along their strands of DNA, which makes rejoining the two strands a lot easier.

[cclix] Technicians, when they want to determine the sequence of DNA, have to heat the DNA in near-boiling water first so they can separate the two strands just to get the sequencing probes inside the DNA so the enzyme can start copying. Your cells accomplish the same job naturally and at a comfortable 37 °C because they have the help of a team of enzymes. / Jell-O (and the agar on Petri dishes) are gels created by long molecules of carbohydrates (sugars). Like DNA, these strands can come back together and form hydrogen bonds with each other in water when the water cools. When one makes Jell-O, the concentration of carbohydrate is so high that the strands form a gel, trapping the water molecules within. Also being long and linear, DNA can form a gel too if its concentration is high enough, much to the annoyance of protein chemists who have to break apart cells to purify proteins.

[cclx] Simply measuring temperature differences can be enough to reveal the natural world. Benjamin Franklin discovered the Gulf Stream by taking temperature readings of the Atlantic while crossing it in a ship. He published the first scientific description of the current – which can affect the weather on two continents – in 1786. 150 years before Franklin – the Pilgrims wouldn't have known about the Gulf Stream, which flows north, and explains why they ended up too far north when they made landfall, settling in New England instead of the Hudson River region. The ship made it back to England in half the time, just 31 days, because upon returning the *Mayflower* had the Gulf Stream in its favor.

[cclxi] Like sending out an e-mail to 2 people, who then forward it to 2 others, and so on so that within 20 rounds, your single e-mail could become over a million.

[cclxii] Our ecology department once sponsored a talk by a biologist from Glacier National Park who was using PCR to DNA-fingerprint grizzly bears. He was pleased his team no longer had to stun them with a tranquilizer gun for tagging. Instead, all they did now was set out barbed wire and when the bears traveled by, a small amount of hair would catch on the wire. They then took the DNA from the bear's hair and amplified it using PCR to DNA fingerprint it. This way, not only could they keep track of the bears, but they also had a good idea which bears were related to one another.

cclxiii The enzyme is called Taq (an abbreviation of the bacteria's name). Taq was *Time*'s first "Molecule of the Year" in 1989. There is a strain of archaen microbe discovered off Puget Sound in a hydrothermal vent that can live and reproduce inside an autoclave at 121 °C. One of the reasons proteins from thermophilic bacteria like *T. aquaticus* can remain so stable at these high temperatures is that they tend to have more "salt bridges" within them. Salt bridges are formed when two amino acids join (bond) with each other within the same protein to help stabilize the protein's overall 3D structure. Salt bridges are similar to disulfide bonds except that here the bond is *ionic*, rather than *covalent*, which means the amino acids in salt bonds are held together because opposite charges attract each other. One amino acid is positive, while the other nearby amino acid in the protein is negative in charge, so they bind to one another...forming this salt bridge not so different than table salt.

cclxiv The HIV virus can become drug resistant because of a similar enzyme it has for duplicating the chromosome, an enzyme with a tendency for making mistakes (mutations). Making mistakes helps in evolution, which is speeded up extremely fast in the case of HIV, effectively making this virus a moving target for the human immune system.

cclxv A well-insulated sleeping bag can keep a person warm even in sub-freezing temperatures because the human body continuously puts out the equivalent of a 100-watt light bulb in heat.

cclxvi *Gamma ray bursts* are the most concentrated sources of energy in the known universe, some the equivalent in brightness to a million trillion suns. They're so intense that astronomers thought at first they had to originate within our own galaxy. / The *rate* of energy release in chemical reactions often has practical significance. A ton of ordinary gasoline, for example, will release more energy than a ton of TNT in the long run, it's just that the TNT releases its energy much faster and can therefore do more damage.

cclxvii If you wind a spring inside a clock, the spring gets heavier, albeit slightly, because some of the energy you put into winding it became additional mass of the spring. / When hydrogen is converted into helium inside the sun, only 1% of the mass of hydrogen turns into energy, yet the sun still manages to convert 4 million tons of hydrogen into energy every second. By comparison, the bomb dropped on Hiroshima converted just 700 mg of matter into pure energy (700 mg is about one-third the mass of a penny). Put another way, more energy is produced by the sun every second than humans have used since the we started using fire. Surprisingly, physicists have created higher temperatures on Earth than occur inside the sun; and here's something else you may find interesting: more energy strikes the Earth from the sun in one hour than all the cities on our planet use in a year.

cclxviii Heat can be used to produce electricity with a device called a *thermal electric converter*. Both Voyager spacecraft rely on heat generated by the decay of plutonium nuclei (rather than solar power since they're too far away from the sun). This decay taking place inside the spacecrafts produce 400 watts of electricity each. / The mummy of Ramesses II was soaked in gamma radiation to kill off fungi & bacteria during its restoration in Paris in the 1970's.

cclxix The Earth's orbit around the sun is elliptical rather than circular so that, when it orbits the sun, its rate of travel is similar to a pendulum. When it's farther away from the sun, the Earth travels slower, like a pendulum does towards the top of its swing, and as it gets closer to the sun, the Earth picks up speed for the same reason the pendulum does when it gets closer to the Earth: both are the result of the inter-conversion between kinetic and potential energy. / The dinosaurs were killed off by kinetic energy released when an asteroid hit the earth, kicking up so much dust into the atmosphere it changed the climate.

cclxx Boyd discovered how explosive the Yellowstone volcano could be. It's now believed that the eruption forming Yellowstone was one of the most explosive events in Earth's entire history. Volcanic ash Boyd identified from Yellowstone's 600,000 year old eruption has been found as far south as the Gulf of Mexico, and as far west as California. It was probably 1000 times more energetic than the Mount St Helens eruption in 1980.

cclxxi J.J. Thompson, discoverer of the electron, must have been a good teacher too. Seven of his students (including his son) went on to win Nobel Prizes of their own. / The inventor of the Petri dish was a lab assistant of Robert Koch.

cclxxii The Lewis & Clark Expedition ate candlefish, roasted whole, while over-wintering at the Pacific in 1805-06. Lewis was partial to candlefish because they were so rich in fat there was no need to make a separate sauce for them. / After Howard Carter knocked a hole in the sealed tomb of King Tut, he first inserted a lit candle to make sure the air was suitable for breathing.

cclxxiii A pottery shop within the ruins of Pompeii had a furnace for making terra cotta lamps, and excavators also found lamp molds.   The eruption that buried Pompeii in August AD 79 produced so much volcanic ash in the air that by 6 p.m. the ones who stayed behind needed lamps to see by as they tried to escape the fallout.   /   King Tut's tomb was stocked with everything he would need after death, including torches. The Egyptians invented candles but had no wicks.   The Romans made tallow candles inside brass molds and also added a wick.   /   It was Benjamin Franklin's dislike for making tallow candles (his father's occupation) that drove him into the printing business.

cclxxiv They were "goddesses of the hearth" and required to remain virgins for 30 years. If they touched a slave, the slave became free.   On the other hand, if a virgin let the fire go out, she might be whipped.

cclxxv Before the trip to Egypt, I had always assumed temples were for the masses, yet it seems this wasn't so in ancient times.   In Egypt, temples were reserved for priests and royalty, whose job it was to curry favor from the gods.   The priests would have discouraged an Egyptian commoner from visiting the hearth of a temple where the statue of the god was by keeping this inner windowless sanctum dark.   /   Howard Carter knew the layout of the Valley of the Kings so well because, before looking for King Tut's tomb, he'd already installed the first electric lighting in nearby tombs.

cclxxvi There were no prisons or police forces and wealthy Romans kept dogs and displayed a "Beware of Dog" sign. They also had iron bars on their windows and a slave guarding their front door at night.   A common inscription on Roman tombstones is "interfectus a latronibus" (killed by robbers).   Archeologists have never found any streetlights in Pompeii, but the ancient city did have "cat's eyes", which were broken bits of white marble incorporated into the sidewalk steppingstones so citizens could see where to walk using reflected moonlight. /   Before reliable gas lighting in the 19th century, Londoners sometimes had to hire someone after the sun went down whose job was to carry a tallow-burning torch.   This "profession" gained a well-deserved reputation for taking their clientele down a dark street, snuffing out the flame, and then robbing them (or worse).   Before the harnessing of electricity, burning things was the only way humanity had of producing light for thousands of years.

cclxxvii Always interested in becoming a better farmer, George Washington had read about Lavoisier's use of the scientific method for improved agriculture and tried to get information from him on it.

cclxxviii Because it was heated, coffee was a safer drink for soldiers in the 1800's.   Florence Nightingale promoted coffee for the same reason.   Union soldiers during the Civil War chewed on coffee beans while marching to absorb the caffeine. Confederate soldiers often had to do without coffee because of Union blockades.   When they could, the two armies would swap tobacco for coffee beans during informal truces along the front lines.

cclxxix They were suspicious that Napoleon and his men never actually converted to Islam. Napoleon eventually had the rebels beheaded and thrown into the Nile.   /   When Constantinople was founded in AD 330, grain from Egypt kept the people fed.

cclxxx The British called them *penny universities* because the price of a cup of coffee was traditionally a penny.   In ancient times, different classes rarely mixed.   One exception was the Roman baths, where all could rub elbows, and not only to take a bath but to see a doctor, visit the library, or even get a haircut. / Coffeehouses in London, like the city's many local neighborhood newspapers, allowed inhabitants to feel as if they were living in a small village (probably like today's cybercafés).   Coffee was slower to catch on in Colonial America due to a tradition of drinking tea.   Following boycotts during both the American Revolution and the War of 1812, coffee increased in popularity as is seen today.

cclxxxi Due to the belief that filth caused plague in the 1600's, Halley's father made enough money producing & selling soap to provide his son with the best education available.   Halley would go on to predict the date of the arrival of his namesake comet, which helped confirm Newton's law of gravity (Thomas Jefferson saw Halley's Comet when he was 15).   To Robert Hooke, coffeehouses were his second home and he demonstrated his latest inventions there including his new microscope.

cclxxxii A popular pastime in the 1800's was to attend a mummy "unrolling" party.   At a promoted event, guests would unwrap the bandages of an Egyptian mummy, or so they assumed.   Mummies were in such demand that supply couldn't keep up and so unscrupulous dealers began disguising the recently deceased as mummies and sending them off to Europe.   To fool their clients, these dealers sped the mummification process

along by first burying the fresh corpse in the desert for several months. In the early 1930's at the Chicago Field Museum, its collection of mummies were x-rayed and some discovered to be frauds, containing only animal bones, assorted human limbs, or a stick. / Thomas Jefferson, a book lover, purchased a set of Diderot's encyclopedias in 1781.
[cclxxxiii] The co-discoverer of oxygen, Joseph Priestly, became friends with Benjamin Franklin after the two met in a coffeehouse in London, finding that each shared a fascination with electricity. / Benjamin Franklin invented a glass street lamp with 4 sides to replace traditional globes so that, if one side broke, the flame wouldn't be extinguished by wind. With the one-piece older style, the entire globe had to be replaced if cracked. / In 1825 Michael Faraday discovered the chemical benzene (C_6H_6) in a gas-fed streetlight in London.
[cclxxxiv] Not just streets but rivers also became lighted, as an aid to navigation. After lights were added to the Mississippi in the late 1800's, steam travel at night became the norm. When Mark Twain returned to the river in 1882 to research his book *Life on the Mississippi*, he described it this way: "As we approached the famous and formidable Plum Point, darkness fell, but that was nothing to shudder about – in these modern times. For now the national government has turned the Mississippi into a sort of two-thousand-mile torchlight procession. In the head of every crossing, and in the foot of every crossing, the government has set up a clear-burning lamp. You are never entirely in the dark, now; there is always a beacon in sight, either before you, or behind you, or abreast. One might almost say that lamps have been squandered there."
[cclxxxv] Thomas Edison, when he designed his electric light bulb, used the gas-burning light as his standard so both would produce the same amount of luminosity.
[cclxxxvi] The Saturn V rocket engine – workhorse of the Apollo moon missions – used kerosene that, upon being combined with liquid oxygen, provided 7 million pounds of thrust during liftoff. And just 130 years before Apollo 11, the steam engine was put to use for the first time on trains. Even as late as the 1830's, very few people had ever traveled faster than a horse could run.
[cclxxxvii] An example of a new idea *without* a corresponding paradigm shift to accompany it was the steam engine, which was apparently invented two thousand years ago by Hero of Alexandria. It wasn't thought to be able to do useful work and so was seen as a novelty, or toy. Paradigm shifts aren't new, though. Ramses the Great had his inscriptions chiseled into the stone monuments rather than raised above the stone as had always been done. Because of this change, it made it harder to remove his name by later pharaohs. Thanks to his paradigm shift, many of his writings can still be read 33 centuries later. Another interesting example of a paradigm shift being too early is what Austrian physician Leopold Auenbrugger did in 1754. He invented percussion – the act of putting one's ear to a patient's body and tapping on it to determine the condition of the internal organs. As a boy, he had learned to tap on his father's wine kegs to determine their level of fullness, and later applied this technique to his patients. He wrote a pamphlet describing it, but was ignored at the time. Eventually, his work was rediscovered in 1808 and became widely accepted after that.
[cclxxxviii] Rubidium can be found in fireworks as it gives off a red color when heated. In 1995, scientists cooled 2,000 rubidium atoms to within ten-millionths of a degree above absolute zero, temporarily forming a new kind of matter called a Bose-Einstein condesate (they earned a Nobel Prize for it). / Rubidium's decay into strontium is one of the ways geologists dated the famous Martian meteorite (which some still believe may harbor traces of life) to 4.5 billion years old. To positively identify it as Martian, they shot energetic electrons at its surface and detected the signature x-ray wavelengths given off by the iron atoms inside. It was the use of EDXA (energy-dispersive x-ray analysis) that revealed Egyptian mummies suffered from silicate pneumoconiosis, a lung disease. / By the time Bunsen discovered rubidium in 1865, there were 60 other elements known to chemists…far more than Aristotle's four, thanks to the Scientific Revolution. / Cesium is used today in atomic clocks. The transition time of cesium's electrons when traversing between two different energy levels near the nucleus are used to define the current value of a second, which is how long it takes for 9,192,631,770 cycles of radiation to be emitted. / By 1890, chemists knew that the sun contained 30 different elements. / Our word chemistry comes from *Khemet*, which is what the ancient Egyptians called Egypt.
[cclxxxix] Even Bartholdi, the French sculptor who designed the Statue of Liberty took out a patent on it (his first proposal had been rejected, one that was for a similar statue at the entrance to the Suez Canal in Egypt). Bartholdi's inspiration for the Statue of Liberty was the

ancient bronze Colossus of Rhodes, which had the same posture, dimensions, and even a burning light to help guide ships into the harbor of Rhodes starting in 292 BC.
ccxc It took two months just to go from Missouri into present day Kansas. During their winter with the Mandan, they had a thermometer that read -45° F at one point, and after the Missouri River froze over, the Mandan could walk across it to trade with the Corps every day.
ccxci The US Calvary preferred Arikara scouts. Custer's favorite scout – Bloody Knife – was half Arikara. Bloody Knife had repeatedly warned Custer of the large number of Indians they were about to encounter and was also killed at the Battle of the Little Big Horn.
ccxcii William Bradford – who also described the first Thanksgiving – told in 1634 of what the "poxe" could do in a short amount of time: "...this spring also, those Indians about their trading house there fell sick of the small poxe and died most miserably...very few escaped." Benjamin Franklin would someday lose a son to smallpox.
ccxciii Traders along the Silk Road also brought back smallpox. / At the Egyptian Museum, I saw the yellow smallpox blisters on the face, abdomen, arms, and neck of Ramesses V, a pharaoh from 1157 BC. Later, smallpox would claim the lives of the Roman Emperor Marcus Aurelius (AD 251) and King Louis XV of France. / Johannes Kepler needed to rely on Tycho Brahe's astronomical observations because of smallpox. The disease had weakened Kepler's own eyesight as a child.
ccxciv The Sioux Chief and medicine man Sitting Bull claimed in a newspaper interview 16 months after his victory at the Little Big Horn that – before he was born – he had already been made aware of smallpox's effects on his people. "I was still in my mother's insides when I began to study all about my people. God gave me the power to see out of the womb. I studied there, in the womb, about many things. I studied about the smallpox that was killing my people – the great sickness that was killing the women and the children."
ccxcv The Spaniards benefited from *herd immunity*, whereby not every member of a population needs previous exposure to gain protection. As long as most have immunity, it acts like a "fire break" to keep the disease away from more susceptible members. The Native Americans, on the other hand, were what epidemiologists would today call a *virgin soil population* because they had no natural immunity; so the disease could spread more rapidly among them.
ccxcvi In December 1775, Washington sent a letter to the Continental Congress expressing his concern that the British under General Howe were about to send smallpox suffers out of Boston to deliberately infect his army. Washington himself was already immune to the disease having gotten it at age 19 in Barbados and had the scars on his face as a reminder of this immunity, which he carried with him the rest of his life. / Revolutionary War soldiers were granted the right by the Continental Congress early in the war to one quart of spruce beer or apple cider per man per day. Washington's soldiers at Valley Forge also got half a cup of rum per day.
ccxcvii Leeuwenhoek had someone write down his final scientific letter while on his deathbed yet during his lifetime never wrote a scientific paper of his own. / Even with Jenner's vaccine, smallpox wasn't eliminated until the 20th century. Abraham Lincoln had smallpox when he traveled from Washington to Gettysburg by train to deliver his famous address in 1863. / Louis Pasteur would later credit Jenner's cowpox discovery with providing the inspiration for his rabies vaccine. Before Pasteur's vaccine, rabies was 100% fatal and greatly feared on the frontier. As Lewis gathered information from the Indians about possible medicinal plants west of the Mississippi, he was also on the lookout for any that might cure rabies. He sent back several pounds of roots to the APS for experiments on rabies. / The first person Pasteur successfully vaccinated against rabies (in 1885) – a boy named Joseph Meister who had been bitten several times by a rabid dog – was the son of a baker in Steige.
ccxcviii Men and women of the Enlightenment believed practical knowledge should benefit as many as possible. Lewis, for example, wasn't above trying out new ideas lacking scientific basis as long as it held the promise of some practical benefit. When Sacagawea was having a difficult childbirth at Fort Mandan, he ground up a rattlesnake's rattle in water after a trapper had assured him it would help. She drank it and gave birth to her son Jean Baptiste soon afterwards.
ccxcix Windmills also supplied water for the boilers of the transcontinental railroad. Sears & Roebuck had a supplement to their catalog specializing in windmills ranchers could order from. Before the *Triple K* got electricity, my grandfather had a ranch hand whose only job was to grease all the windmills. Wind is formed on Earth due to solar energy's uneven

heating of our planet's surface.   When my grandfather was born the American farmer fed an average of just 7 – today he feeds more than 100. Now is the first time in history that just 2% of the human population is able to feed the other 98%.

ccc The APS's mission was to pursue equally "all philosophical Experiments that let Light into the Nature of Things, tend to increase the Power of Man over Matter, and multiply the Conveniences or Pleasures of Life". Washington, Jefferson, Benjamin Rush, and Meriwether Lewis were all members (meetings were sometimes held in Franklin's own dining room). Other members someday would include Louis Pasteur and Charles Darwin. The APS had a recipe for brewing ale from pumpkins (1771).   Like Lewis but 2000 years earlier, Alexander the Great also sent back samples of minerals and other curiosities he found on his travels to his former teacher, Aristotle.   Napoleon founded a scientific institute while in Egypt and one of the issues it addressed was, perhaps not surprisingly, better ways to bake bread.

ccci While searching the banks of the Yellowstone River in Montana, William Clark found the first dinosaur fossil in North America.   Seventy years later, in the summer of 1876 while the Battle of the Little Bighorn was taking place just 100 miles away, famed fossil hunter Edward Cope was also in Montana, discovering a new species of dinosaur he would soon call *Monoclonius*.

cccii Educated in Scotland, Dr. Rush trained Lewis on medicine and was a strong proponent of "bleeding" as a treatment for various ailments, erroneously believing a person could safely lose up to 80% of their blood. Lewis later bled Sacagawea in Montana when she became sick but soon realized it wasn't doing her any good. Laudanum did help her sleep and she felt better the next day.   Like many physicians of his time, Rush also believed in spontaneous combustion; for example that a heavy drinker was more likely to suddenly burst into flames than someone who didn't drink alcohol.

ccciii Which is why the journals wouldn't be published for 100 years. They might have been out sooner except that Lewis became discouraged after hearing he was to be scooped by two of his own men, a sergeant and a private. These two did have plans to publish accounts of the expedition ahead of Lewis but only one of them succeeded.

www.ingramcontent.com/pod-product-compliance
Lightning Source LLC
Chambersburg PA
CBHW051209170526
45166CB00005B/1820